T0304765

Fundamentals of Structural Engineering

This book provides an introduction to the principles of structural engineering using a problem-based approach. It covers the basic concepts of structural analysis and design, including statics, strength of materials, and mechanics of materials. The text emphasizes the application of these principles to real-world structural engineering problems and includes numerous example problems and case studies to illustrate key concepts. The problem-based approach helps students develop their problem-solving skills, critical thinking abilities, and intuition for structural engineering. *Fundamentals of Structural Engineering: A Problem-Based Approach* is designed for undergraduate students studying structural engineering or related fields.

- Covers all the key concepts in structural engineering, including statics, strength of materials, mechanics of materials, load estimation, and analysis techniques.
- Utilizes a problem-based approach that helps students understand and apply the principles of structural engineering in a practical, hands-on way.
- Includes numerous worked examples, practice problems, and case studies that provide students with a clear understanding of how the concepts they have learned can be applied to real-world structural engineering problems.

Fundamentals of
Structural Engineering
A Problem-Based Approach

Tanvir Mustafy

CRC Press
Taylor & Francis Group
Boca Raton London New York

CRC Press is an imprint of the
Taylor & Francis Group, an **informa** business

Designed cover image: Shutterstock

First edition published 2025
by CRC Press
2385 NW Executive Center Drive, Suite 320, Boca Raton FL 33431

and by CRC Press
4 Park Square, Milton Park, Abingdon, Oxon, OX14 4RN

CRC Press is an imprint of Taylor & Francis Group, LLC

© 2025 Tanvir Mustafy

Reasonable efforts have been made to publish reliable data and information, but the author and publisher cannot assume responsibility for the validity of all materials or the consequences of their use. The authors and publishers have attempted to trace the copyright holders of all material reproduced in this publication and apologize to copyright holders if permission to publish in this form has not been obtained. If any copyright material has not been acknowledged please write and let us know so we may rectify in any future reprint.

Except as permitted under U.S. Copyright Law, no part of this book may be reprinted, reproduced, transmitted, or utilized in any form by any electronic, mechanical, or other means, now known or hereafter invented, including photocopying, microfilming, and recording, or in any information storage or retrieval system, without written permission from the publishers.

For permission to photocopy or use material electronically from this work, access www.copyright.com or contact the Copyright Clearance Center, Inc. (CCC), 222 Rosewood Drive, Danvers, MA 01923, 978–750–8400. For works that are not available on CCC please contact mpkbookspermissions@tandf.co.uk

Trademark notice: Product or corporate names may be trademarks or registered trademarks and are used only for identification and explanation without intent to infringe.

ISBN: 978-1-032-63805-8 (hbk)
ISBN: 978-1-032-63806-5 (pbk)
ISBN: 978-1-032-63807-2 (ebk)

DOI: 10.1201/9781032638072

Typeset in Times LT Std
by Apex CoVantage, LLC

Contents

About the Author

Dr. Tanvir Mustafy received his B.Sc. degree in Civil Engineering from BUET, Bangladesh, in 2011, an M.Sc. degree in structural engineering from the University of Alberta, Canada, in 2013, and a Ph.D. Degree (Computational Mechanics) from University of Montreal, Canada, in 2019. He also did a Postdoc from Imperial College London in the same year. His research and teaching interests include the theory and application of machine learning, structural engineering, earthquake engineering, advanced finite element modeling, dynamics of structures, data analysis, and injury biomechanics. Dr. Mustafy currently serves as an Associate Professor in the Department of Civil and Environmental Engineering at the North South University (NSU), Bangladesh. Before joining NSU, he worked as an Associate Professor in the Department of Civil Engineering at the Military Institute of Science and Technology (MIST), Bangladesh. Prior to that, Dr. Mustafy worked as a member of a prestigious scientist group led by one of the most renowned Research Chairs in Canada during his doctoral period. He traveled to France as a visiting scholar and spent three months working at Aix-Marseille University. Dr. Mustafy also received the prestigious Professional Structural Engineer (SEng) by BUET, ICC (USA), RAJUK, URP and IEB in 2023.

1 Force–Equilibrium

1.1 INTRODUCTION

A **force** is a push or pull that can cause an object to accelerate. Forces can be categorized as contact forces (forces that act when two objects are in contact) and non-contact forces (forces that act at a distance, such as gravity and magnetism).

Equilibrium refers to a state in which the net force acting on an object is zero. In other words, the object is not accelerating. This can happen when the forces acting on the object are balanced, meaning, that the force in one direction is equal in magnitude but opposite in direction to the force in the other direction.

A **free-body diagram** is a sketch of an object with all the forces acting on it represented by vectors. It is used to help identify and analyze the forces acting on an object and to determine the net force and the forces in the x and y direction.

Resultants and components refer to the relationship between the forces acting on an object and the net force. The *net force* is the vector sum of all the forces acting on an object and is also known as the "resultant" force. Forces can be resolved into their horizontal and vertical components, which are the projections of the force onto the x- and y-axes.

Together, these concepts can be used to understand and analyze the motion of objects and the forces acting on them. They are particularly useful in fields such as physics, engineering, and mechanics, where understanding the behavior of objects under different conditions is essential.

1.2 CONCURRENT FORCES

Concurrent forces are forces that act on an object in the same direction or plane. They are also known as "coincident" or "collinear" forces. When forces are concurrent, they can be combined using vector addition to find the net force, also called the resulting force, acting on the object.

For example, if two forces of 10 N and 15 N act on an object in the same direction, the net force acting on the object is $25\,\mathrm{N}\,(10\,\mathrm{N}+15\,\mathrm{N})$. The direction of the net force is the same as the direction of the individual forces.

When forces are concurrent but in opposite direction, they are called "opposing force," and their net force is the difference of the magnitude.

It is important to note that concurrent forces are different from "parallel forces," which act on an object in the same direction but do not share a common point of application. In the case of parallel forces, the forces must be resolved into their components before the net force can be found.

Example 1

A particle is under the influence of several forces shown at Figure 1.1. Find the magnitude and direction of resultant force of the system.

Solution

$$\rightarrow + \sum F_x = 170*\cos30 + 220*\cos50 - 200 - 180*\cos45 + 0 - 300*\cos40$$
$$= 147.22 + 141.41 - 200 - 127.28 + 0 - 229.81$$
$$= -268.45\,\mathrm{N}\,(\leftarrow)$$

DOI: 10.1201/9781032638072-1

$$\uparrow + \sum F_y = 170 * \sin 30 - 220 * \sin 50 - 180 * \sin 45 + 0 + 225 + 300 * \sin 40$$
$$= 85 - 168.53 - 127.28 + 225 + 192.84$$
$$= 207.03 \, N \, (\uparrow)$$

$$R = \sqrt{\left(\sum F_x\right)^2 + \left(\sum F_y\right)^2}$$
$$= \sqrt{(268.45)^2 + (207.03)^2}$$
$$= 339.01 N (\nwarrow)$$

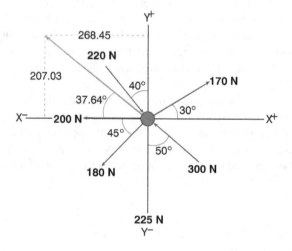

$$\theta = \tan^{-1} \frac{\sum F_y}{\sum F_x} = \frac{207.03}{268.45}$$
$$= 37.64°$$

Answer: 339.01 N, 37.64°

Example 2

Determine the magnitude and direction of the resultant force of the system shown at Figure 1.3.

Solution

$$\rightarrow + \sum F_x = 350 \times \frac{4}{5} - 180 \times \frac{3}{\sqrt{34}} = 187.4 \, N \, (\rightarrow)$$
$$\uparrow + \sum F_y = 350 \times \frac{3}{5} + 180 \times \frac{5}{\sqrt{34}} = 364.3 \, N \, (\uparrow)$$

$$R = \sqrt{\left(\sum F_x\right)^2 + \left(\sum F_y\right)}$$
$$= \sqrt{(187.4)^2 + (364.3)^2}$$
$$= 409.6\,N\,(\nearrow)$$

Answer: 409.6 N, 62.78°

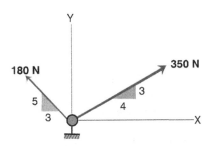

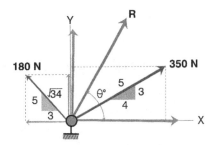

1.3   RESULTANT OF COPLANAR FORCES

The resultant of coplanar forces is the vector sum of all the forces acting on an object in a specific plane. In other words, it is the single force that can replace all the other forces and produce the same effect on the object's motion.

To find the resultant of coplanar forces, one must first resolve each force into its horizontal and vertical components. Then, the horizontal components are added together and the vertical components are added together to find the horizontal and vertical components of the resultant force. Finally, the horizontal and vertical components are combined using the Pythagorean theorem to find the magnitude and direction of the resultant force.

Alternatively, you can use the parallelogram method to find the resultant of coplanar forces graphically. The parallelogram method involves drawing the forces as vectors and connecting the tail of one force vector to the head of the next, forming a parallelogram. The diagonal of the parallelogram from the tail of the first force vector to the head of the last force vector is the resultant force.

It is worth noting that the concept of resultant force only makes sense when there are more than one force acting on an object in a specific plane.

- **Parallel Forces**

    The resultant of parallel coplanar forces is the single force that can replace all the other parallel forces and produce the same effect on the object's motion.

    Parallel coplanar forces are forces that act on an object in the same direction and in the same plane but do not share a common point of application. In other words, they are forces that have the same direction but different points of application. Since they are parallel, they have the same direction, so the magnitude of the net force is the sum of the magnitudes of the individual forces.

To find the magnitude of the net force, one simply adds the magnitudes of the individual forces. The direction of the net force will be the same as the direction of the individual forces since they are parallel. It is worth noting that for parallel forces, no matter how many forces are acting on an object, their net force can be found by simply adding their magnitudes.

Example 3 (Parallel Forces)

Determine the magnitude of the resultant force and location of it from A at Figure 1.4.

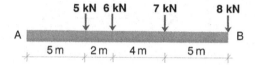

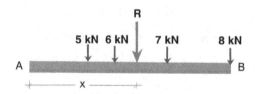

Solution

Add horizontal components.

$$\rightarrow + \sum F_x = 0 + 0 + 0 + 0 = 0$$

Add vertical components.

$$\uparrow + \sum F_y = -5 - 6 - 7 - 8 = -26\,\text{kN}\,(\downarrow)$$

Find resultant.

$$R = \sqrt{\left(\sum F_x\right)^2 + \left(\sum F_y\right)^2}$$
$$= \sqrt{(0)^2 + (-26)^2}$$
$$= +26\,\text{kN}\,(\downarrow)$$

Find location of resultant.

Applying **Varignon's theorem** at point A, moment of resultant force at A = sum of moments of all components at A.

$$26 \times x = 5 \times 5 + 6 \times 7 + 7 \times 11 + 8 \times 16$$
$$\Rightarrow x = 10.48\,\text{m}$$

Answer: 26 kN, 10.48 m

- **Non-Parallel Forces**

    The resultant of non-parallel coplanar forces is the single force that can replace all the other non-parallel forces and produce the same effect on the object's motion.

    Non-parallel coplanar forces are forces that act on an object in the same plane but not in the same direction. Since they are not parallel, they do not have the same direction, so the magnitude and direction of the net force cannot be found by simply adding the magnitudes of the individual forces.

    To find the magnitude and direction of the net force, one must use vector addition to find the vector sum of all the forces. This can be done either algebraically, by resolving each force into its horizontal and vertical components and then using the Pythagorean theorem to find the magnitude and direction of the net force, or graphically, by using the head-to-tail method or the polygon method.

    The head-to-tail method involves drawing the forces as vectors and connecting the head of one force vector to the tail of the next, forming a polygon. The diagonal of the polygon from the tail of the first force vector to the head of the last force vector is the net force, also called the resultant force.

    It is worth noting that the concept of resultant force only makes sense when there are more than one force acting on an object in a specific plane.

Example 4 (Non-Parallel Forces)

Determine the magnitude, angle of the resultant force, and location of it from A at Figure 1.5.

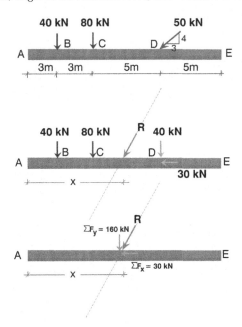

Solution

Add horizontal components.

$$\rightarrow + \sum F_x = 0 + 0 - 30 = -30\,\text{kN}\,(\leftarrow)$$

Adding vertical components.

$$\uparrow + \sum F_y = -40 - 80 - 40 = -160\ \text{kN} (\downarrow)$$

Find resultant.

$$R = \sqrt{\left(\sum F_x\right) + \left(\sum F_y\right)}$$
$$= \sqrt{(-30)^2 + (-160)^2}$$
$$= 162.8\ \text{kN}$$

Find angle of resultant.

$$\theta = \tan^{-1}\frac{\sum F_y}{\sum F_x} = \tan^{-1}\frac{160}{30}$$
$$= 79.4°$$

Find location of resultant by applying **Varignon's theorem**.
Moment of resultant force at A = sum of moments of all components at A.

$$160 \times x = 40 \times 3 + 80 \times 6 + 40 \times 11$$
$$\Rightarrow x = 6.5\ \text{m}$$

Answer: $162.8\ \text{kN}, 79°, 10.48\ \text{m from A}$

Example 5 (Non-Parallel Forces)

Determine the magnitude, angle of the resultant force, and location of it from A at Figure 1.6.

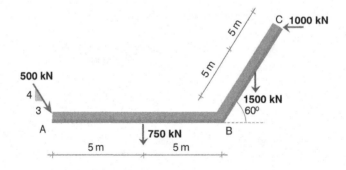

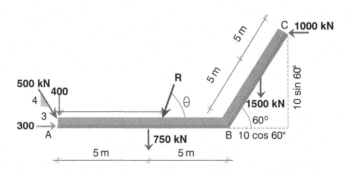

Solution

Add horizontal components.

$$\rightarrow + \sum F_x = 300 - 1000 = -700 \, kN \, (\leftarrow)$$

Add vertical components.

$$\uparrow + \sum F_y = -400 - 750 - 1500 = -2650 \, kN \, (\downarrow)$$

Find resultant.

$$R = \sqrt{\left(\sum F_x\right)^2 + \left(\sum F_y\right)^2}$$
$$= \sqrt{(-700)^2 + (-2650)^2}$$
$$= 2740.9 \, kN$$

Find angle of resultant.

$$\theta = \tan^{-1} \frac{\sum F_y}{\sum F_x} = \frac{2650}{700}$$
$$= 75.2°$$

Find location of resultant by applying **Varignon's theorem**.
   Moment of resultant force at A = sum of moments of all components at A.

$$2650 \times x = 750 \times 5 + 1500 * (5 + 5 + 5\cos 60) - 1000 \times 10 \, sn60$$
$$\Rightarrow x = 5.22 \, m$$

Answer: 2740.9 kN, 75.2°, 5.22 m from A

Example 6 (Non-Parallel Forces)

Determine the magnitude, angle of the resultant force, and location of it from O at Figure 1.7.

Solution

Add horizontal components.

$$\rightarrow + \sum F_x = -200\sin 30 - 400 + 100\cos 30$$
$$= 413.4 \, kN \, (\leftarrow)$$

Add vertical components.

$$\uparrow + \sum F_y = -200\cos 30 - 500 + 100\sin 30$$
$$= -623.2 \, kN \, (\downarrow)$$

Find resultant.

$$R = \sqrt{\left(\sum F_x\right)^2 + \left(\sum F_y\right)^2}$$
$$= \sqrt{(413.4)^2 + (-623.2)^2}$$
$$= 747.8 \, kN$$

Find angle of resultant.

$$\theta = \tan^{-1}\frac{\sum F_y}{\sum F_x} = \frac{623.2}{413.4}$$
$$= 56.4°$$

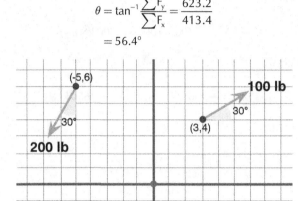

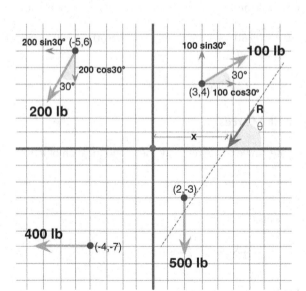

Find location of resultant by applying **Varignon's theorem**.
    Moment of resultant force at O = sum of moments of all components at O.

$$623.2 \times x = (100\cos 30)*4 - (100\sin 30)*3 - (200\cos 30)*5$$
$$- (200\sin 30)*6 + (400 \times 7) + (500 \times 2)$$
$$\Rightarrow x = 4.06\,m$$

Answer: 747.8 kN, 35.5°, 4.06 m from O

1.4   EQUILIBRIUM OF CONCURRENT FORCES

Equilibrium of concurrent forces refers to a state in which the net force acting on an object is zero. In other words, the object is not accelerating. This can happen when the forces acting on the object are balanced, meaning, that the force in one direction is equal in magnitude but opposite in direction to the force in the other direction.

For example, if a 10 N force is acting to the left and a 10 N force is acting to the right, the net force acting on the object is zero and the object is in a state of equilibrium.

It is also possible for an object to be in equilibrium when forces that are not concurrent are acting on it, as long as the net force is zero, which is called "conditions of equilibrium."

The conditions of equilibrium are as follows:

- The vector sum of all the forces acting on an object must be zero.
- The vector sum of all the torques (moments) acting on an object must be zero.

If an object is in equilibrium, it is not necessarily at rest. It can be in motion at a constant velocity, which is known as dynamic equilibrium.

It is also important to note that equilibrium is a condition, not a state; an object may be in equilibrium at one instant but not at the next if an external force is applied to it.

Example 7

Determine the surface reactions of the cylindrical drum which weighs 100 N at Figure 1.8.

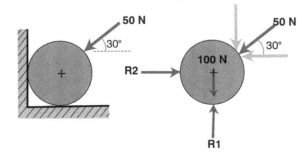

Solution

$$\sum F_x = 0 \rightarrow +$$
$$R_2 - 50\cos30 = 0$$
$$R_2 = 43.3\,N(\rightarrow)$$
$$\sum F_y = 0 \uparrow +$$
$$R_1 - 100 - 50\sin30 = 0$$
$$R_1 = 125\,N\,(\uparrow)$$

Answer: 43.3 N, 125 N

Example 8

Determine the surface reactions of the cylindrical drum which weighs 100 N at Figure 1.9.

Solution

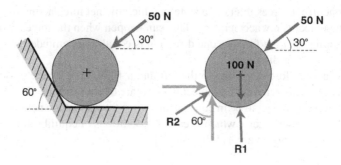

$$\sum F_x = 0 \rightarrow +$$
$$R_2\sin60 - 50\cos30 = 0$$
$$R_2 = 50\,N\,(\nearrow)$$
$$\sum F_y = 0\uparrow +$$
$$R_2\cos60 + R_1 - 100 - 50\sin30 = 0$$
$$\Rightarrow 50\cos60 + R_1 - 100 - 50\sin30 = 0$$
$$R_1 = 100\,N\,(\uparrow)$$

Answer: 50 N, 100 N

Example 9

Determine the surface reactions of the cylindrical drum which weighs 100 N at Figure 1.10.

Solution

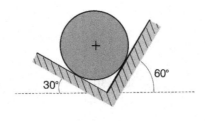

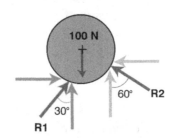

$$\sum F_x = 0 \rightarrow +$$
$$R_2\sin30 - R_1\sin60 = 0$$
$$\sum F_y = 0 \uparrow +$$
$$R_2\cos60 + R_1\cos60 - 100 = 0$$

Simultaneously solving the equations, we find $R_2 = 86.6$ N
   and $R_1 = 50$ N.

Answer: 86.6 N, 50 N

Example 10

Determine the surface reactions of the cylindrical drums at Figure 1.11. Each drum weighs 50 N.

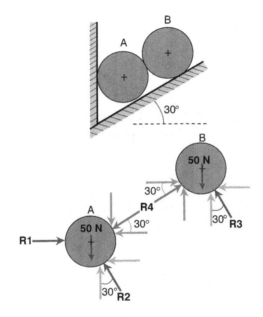

Solution

Free body of cylinder B.

$$\sum F_x = 0 \rightarrow +$$
$$R_4\cos30 - R_3\sin30 = 0$$
$$\sum F_y = 0 \uparrow +$$
$$R_4\sin60 + R_3\cos30 - 50 = 0$$

Simultaneously solving the equations, we find $R_3 = 43.3$ N and $R_4 = 25$ N.

Free body of cylinder A.

$$\sum F_y = 0 \uparrow +$$
$$-50 + R_2\cos30 - R_4\sin30 = 0$$

$$\Rightarrow -50 + R_2\cos30 - 25\sin30 = 0$$
$$R_2 = 72.2\,\text{N}$$
$$\sum F_x = 0 \rightarrow +$$
$$R_1 - R_2\sin30 - R_4\cos30 = 0$$
$$\Rightarrow R_1 - 72.2\sin30 - 25\cos30 = 0$$
$$R_1 = 57.7\,\text{N}$$

Answer: $R_1 = 57.7\,\text{N}, R_2 = 72.2\,\text{N}, R_3 = 43.3\,\text{N},$ and $R_4 = 25\,\text{N}$

Example 11

Determine the surface reaction under block *A*, which weighs 500 N; also determine tension in the cable at Figure 1.12.

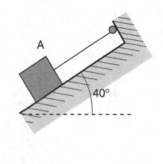

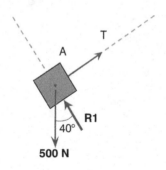

Solution

Free body of block A.

$$\sum F_x = 0 \rightarrow +$$
$$T - 500\sin40 = 0$$
$$\Rightarrow T = 321.4\,\text{N}$$
$$\sum F_y = 0 \uparrow +$$
$$R_1 - 500\cos40 = 0$$
$$\Rightarrow R_1 = 383\,\text{N}$$

Answer: $T = 321.4\,\text{N},$ and $R_1 = 383\,\text{N}$

Example 12

Two blocks, A and B, resting on two inclined planes, are connected together by a cable over a frictionless pulley.

Determine (1) the angle θ required to keep both blocks in equilibrium (2) surface reactions under both blocks and (3) tension in the cable, given that block A weighs 500 N and block B weighs 400 N at Figure 1.13.

Solution

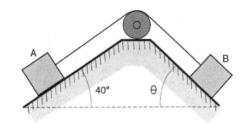

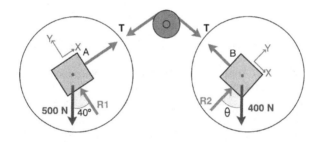

Free body of block A.

$$\sum F_x = 0 \rightarrow +$$
$$T - 500\sin40 = 0$$
$$\Rightarrow T = 321.4\,N$$
$$\sum F_y = 0\uparrow +$$
$$R_1 - 500\cos40 = 0$$
$$\Rightarrow R_1 = 383\,N$$

Free body of block B.

$$\sum F_x = 0 \rightarrow +$$
$$\Rightarrow -T + 400\,\sin\theta = 0$$
$$\Rightarrow -321.4 + 400\,\sin\theta = 0$$
$$\Rightarrow \theta = 53.46°$$

$$\sum F_y = 0\uparrow +$$
$$R_2 - 400\cos\theta = 0$$
$$\Rightarrow R_2 - 400\cos53.46 = 0$$
$$\Rightarrow R_2 = 238.2\,N$$

Answer: T = 321.4 N, R_1 = 383 N and R_2 = 238.2 N

1.5   EQUILIBRIUM OF COPLANAR FORCES

The equilibrium of coplanar forces refers to a state in which the net force acting on an object in a specific plane is zero. In other words, the object is not accelerating in that plane. This can happen when the forces acting on the object in that plane are balanced, meaning, that the force in one direction is equal in magnitude but opposite in direction to the force in the other direction.

The concept of coplanar forces is closely related to the concept of "resultant force", which is the vector sum of all the forces acting on an object in a specific plane. When the net force acting on an object is zero, the object is in equilibrium.

For example, if two forces of 10 N and 15 N act on an object in the same direction, the net force acting on the object is $25\,\mathrm{N}\,(10\,\mathrm{N}+15\,\mathrm{N})$, and the object is not in equilibrium. But if two forces of 10 N and −15 N act on an object in the same direction, the net force acting on the object is $-5\,\mathrm{N}\,(10\,\mathrm{N}+-15\,\mathrm{N})$, which is zero, and the object is in equilibrium.

It is also important to note that just like equilibrium of concurrent forces, equilibrium of coplanar forces is also a condition, not a state; an object may be in equilibrium at one instant but not at the next if an external force is applied to it in that plane.

Equilibrium of coplanar forces is closely related to the concept of "conditions of equilibrium", which states that for an object to be in equilibrium, the vector sum of all the forces acting on it must be zero and the vector sum of all the torques (moments) acting on it must be zero.

Example 13

Determine reactions at A and C support at Figure 1.14. Also determine cable tension in BC, given that the member AB weighs 20 kN and the length of AB is 5 m.

Solution

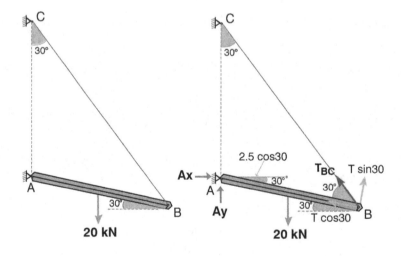

Free body of member AB.

$$\sum M_A = 0 + \circlearrowleft$$
$$20*(2.5\cos30) - (T_{BC}\sin30)*5 = 0$$
$$\Rightarrow T_{BC} = 17.32\,N$$

Free body of pin C (Figure 1.14).

$$\sum F_x = 0 \rightarrow +$$
$$C_x + T_{BC}\sin30 = 0$$
$$\Rightarrow C_x + 17.32\sin30 = 0$$
$$\Rightarrow C_x = -8.66\,N\,(\leftarrow)$$
$$\sum F_y = 0 \uparrow +$$
$$C_y - T_{BC}\cos30 = 0$$
$$\Rightarrow C_y - 17.32\cos30 = 0$$
$$\Rightarrow C_y = 15\,N\,(\uparrow)$$

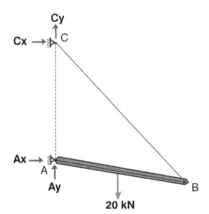

Free body of the whole system.

$$\sum F_x = 0 \rightarrow +$$
$$A_x + C_x = 0$$
$$\Rightarrow A_x + (-8.66) = 0$$
$$\Rightarrow A_x = 8.66\,N\,(\rightarrow)$$
$$\sum F_y = 0 \uparrow +$$
$$A_y + C_y - 20 = 0$$
$$\Rightarrow A_y + 15 - 20 = 0$$
$$\Rightarrow A_y = 5\,N\,(\uparrow)$$

Answer: $A_x = 8.66\,N, A_y = 5\,N, C_x = -8.66\,N, C_y = 15\,N$ and $T_{BC} = 17.32\,N$

Example 14

A cylindrical drum *DE* is supported by a mast *AB*, cable *BC*, and wall *ACD* at Figure 1.15. The drum weighs 300 N and has 1000 mm of radius. Determine support reactions at *A* and *C*, surface reaction at *D*, and tension in cable *BC*, given that the member *AB* has negligible weight, and its length is 4 m.

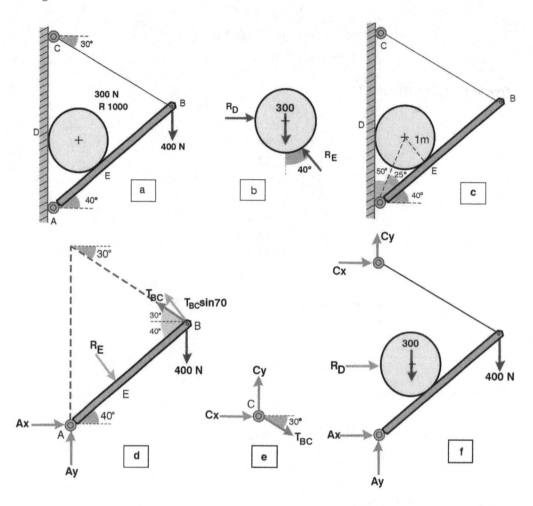

Free body of cylinder DE (Figure 1.15b).

$$\sum F_y = 0 \uparrow +$$
$$R_E\cos40 - 300 = 0$$
$$\Rightarrow R_E = 391.62 \,(\nwarrow)$$
$$\sum F_x = 0 \rightarrow +$$
$$R_D - R_E\sin40 = 0$$
$$\Rightarrow R_D - 391.62\sin40 = 0$$
$$\Rightarrow R_D = 251.73\,N\,(\rightarrow)$$

Find length of AE (Figure 1.15c).

$$\tan25 = \frac{1}{AE}$$
$$\Rightarrow AE = \frac{1}{\tan25} = 2.144\,m$$

Free body of mast AB (Figure 1.15d).

$$\sum M_A = 0 + \circlearrowleft$$

$$R_E * AE + 400 * (AB\cos40) - (T_{BC}\sin70) * AB = 0$$
$$\Rightarrow 391.62 * (2.144) + 400 * (5\cos40) - (T_{BC}\sin70) * 5 = 0$$
$$\Rightarrow T_{BC} = 504.79\,N$$

Free body of pin C (Figure 1.15e).

$$\sum F_x = 0 \rightarrow +$$
$$C_X + T_{BC}\cos30 = 0$$
$$\Rightarrow C_X + 504.79\cos30 = 0$$
$$\Rightarrow C_X = -437.16\,N\,(\leftarrow)$$
$$\sum F_y = 0 \uparrow +$$
$$C_Y - T_{BC}\sin30 = 0$$
$$\Rightarrow C_Y - 504.79\sin30 = 0$$
$$\Rightarrow C_Y = 252.39\,N\,(\uparrow)$$

Free body of the whole system (Figure 1.15f).

$$\sum F_x = 0 \rightarrow +$$
$$A_X + R_D + C_X = 0$$
$$\Rightarrow A_X + 251.73 + (-437.16) = 0$$
$$\Rightarrow A_X = 185.43\,N\,(\rightarrow)$$
$$\sum F_y = 0 \uparrow +$$
$$A_Y + C_Y - 300 - 400 = 0$$
$$\Rightarrow A_Y + 252.39 - 300 - 400 = 0$$
$$\Rightarrow A_Y = 447.61N\,(\uparrow)$$

1.6 CABLE ANALYSIS

Cable analysis with point load on cable and distributed load on cable involves determining the stresses and deformations in a cable under the influence of external loads. The cable can be modeled as a slender, flexible beam that is supported at its ends and subjected to various types of loads.

1. Point Load on Cable
   A *point load* on a cable is a concentrated load that is applied at a specific point along the cable's length. The load can be represented as a single force acting on the cable.
   To analyze a cable with a point load, the following steps can be followed:
   - Draw the free-body diagram of the cable, showing all the forces acting on the cable, including the point load and any other external loads.
   - Determine the reaction forces at the cable supports using the equilibrium equations.
   - Determine the internal forces in the cable using the equilibrium equations and the compatibility equation, which relates the internal forces to the cable's deformation.
   - Determine the cable's deformation and stress using the equations of compatibility and elasticity, which relate the deformation and stress to the internal forces.

2. Distributed Load on Cable

A *distributed load* on a cable is a load that is spread over the cable's length. The load can be represented as a continuous force per unit length along the cable.

To analyze a cable with a distributed load, the following steps can be followed:

- Draw the free-body diagram of an element of the cable, showing all the forces acting on the element, including the distributed load and any other external loads.
- Determine the internal forces in the element using the equilibrium equations and the compatibility equation.
- Integrate the internal forces over the length of the cable to obtain the internal forces throughout the cable.
- Determine the cable's deformation and stress using the equations of compatibility and elasticity, which relate the deformation and stress to the internal forces.

In both cases, it is important to note the assumptions made during the analysis, such as neglecting cable mass and damping and ensuring that the cable material properties are accurately represented in the analysis.

Example 15 (Point Load on Cable)

Determine vertical distance of *B* from level of *A*, tension in cable *AB*, *BC*, *DE*, and support reactions at *A* and *E* at Figure 1.16.

Solution

Whole free-body diagram.

$$\sum M_E = 0 + \circlearrowleft$$
$$-A_x * 20 + A_Y * 60 - 6 * 40 - 12 * 30 - 4 * 15 = 0$$
$$\Rightarrow -20A_x + 60A_Y = 660$$

Free body diagram of ABC.

$$\sum M_C = 0 + \circlearrowleft$$
$$A_x * 5 + A_Y * 30 - 6 * 10 = 0$$
$$\Rightarrow 5A_x + 30A_Y = 60$$

Solving the equations simultaneously, $A_x = -18\,\text{kip}$ and $A_Y = 5\,\text{kip}$.

Free-body diagram of AB.

$$\sum M_B = 0 + \circlearrowleft$$
$$A_x * y_B + A_Y * 20 = 0$$
$$\Rightarrow -18 * y_B + 5 * 20 = 0$$
$$\Rightarrow y_B = 5.56\,\text{ft}$$

Whole free-body diagram.

$$\sum F_x = 0 \rightarrow +$$
$$A_x + E_X = 0$$
$$\Rightarrow -18 + E_X = 0$$
$$\Rightarrow E_X = 18\,\text{kip}$$
$$\sum F_y = 0 \uparrow +$$
$$A_Y - 6 - 12 - 4 + E_Y = 0$$
$$\Rightarrow 5 - 22 + E_Y = 0$$
$$\Rightarrow E_Y = 17\,\text{kip}$$

From geometry, we find:

$$\theta_A = \tan^{-1}\frac{5.56}{20} = 15.53°$$
$$\theta_B = \tan^{-1}\frac{5.56 - 5}{10} = 3.21°$$

Joint B:

$$\sum F_x = 0 \rightarrow +$$
$$\Rightarrow -T_{AB}\cos\theta_A + T_{BC}\cos\theta_B = 0$$
$$\sum F_y = 0 \uparrow +$$
$$\Rightarrow T_{AB}\sin\theta_A + T_{BC}\sin\theta_B - 6 = 0$$

Substituting θ_A and θ_B and then solving simultaneously, $T_{AB} = 18.7\,\text{kip}$ and $T_{BC} = 18.0\,\text{kip}$

Joint E:

$$\sum F_x = 0 \rightarrow +$$
$$\Rightarrow -T_{DE}\cos\theta_E + E_x = 0$$
$$\Rightarrow -T_{DE}\cos\theta_E = 18$$
$$\sum F_y = 0 \uparrow +$$
$$\Rightarrow -T_{DE}\sin\theta_E + E_Y = 0$$
$$\Rightarrow -T_{DE}\sin\theta_E = 17$$

Solving, we find:

$$T_{DE} = \sqrt{18^2 + 17^2} = 24.8\,\text{kip}$$

Example 16 (Distributed Load on Cable)

The cable *ABC* is parabolic in shape and carries a distributed load $w = 0.6$ k/ft at Figure 1.17. Determine tension at point *B*, which is the lowest point on the cable. Also, determine tension at *A* and *C* of the cable and reactions at *A* and *C* support.

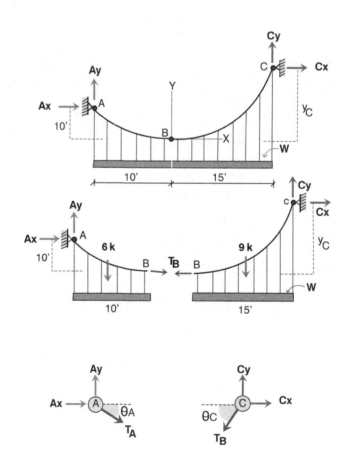

Solution

Determine y_c.

From geometric properties of parabola,

$$\frac{y_c}{y_a} = \frac{x_c^2}{x_a^2};$$

$$y_c = y_a \left(\frac{x_c}{x_a}\right)^2 = 10\left(\frac{15}{10}\right)^2 = 22.5\ ft$$

Free-body diagram of BC.

$$\sum M_c = 0 \circlearrowleft +$$

$$T_B \times y_c - 9 \times 7.5 = 0$$
$$T_B \times 22.5 - 9 \times 7.5 = 0$$
$$T_B = 3\ kip$$

$$\sum F_x = 0 \rightarrow^+$$

$$-T_B + C_x = 0$$
$$-3 + C_x = 0;$$
$$C_x = 3\ kip\,(\rightarrow)$$

$$\sum F_y = 0 \uparrow^+$$

$$-9 + C_y = 0;$$
$$C_y = 9\ kip\,(\uparrow)$$

Free-body diagram of AB.

$$\sum F_x = 0 \rightarrow^+$$

$$T_B + A_x = 0$$
$$3 + A_x = 0;$$
$$A_x = -3\ kip\,(\leftarrow)$$

$$\sum F_y = 0 \uparrow^+$$

$$-6 + A_y = 0;$$
$$A_y = 6\ kip\,(\uparrow)$$

Free-body diagram of joint A and joint C.

$$T_A = \sqrt{A_x^2 + A_y^2} = \sqrt{3^2 + 6^2} = 6.71\,\text{kip}$$
$$T_B = \sqrt{C_x^2 + C_y^2} = \sqrt{3^2 + 9^2} = 9.49\,\text{kip}$$

1.7   BEAM ANALYSIS

Example 17 (Determine Reaction of the Beam)

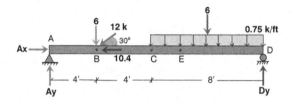

Solution

$$\sum M_D = 0 \circlearrowleft^+$$

$$A_y \times 16 - 6 \times 12 - 6 \times 4 = 0;$$
$$A_y = 6\,\text{kip}\,(\uparrow)$$

$$\sum F_y = 0 \uparrow^+$$

$$A_y - 6 - 6 + D_y = 0;$$
$$D_y = 6\,\text{kip}\,(\uparrow)$$

$$\sum F_x = 0 \rightarrow^+$$

$$A_x - 10.4 = 0$$
$$A_x = 10.4\,\text{kip}\,(\rightarrow)$$

Example 18 (Determine Reaction of the Beam)

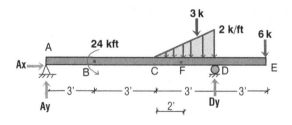

Solution

$$\Sigma M_D = 0 \circlearrowleft^+$$
$$A_y \times 9 - 24 - 3 \times 1 + 6 \times 3 = 0$$
$$A_y = 1 \, kip \, (\uparrow)$$

$$\Sigma F_y = 0 \uparrow^+$$
$$A_y - 3 + D_y - 6 = 0$$

$$D_y = 8 \, kip \, (\uparrow)$$
$$\Sigma F_x = 0 \rightarrow^+$$
$$A_x = 0;$$

Example 19 (Determine Reaction of the Beam)

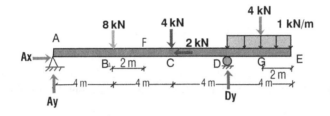

Solution

$$\Sigma M_D = 0 \circlearrowleft^+$$
$$A_y \times 12 - 8 \times 8 - 4 \times 4 + 4 \times 2 = 0;$$

$$A_y = 6 \, kip \, (\uparrow)$$
$$\Sigma F_y = 0 \uparrow^+$$

$$A_y - 8 - 4 + D_y - 4 = 0;$$
$$D_y = 10 \, kip \, (\uparrow)$$

$$\Sigma F_x = 0 \rightarrow^+ A_x - 2 = 0;$$
$$A_x = 2 \, kip \, (\rightarrow)$$

Example 20 (Determine Reaction of the Beam)

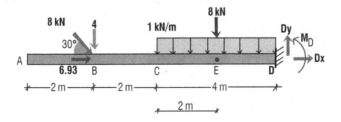

Solution

$$\sum M_D = 0 \circlearrowright^+$$
$$-4 \times 6 - 8 \times 2 - M_D = 0;$$
$$M_D = -40 \text{ kip} (\circlearrowright)$$

$$\sum F_y = 0 \uparrow^+$$
$$-4 - 8 + D_y = 0$$
$$D_y = 12 \text{ kip} (\uparrow)$$

$$\sum F_x = 0 \rightarrow^+$$
$$6.93 + D_x = 0;$$

$$D_x = -6.93 \text{ kip} (\leftarrow)$$

1.8   CENTER OF GRAVITY (GEOMETRIC SHAPE)

The *center of gravity* is the point in a body or object where the weight of the body is considered to be concentrated. In other words, it is the point where the object can be balanced perfectly.

The center of gravity of geometric shapes can be calculated using the following formulas:

1. *Center of gravity of a point.* The center of gravity of a single point is simply the location of that point.
2. *Center of gravity of a line.* The center of gravity of a straight line is located at its midpoint.
3. *Center of gravity of a triangle.* The center of gravity of a triangle is located at the intersection of its medians, which are the lines drawn from each vertex to the midpoint of the opposite side. The center of gravity is one-third of the distance from each vertex to the opposite side.
4. *Center of gravity of a rectangle.* The center of gravity of a rectangle is located at its geometric center, which is the point where the two diagonals intersect. The geometric center is halfway between the midpoint of each side.
5. *Center of gravity of a circle.* The center of gravity of a circle is located at its geometric center, which is also the center of the circle.
6. *Center of gravity of a semicircle.* The center of gravity of a semicircle is located along its axis of symmetry, which is the line passing through the center of the circle and perpendicular to its diameter. The center of gravity is located at a distance of $4r/3\pi$ from the base of the semicircle, where r is the radius of the circle.

These formulas can be used to determine the center of gravity of more complex shapes by breaking them down into simpler geometric shapes and calculating the center of gravity of each part. The overall center of gravity can then be determined using the weighted average method.

The center of gravity of an area is the point at which the entire area can be balanced. It is also called the *centroid* or *geometric center* of the area. The center of gravity of an area can be calculated using the following formula:

$$x = \frac{\{\Sigma(xi * Ai)\}}{A}$$
$$y = \frac{\{\Sigma(yi * Ai)\}}{A}$$

where x and y are the coordinates of the center of gravity along the x- and y-axes, respectively; xi and yi are the coordinates of each element of the area; Ai is the area of each element; and A is the total area of the object.

This formula can be used to find the center of gravity of any 2D shape, regardless of its complexity. The shape can be divided into smaller elements, such as triangles or rectangles, and the coordinates and areas of each element can be calculated. The overall center of gravity can then be calculated using the weighted average method, where the location of the center of gravity is calculated based on the area and position of each individual element.

Example 21

Determine the location of the centroid of the triangular shape of width b and height h.

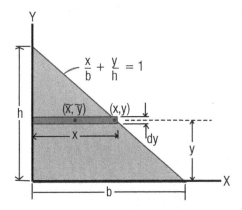

Solution

1. From equation of hypotenuse of the triangle, we find:

$$\frac{x}{b}+\frac{y}{h}=1;$$

$$x=b\left(1-\frac{y}{h}\right)$$

2. Consider a differential element, gray rectangle, of width x and height dy. The area of the differential element (dA) and coordinate of its centroid $(\tilde{x};\tilde{y})$ can be expressed as:

$$dA=xdy;$$

$$\tilde{x}=\frac{x}{2};$$

$$\tilde{y}=y$$

3. Now, determine $\int dA$.

$$\Sigma dA=\int_0^h xdy$$

$$=\int_0^h b\left(1-\frac{y}{h}\right)dy$$

$$=b\int_0^h\left(1-\frac{y}{h}\right)dy$$

$$= b \left[y - \frac{y^2}{2h} \right]_0^h$$

$$= b \left(h - \frac{h^2}{2h} - 0 + 0 \right)$$

$$= b \left(h - \frac{h}{2} \right) = \frac{bh}{2}$$

4. Then, determine $\int \tilde{y} dA$.

$$\int \tilde{y} dA = \int_0^h \tilde{y} (x dy)$$

$$= \int_0^h y \left[b \left(1 - \frac{y}{h} \right) dy \right]$$

$$= b \int_0^h \left(y - \frac{y^2}{h} \right) dy$$

$$= b \left[\frac{y^2}{2} - \frac{y^3}{3h} \right]_0^h$$

$$= b \left(\frac{h^2}{2} - \frac{h^3}{3h} - 0 + 0 \right)$$

$$= b \left(\frac{h^2}{2} - \frac{h^2}{3} \right) = \frac{bh^2}{6}$$

5. From the formula of a centroid, we know:

$$\bar{y} = \frac{\int \tilde{y} dA}{\int dA} = \frac{bh^2 / 6}{bh / 2} = \frac{h}{3}$$

6. Similarly, it can be proved that:

$$\bar{x} = \frac{b}{3}$$

Answer: $\bar{x} = \dfrac{b}{3}$, and $\bar{y} = \dfrac{h}{3}$

Example 22

Determine the location of the centroid of the circular quadrant or radius R.

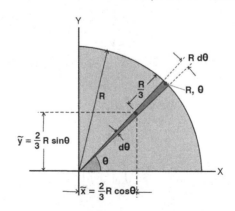

Solution

1. Any point on the perimeter of the circle can be expressed as (R, θ) in polar coordinate system.
2. Consider a differential element, gray triangle, of height R and base $R\, d\theta$. The centroid of this triangle is 2R/3 distance away from origin. Now:

$$dA = \frac{1}{2}(R)(Rd\theta);$$

$$\tilde{x} = \frac{2}{3}R\cos\theta;$$

$$\tilde{y} = \frac{2}{3}R\sin\theta$$

3. Now, determine $\int dA$.

$$\int dA = \int_0^{\pi/2} \frac{1}{2}R^2 d\theta$$

$$= \frac{R^2}{2}\int_0^{\pi/2} d\theta$$

$$= \frac{R^2}{2}[\theta]_0^{\pi/2}$$

$$= \frac{R^2}{2}\left(\frac{\pi}{2} - 0\right)$$

$$= \frac{\pi R^2}{4}$$

4. Then, determine $\int \tilde{y} dA$.

$$\int \tilde{x} dA = \int_0^{\pi/2}\left(\frac{2}{3}R\cos\theta\right)\left(\frac{1}{2}R^2 d\theta\right)$$

$$= \frac{R^3}{3}\int_0^{\pi/2}\cos\theta d\theta$$

$$= \frac{R^3}{3}[\sin\theta]_0^{\pi/2}$$

$$= \frac{R^3}{3}\left(\sin\frac{\pi}{2} - \sin 0\right)$$

$$= \frac{R^3}{3}(1-0) = \frac{R^3}{3}$$

5. From the formula of a centroid, we know:

$$\overline{x} = \frac{\int \tilde{x} dA}{\int dA} = \frac{R^3/3}{\pi R^2/4} = \frac{4R}{3\pi}$$

6. Similarly, it can be proved that:

$$\overline{y} = \frac{4R}{3\pi}$$

Answer: $\overline{x} = \dfrac{4R}{3\pi}$, *and* $\overline{y} = \dfrac{4R}{3\pi}$

1.9 CENTER OF GRAVITY (COMPOSITE SHAPE)

The *center of gravity* (CG), also known as the *center of mass*, of a composite shape is the point at which the entire weight of the object is considered to be concentrated. For a composite shape, the CG can be determined by finding the weighted average of the CGs of its individual components.

To find the CG of a composite shape, follow these steps:

1. Break the composite shape down into its individual components.
2. Determine the CG of each individual component.
3. Calculate the weight of each individual component using its density and volume or mass and acceleration due to gravity.
4. Multiply the weight of each component by the distance between its CG and a reference point, such as the origin or any fixed point on the object.
5. Add up the products from step 4 for all components.
6. Divide the result from step 5 by the total weight of the composite shape.

The resulting value will give the coordinates of the CG of the composite shape. It is important to note that the CG of a composite shape may not be located within the physical boundaries of the object.

In addition to composite shapes, the CG can be determined for a variety of objects, including regular and irregular shapes, rigid bodies, and fluids. For simple regular shapes, such as a sphere or cube, the CG is located at the geometric center of the object. However, for irregular shapes or objects with non-uniform density, the CG can be more difficult to determine. Applications of the concept of CG include aircraft design, where engineers must ensure that the CG is located within a specific range to ensure safe and stable flight. Similarly, in automotive design, the CG of a vehicle must be carefully considered to ensure stability and handling.

Example 23

Determine the location of the centroid of the following *L* section.

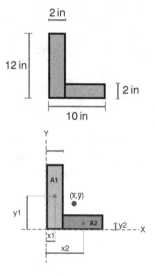

Solution

At first, divide the L section into two rectangular regions, A1 and A2. Consider the coordinate system as shown in the figure.

Determine areas.

$$A_1 = 12 \times 2 = 24 \text{ in}^2; \ A_2 = 8 \times 2 = 16 \text{ in}^2$$

Find x coordinate of the centroid of each area.

$$x_1 = \frac{2}{2} = 1 \text{ in}; \ x_2 = 2 + \frac{8}{2} = 6 \text{ in}$$

Find the y coordinate of the centroid of each area.

$$y_1 = \frac{12}{2} = 6 \text{ in}; \ y_2 = \frac{2}{2} = 1 \text{ in}$$

Determine the location of the centroid of the L section.

$$\bar{x} = \frac{\sum Ax}{\sum A} = \frac{A_1 x_1 + A_2 x_2}{A_1 + A_2} = \frac{(24 \times 1) + (16 \times 6)}{24 + 16} = 3 \text{ in}$$

$$\bar{y} = \frac{\sum Ay}{\sum A} = \frac{A_1 y_1 + A_2 y_2}{A_1 + A_2} = \frac{(24 \times 6) + (16 \times 1)}{24 + 16} = 4 \text{ in}$$

Answer: $\bar{x} = 3$ in, and $\bar{y} = 4$ in

Example 24

Determine the location of the centroid of the circular quadrant or radius R.

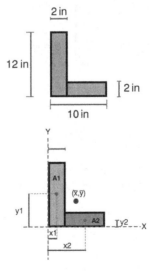

Solution

segment	$A(\text{in}^2)$	$x(\text{in})$	$Ax(\text{in}^3)$	$y(\text{in})$	$Ay(\text{in}^3)$
1	$12 \times 2 = 24$	$2/2 = 1$	24	$12/2 = 6$	144
2	$8 \times 2 = 16$	$2 + 8/2 = 6$	96	$2/2 = 1$	16
	$\Sigma A = 40$		$\Sigma Ax = 120$		$\Sigma Ay = 160$

$$\bar{x} = \frac{\Sigma Ax}{\Sigma A} = \frac{120}{40} = 3 \text{ in}$$

$$\bar{y} = \frac{\Sigma Ay}{\Sigma A} = \frac{160}{40} = 4 \text{ in}$$

Answer: $\bar{x} = 3$ *in, and* $\bar{y} = 4$ *in*

Example 25

Determine the location of the centroid of the following T section.

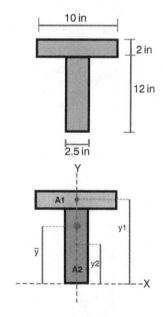

Solution

At first, divide T section into two rectangular regions, A_1 and A_2. Consider the coordinate system as shown in the figure. Due to the symmetry about the y-axis, $\bar{x} = 0$. So we need to determine only $\bar{y}$.

Determine areas.

$$A_1 = 10 \times 2 = 20 \text{ in}^2; \ A_2 = 12 \times 2.5 = 30 \text{ in}^2$$

Find the y coordinate of the centroid of each area.

$$y_1 = 12 + \frac{2}{2} = 13 \text{ in}; \ y_2 = \frac{12}{2} = 6 \text{ in}$$

Determine the location of the centroid of the T section.

$$\bar{y} = \frac{\sum Ay}{\sum A} = \frac{A_1 y_1 + A_2 y_2}{A_1 + A_2}$$

$$= \frac{(20 \times 13) + (30 \times 6)}{20 + 30} = 8.8 \text{ in}$$

Answer: $\bar{x} = 0$ in, and $\bar{y} = 8.8$ in

Example 26

Determine the location of the centroid of the following shape.

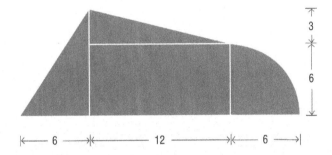

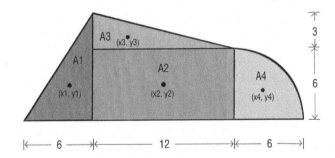

Solution

Determine the area A of each subdivision.

$$A_1 = \frac{1}{2} \times 6 \times 9 = 27 \text{ in}^2$$

$$A_2 = 12 \times 6 = 72 \text{ in}^2$$

$$A_3 = \frac{1}{2} \times 12 \times 3 = 18 \text{ in}^2$$

$$A_4 = \frac{\pi \times 6^2}{4} = 28.27 \text{ in}^2$$

Find the x coordinate of the centroid of each subdivision.

$$x_1 = \frac{2}{3} \times 6 = 4 \text{ in}$$

$$x_2 = 6 + \frac{12}{2} = 12 \text{ in}$$

$$x_3 = 6 + \frac{1}{3} \times 12 = 10 \text{ in}$$

$$x_4 = 6 + 12 + \frac{4 \times 6}{3\pi} = 20.54 \text{ in}$$

Find the y coordinate of the centroid of each subdivision.

$$y_1 = \frac{1}{3} \times 9 = 3 \text{ in}$$

$$y_2 = \frac{6}{2} = 3 \text{ in}$$

$$y_3 = 6 + \frac{1}{3} \times 3 = 7 \text{ in}$$

$$y_4 = \frac{4 \times 6}{3\pi} = 2.54 \text{ in}$$

Determine the $\bar{x}$ and $\bar{y}$ of the shape.

$$\bar{x} = \frac{\sum Ax}{\sum A}$$
$$= \frac{(27 \times 4) + (72 \times 12) + (18 \times 10) + (28.27 \times 20.54)}{27 + 72 + 18 + 28.27} = 11.93 \text{ in}$$

$$\bar{y} = \frac{\sum Ay}{\sum A}$$
$$= \frac{(27 \times 3) + (72 \times 3) + (18 \times 7) + (28.27 \times 2.54)}{27 + 72 + 18 + 28.27} = 3.41 \text{ in}$$

Answer: $\bar{x} = 11.93 \text{ in}$, and $\bar{y} = 3.41 \text{ in}$

Example 27

Determine the location of the centroid of the following shape.

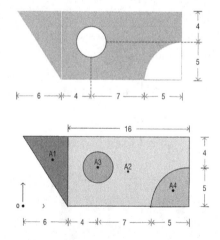

Solution

Determine the area A of each subdivision.

$$A_1 = \frac{1}{2} \times 6 \times 9 = 27 \text{ in}^2$$
$$A_2 = 16 \times 9 = 144 \text{ in}^2$$
$$A_3 = \pi \times 2^2 = 12.56 \text{ in}^2$$
$$A_4 = \frac{\pi \times 5^2}{4} = 19.63 \text{ in}^2$$

Find the x coordinate of the centroid of each subdivision.

$$x_1 = \frac{2}{3} \times 6 = 4 \text{ in}$$
$$x_2 = 6 + \frac{16}{2} = 14 \text{ in}$$
$$x_3 = 6 + 4 = 10 \text{ in}$$
$$x_4 = 6 + 16 + \frac{4 \times 5}{3\pi} = 19.87 \text{ in}$$

Find the y coordinate of the centroid of each subdivision.

$$y_1 = \frac{2}{3} \times 9 = 6 \text{ in}$$
$$y_2 = \frac{9}{2} = 4.5 \text{ in}$$
$$y_3 = 5 \text{ in}$$
$$y_4 = \frac{4 \times 5}{3\pi} = 2.12 \text{ in}$$

Determine the $\bar{x}$ and $\bar{y}$ of the shape.

$$\bar{x} = \frac{\Sigma Ax}{\Sigma A}$$
$$= \frac{(27 \times 4) + (144 \times 14) - (12.56 \times 10) - (19.63 \times 19.87)}{27 + 144 - 12.56 - 19.63} = 11.58 \text{ in}$$
$$\bar{y} = \frac{\Sigma Ay}{\Sigma A}$$
$$= \frac{(27 \times 6) + (144 \times 4.5) + (12.56 \times 5) + (19.63 \times 2.12)}{27 + 144 - 12.56 - 19.63} = 5.08 \text{ in}$$

Answer: $\bar{x} = 11.58$ *in, and* $\bar{y} = 5.08$ *in*

1.10 MOMENT OF INERTIA (COMPOSITE SHAPE)

The moment of inertia of a composite shape can be calculated by adding the moments of inertia of its individual parts.

The moment of inertia of an object describes its resistance to rotational motion. It is the sum of the products of the mass of each particle in the object and the square of its distance from the axis of rotation.

For a composite shape, the moment of inertia can be calculated by dividing the object into simpler shapes and summing their individual moments of inertia. This can be expressed mathematically as:

$$I\_total = \Sigma I\_i$$

where I_total is the moment of inertia of the composite shape, and ΣI_i is the sum of the moments of inertia of its individual parts.

For example, consider a rectangular plate with a circular hole in the center. The moment of inertia of the plate can be calculated by dividing it into a rectangle and a circle and summing their individual moments of inertia. The moment of inertia of a rectangular plate of width w and height h about an axis perpendicular to its face and passing through its center is given by:

$$I\_rectangle = (1/12) * m * (w^2 + h^2)$$

where m is the mass of the rectangle.

The moment of inertia of a circular hole of radius r about an axis passing through its center is given by:

$$I\_circle = (1/2) * m * r^2$$

where m is the mass of the circle.

Therefore, the moment of inertia of the composite shape can be calculated as:

$$I\_total = I\_rectangle + I\_circle$$

Note that the axis of rotation must be specified when calculating the moment of inertia of a composite shape.

Example 28

Determine the centroidal moment of inertia of the rectangular section.

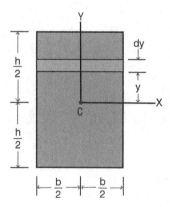

Solution

From the definition of *moment of inertia,* we know:

$$T_x = \int y^2 dA$$

$$= \int_{-h/2}^{+h/2} y^2 (bdy)$$

$$= b \int_{-h/2}^{+h/2} y^2 dy$$

$$= b \left[\frac{y^3}{3} \right]_{-h/2}^{+h/2}$$

$$= \frac{b}{3} \left[\frac{h^3}{8} - \frac{-h^3}{8} \right]$$

$$= \frac{bh^3}{12}$$

Similarly, it can be shown that:

$$\bar{I}_y = \frac{hb^3}{12}$$

Example 29

Determine the centroidal moment of inertia of the angle section.

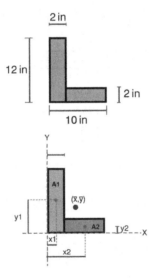

Solution

It has been solved earlier that:

$$\bar{x} = 3\ in,\ and\ \bar{y} = 4\ in$$

Now:

$$T_x = \Sigma(T_x + Ad^2)$$

$$= \left[\frac{2 \times 12^3}{12} + (12 \times 2) \times (6-4)^2\right] + \left[\frac{8 \times 2^3}{12} + (8 \times 2) \times (4-1)^2\right]$$

$$= 533 \text{ in}^4$$

$$T_y = \Sigma(T_y + Ad^2)$$

$$= \left[\frac{12 \times 2^3}{12} + (12 \times 2) \times (3-1)^2\right] + \left[\frac{2 \times 8^3}{12} + (8 \times 2) \times (6-3)^2\right]$$

$$= 333 \text{ in}^4$$

Example 30

Determine the centroidal moment of inertia of the angle section.

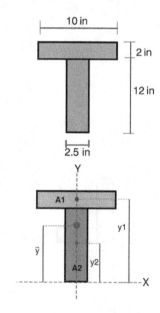

Solution

It has been solved earlier that:

$$\bar{x} = 0 \text{ in, and } \bar{y} = 8.8 \text{ in}$$

Now:

$$\bar{I}_x = \Sigma(\bar{I}_x + Ad^2)$$

$$= \left[\frac{10 \times 2^3}{12} + (10 \times 2) \times (13-8.8)^2\right] + \left[\frac{2.5 \times 12^3}{12} + (2.5 \times 12) \times (8.8-6)^2\right] = 955 \text{ in}^4$$

$$\bar{I}_y = \Sigma(\bar{I}_y + Ad^2)$$

$$= \left[\frac{2 \times 10^3}{12}\right] + \left[\frac{12 \times 2.5^3}{12}\right] = 182 \text{ in}^4$$

2 Stress and Strain

Topics to be covered in this chapter are as follows:

- Shear force and bending moment diagrams and implications in design of statically deter-minate beams and frames.
- Fundamental concepts of stress and strain.
- Mechanical properties of materials.
- Stresses and strains in members subject to tension, compression, shear and temperature changes.
- Joints welded and riveted.

2.1 REVIEW OF BEAM REACTIONS

Beam reactions are the forces and moments that occur at the supports of a beam in order to balance the external loads applied to the beam. The reactions can be calculated using the principles of stat-ics, which state that for a system to be in equilibrium, the sum of the forces and moments acting on it must be zero.

There are generally two types of beam supports: fixed supports and pinned supports. *Fixed sup-ports* are rigid connections that prevent both translation and rotation of the beam at the support, while *pinned supports* allow rotation but prevent translation.

To calculate the reactions at the supports of a beam, the following steps can be followed:

1. Draw a free-body diagram of the beam showing all the external loads applied to it, includ-ing point loads, distributed loads, and moments.
2. Identify the type of support at each end of the beam and the direction of the reactions (upward, downward, clockwise, or counterclockwise).
3. Apply the equations of equilibrium to the free-body diagram to calculate the unknown reactions. There are three equations of equilibrium: $\Sigma Fx = 0, \Sigma Fy = 0,$ and $\Sigma M = 0$ where ΣFx is the sum of the forces in the x-direction, ΣFx is the sum of the forces in the y-direction, and ΣM is the sum of the moments about a point.
4. Solve the equations simultaneously to find the values of the unknown reactions.

For example, consider a simply supported beam with a point load of 10 kN at its midpoint. The beam is 4 m long and has a uniform cross section of 0.1 m × 0.1 m. The reactions at the supports can be calculated as follows:

1. Draw a free-body diagram of the beam.

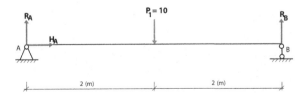

2. Since the beam is simply supported, there are two reactions: an upward force at the left support and a downward force at the right support.

DOI: 10.1201/9781032638072-2   37

3. Apply the equations of equilibrium to the free-body diagram:

$\Sigma Fx = 0 : H_A = 0$

$\Sigma Fy = 0 : R_A + R_B - 10 = 0$

$\Sigma M_A = 0$ (about point A, the left support): $R_B * 4 - 10 * 2 = 0$. Solve the equations simultaneously to find the values of the reactions:

RA = 5 kN (↑)

RB = 5 kN (↓)

4. Therefore, the reactions at the supports of the beam are 5 kN upward at the left support and 5 kN downward at the right support.

Example 1

Determine the reaction of the beam.

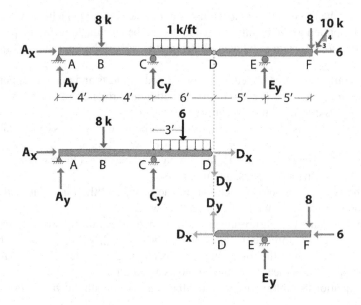

Solution

Segment *DF*.

$$\Sigma M_D = 0 \circlearrowright^+$$
$$8 \times 10 - E_y \times 5 = 0$$
$$E_y = 16 \; kip \; (\uparrow)$$
$$\Sigma F_y = 0 \uparrow^+$$
$$D_y + 16 - 8 = 0$$
$$D_y = -8 \; kip \; (\downarrow)$$
$$\Sigma F_x = 0 \rightarrow^+$$
$$-D_x - 6 = 0;$$
$$D_x = -2 \; kip \; (\rightarrow)$$

Segment AD.

$$\sum M_C = 0 \circlearrowleft^+$$
$$A_y \times 8 - 8 \times 4 + 6 \times 3 + (-8) \times 6 = 0;$$
$$A_y = 7.75 \; kip \; (\uparrow)$$

$$\sum F_y = 0 \uparrow^+$$
$$7.75 - 8 + C_y - 6 - (-8) = 0;$$
$$C_y = -1.75 \; kip \; (\downarrow)$$

$$\sum F_x = 0 \rightarrow^+$$
$$A_x - D_x = 0$$
$$A_x = 6 \; kip \; (\rightarrow)$$

Example 2

Determine the reaction of the beam.

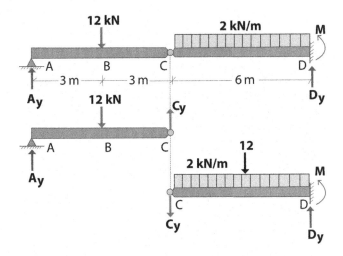

Solution

Segment AC.

$$\sum M_C = 0 \circlearrowleft^+$$
$$A_y \times 6 - 12 \times 3 = 0$$
$$A_y = 6 \; kip \; (\uparrow)$$

$$\sum F_y = 0 \uparrow^+$$
$$A_y - 12 + C_y = 0$$
$$C_y = 6 \; kip \; (\uparrow)$$

Segment *CD*.

$$\Sigma M_D = 0 \circlearrowleft^+$$
$$-C_y \times 6 - 12 \times 3 - M = 0;$$
$$M = -72 \, kip \, (\curvearrowright)$$

$$\Sigma F_y = 0 \uparrow^+$$
$$-C_y - 12 + D_y = 0;$$
$$D_y = 18 \, kip \, (\uparrow)$$

Example 3

Determine the reaction of the beam.

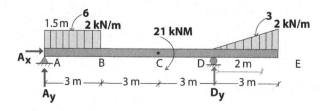

Solution

$$\Sigma M_D = 0 \circlearrowleft^+$$
$$A_y \times 9 - 6 \times 7.5 + 21 + 3 \times 2 = 0;$$
$$A_y = 2 \, kip \, (\uparrow)$$

$$\Sigma F_y = 0 \uparrow^+$$
$$A_y - 6 + D_y - 3 = 0;$$
$$D_y = 7 \, kip \, (\uparrow)$$

$$\Sigma F_x = 0 \rightarrow^+$$
$$A_x = 0$$

2.2   ANALYSIS OF INTERNAL FORCES

Analysis of internal forces is an essential concept in the field of mechanics and engineering. It is the study of the forces that act within a structure, such as a beam or a truss, and how these forces affect the structural integrity of the object. There are two main types of internal forces: normal forces and shear forces. *Normal forces* are forces that act perpendicular to the cross-sectional area of the structure, while *shear forces* are forces that act parallel to the cross-sectional area.

To analyze the internal forces in a structure, engineers use a variety of techniques, including the method of joints, the method of sections, and the graphical method. These techniques involve breaking the structure down into smaller parts and analyzing the forces acting on each part. The method of joints involves analyzing the forces acting on each joint in the structure, while the method of sections involves cutting the structure into sections and analyzing the forces acting on each section. The graphical method involves constructing a free-body diagram of the structure and analyzing the forces acting on each part. Once the internal forces have been analyzed, engineers can determine the stresses and strains within the structure. This information is essential for designing safe and efficient structures, as well as for determining the materials and dimensions needed for a particular project.

Let us consider a simple theoretical example of analyzing internal forces in a beam.

Suppose we have a beam that is 10 m long and has a rectangular cross section of 0.1 m × 0.2 m. The beam is supported on each end by a fixed support, and a force of 1,000 N is applied at the center of the beam.

To analyze the internal forces in the beam, we can use the method of sections. We can cut the beam into two sections at the center and analyze the forces acting on each section separately. First, let us consider the left section of the beam. We can draw a free-body diagram of the left section showing the external forces acting on the section and the internal forces within the section.

The external forces acting on the section are the support reactions at the left end of the beam, and the internal forces within the section are the normal force and the shear force. Using the equations of equilibrium, we can determine that the normal force is zero and the shear force is 500 N, acting to the right. Next, let us consider the right section of the beam. We can draw a free-body diagram of the right section showing the external forces acting on the section and the internal forces within the section.

The external forces acting on the section are the support reactions at the right end of the beam, and the internal forces within the section are the normal force and the shear force. Using the equations of equilibrium, we can determine that the normal force is zero and the shear force is 500 N, acting to the left. Thus, we have analyzed the internal forces in the beam and determined that there is a shear force of 500 N acting in each section of the beam. This information can be used to determine the stresses and strains within the beam and ensure that it is designed to withstand the applied load.

2.2.1 Axial Force

Axial force is an internal force that acts along the axis of an object, such as a beam, column, or shaft. It is also known as axial load or compression/tension force. Axial force can either be compressive or tensile, depending on whether it is pushing or pulling on the object. For example, a rope pulling on a column would create a tensile axial force, while a weight pushing down on a column would create a compressive axial force.

The magnitude of the axial force is equal to the external load applied to the object, and it is distributed uniformly across the cross-sectional area of the object. The axial force creates internal stresses within the object, which can cause deformation or failure if they exceed the material's strength. To analyze the axial force in an object, engineers use the equations of equilibrium and the principles of mechanics to determine the external loads acting on the object, the internal forces within the object, and the resulting stresses and strains. This information is essential for designing safe and efficient structures, as well as for determining the materials and dimensions needed for a particular project.

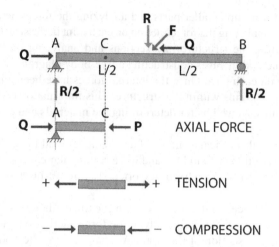

2.2.2 SHEAR FORCE

Shear force is an internal force that acts parallel to the cross-sectional area of an object, such as a beam or a shaft. It is also known as transverse force or lateral force. Shear force occurs when an external load is applied perpendicular to the longitudinal axis of the object, causing a portion of the object to experience a parallel force that slides or shears along an adjacent section of the object. For example, when a beam is supported on both ends and a load is applied at a point along the beam's length, shear force will be generated in the beam.

The magnitude of the shear force varies along the length of the object, and it is generally greatest at the location of the external load. Shear force creates internal stresses within the object, which can cause deformation or failure if they exceed the material's strength. To analyze the shear force in an object, engineers use the equations of equilibrium and the principles of mechanics to determine the external loads acting on the object, the internal forces within the object, and the resulting stresses and strains. This information is essential for designing safe and efficient structures, as well as for determining the materials and dimensions needed for a particular project.

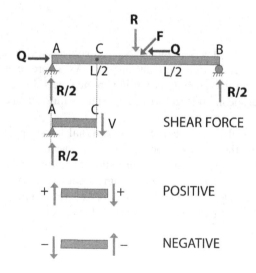

2.2.3 BENDING MOMENT

Bending moment is an internal force that results from an external load applied perpendicular to the longitudinal axis of an object, such as a beam or a bridge. It is also known as flexural moment. Bending moment occurs when an external load is applied to an object, causing the object to bend or deform. As the object bends, internal stresses are created within the object, resulting in the formation of a moment that resists the bending. This moment is known as the bending moment and is calculated using the principles of statics and mechanics.

The magnitude of the bending moment varies along the length of the object, and it is generally greatest at the location where the external load is applied. The bending moment creates both compressive and tensile stresses within the object, which can cause deformation or failure if they exceed the material's strength. To analyze the bending moment in an object, engineers use the equations of equilibrium and the principles of mechanics to determine the external loads acting on the object, the internal forces within the object, and the resulting stresses and strains. This information is essential for designing safe and efficient structures, as well as for determining the materials and dimensions needed for a particular project.

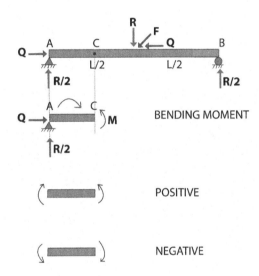

Example 4

Determine the following items: (1) axial force and shear force at just left of *B*, (2) bending moment at *B*, and (3) axial force, shear force, and bending moment at *E*.

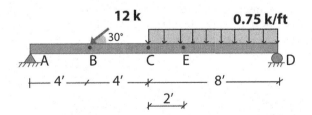

Solution

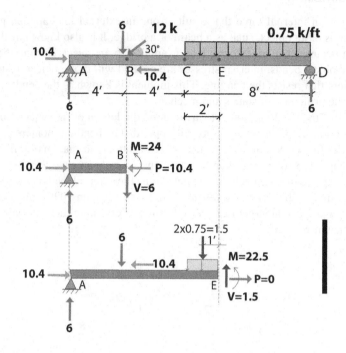

SegmentAB

$$\Sigma F_x = 0 \rightarrow \quad 10.4 + P = 0 \quad P = -10.4\,k\,(\rightarrow)$$
$$\Sigma F_y = 0 \uparrow^+ \quad 6 - V = 0 \quad V = +6\,k\,(\downarrow)$$
$$\Sigma M_B = 0 \circlearrowright^+ \quad 6 \times 4 - M = 0 \quad M = +24\,kft\,(\circlearrowright)$$

SegmentAE

$$\Sigma F_x = 0 \rightarrow + \quad 10.4 - 10.4 + P = 0 \quad\quad\quad\quad P = 0$$
$$\Sigma F_y = 0 \uparrow^+ \quad 6 - 6 - 1.5 - V = 0 \quad\quad V = -1.5\,k\,(\uparrow)$$
$$\Sigma M_E = 0 \circlearrowright^+ \quad 6 \times 10 - 6 \times 6 - 1.5 \times 1 - M = 0$$
$$M = 22.5\,kft\,(\circlearrowright)$$

Example 5

Determine the following items: (1) axial force and shear force at *B*, (2) bending moment at just left of *B*, and (3) axial force, shear force, and bending moment at *F*.

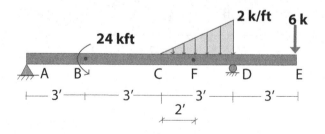

Solution

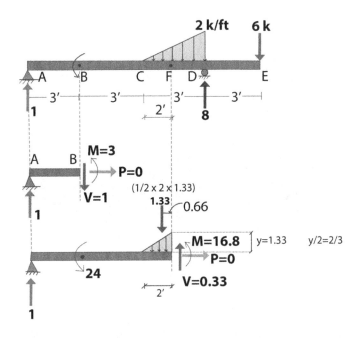

SegmentAB

$\Sigma F_x = 0 \rightarrow +$ $P = 0$ $P = 0 \, k$

$\Sigma F_y = 0 \uparrow^+$ $1 - V = 0$ $V = 1 \, k \, (\downarrow)$

$\Sigma M_B = 0 \circlearrowleft^+$ $1 \times 3 - M = 0$ $M = +3 \, kft \, (\circlearrowleft)$

SegmentAE

$\Sigma F_x = 0 \rightarrow +$ $P = 0$ $P = 0$

$\Sigma F_y = 0 \uparrow^+$ $1 - 1.33 - V = 0$ $V = -0.33 \, k \, (\uparrow)$

$\Sigma M_E = 0 \circlearrowleft^+$ $1 \times 8 - 24 - 1.33 \times 0.66 - M = 0$

                                  $M = 16.8 \, kft$

Example 6

Determine the following items: (1) axial force and shear force at just left of F, (2) bending moment at F, and (3) axial force, shear force, and bending moment at G.

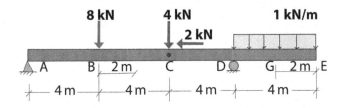

Solution

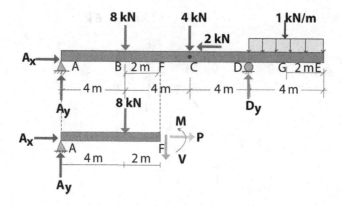

Reactions:

$$\sum M_A = 0 \circlearrowright^+$$
$$8 \times 4 + 4 \times 8 - Dy \times 12 + 4 \times 14 = 0$$
$$Dy = 10(\uparrow)$$
$$\sum Fy = 0 \uparrow^+$$
$$A_y - 8 - 4 + 10 - 4 = 0$$
$$Ay = 6(t)$$
$$\sum F_x = 0 \to +$$
$$A_x - 2 = 0$$
$$A_x = 2(\to)$$

Section AF
$$\sum F_x = 0 \to +$$
$$2 + P = 0$$
$$P = -2(\leftarrow)$$
$$\sum Fy = 0 \uparrow^+$$
$$6 - 8 - V = 0$$
$$V = -2(\uparrow)$$
$$\sum M_F = 0 \circlearrowright^+$$
$$6 \times 4 - 8 \times 2 - M = 0$$
$$M = 8(\circlearrowleft)$$

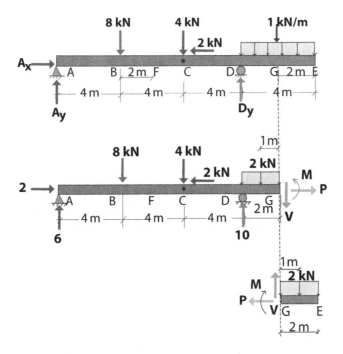

Section GE

$\Sigma F_x = 0 \rightarrow$ $\quad\quad\quad\quad\quad$ $P = 0$ $\quad$ $P = 0k$

$\Sigma F_y = 0 \uparrow^+$ $\quad\quad\quad$ $V - 2 = 0$ $\quad$ $V = 2k(\uparrow)$

$\Sigma M_G = 0 \circlearrowleft^+$ $\quad$ $2 \times 1 + M = 0$ $\quad$ $M = -2kft(\circlearrowleft)$

Reactions:

Section AG

$\Sigma F_x = 0 \rightarrow$

$2 - 2 + P = 0$

$\quad\quad P = 0$

$\Sigma F_y = 01^+$

$6 - 8 - 4 + 10 - 2 - V = 0$

$\quad\quad V = 2(\downarrow)$

$\Sigma M_G = 0 \circlearrowleft^+$

$6 \times 14 - 8 \times 10 - 4 \times 6 + 10 \times 2$

$\quad\quad - 2 \times 1 - M = 0$

$\quad\quad M = -2(\circlearrowleft)$

Example 7

Determine the support reactions of the beam, and then determine the axial force, shear force, and bending moment at G.

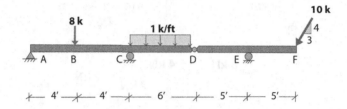

Solution

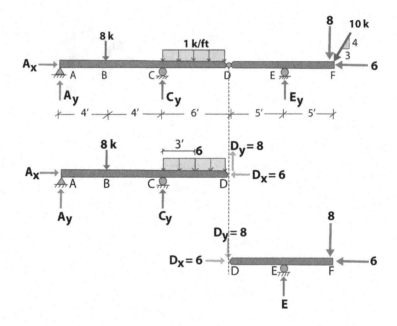

Segment DF

$$\Sigma F_x = 0 \rightarrow \qquad -D_x - 6 = 0 \qquad D_x = -6 \ k\,(\rightarrow)$$
$$\Sigma M_D = 0 \circlearrowleft^+ \quad -E_y * 5 + 8 * 10 = 0 \quad E_y = +16 \ k\,(\uparrow)$$
$$\Sigma F_y = 0 \uparrow^+ \qquad D_y + 16 - 8 = 0 \qquad D_y = -8 \ k\,(\downarrow)$$

Segment AD

$$\Sigma F_x = 0 \rightarrow + \qquad\qquad A_x - 6 = 0 \qquad\qquad A_x = 6 \ k\,(\rightarrow)$$
$$\Sigma M_A = 0 \circlearrowleft^+ \quad 8 * 4 - C_y * 8 + 6 * 11 - 8 * 14 = 0 \quad C_y = -1.75 \ k\,(\downarrow)$$
$$\Sigma F_y = 0 \uparrow^+ \qquad A_y - 8 - 1.75 - 6 + 8 = 0$$
$$A_y = 7.75 \ k\,(\uparrow)$$

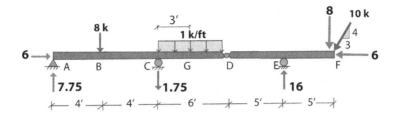

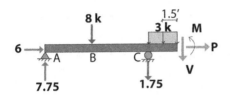

Segment AG

$\Sigma F_x = 0 \rightarrow +$ $6 + P = 0$ $P = -6 (\leftarrow)$

$\Sigma F_y = 0 \uparrow^+$ $7.75 - 1.338 - 1.75 - 3 - V = 0$ $V = -5 \ k (\uparrow)$

$\Sigma M_G = 0 \circlearrowleft^+$ $7.75 \times 11 - 8 \times 7 - 1.75 \times 3 - 3 \times 1.5 - M = 0$

                                             $M = 19.5 \ kft (\circlearrowleft)$

Example 8

Determine the support reactions of the beam, and then determine the axial force, shear force, and bending moment at *E*.

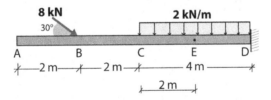

Solution

Reactions:

$\Sigma F_x = 0 \rightarrow$ $6.93 + D_x = 0$ $D_x = -6.93 \ k (\leftarrow)$

$\Sigma M_D = 0 \circlearrowleft^+$ $-4*6 - 8*2 - M_D = 0$ $M_D = -40 \ kft (\circlearrowleft)$

$\Sigma F_y = 0 \uparrow^+$ $-4 - 8 + D_y = 0$ $D_y = +12 \ k (\uparrow)$

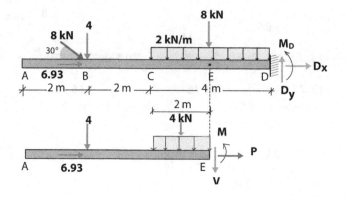

Section AE

$$\Sigma F_x = 0 \rightarrow + \qquad 6.93 + P = 0 \qquad P = -6.93 (\leftarrow)$$
$$\Sigma F_y = 0 \uparrow^+ \qquad -4 - 4 - V = 0 \qquad V = -8 \, k (\uparrow)$$
$$\Sigma M_E = 0 \circlearrowleft^+ \quad -4 \times 4 - 4 \times 2 - M = 0$$
$$M = -24 \, kft (\circlearrowleft)$$

Example 9

Determine the support reactions of the beam, and then determine the bending moment at A, C, and E.

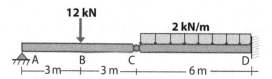

Solution

Section AC

$$\Sigma M_A = 0 \circlearrowleft^+ \quad 12*3 + C_y*6 = 0 \quad C_y = 6k(\uparrow)$$
$$\Sigma F_y = 0 \uparrow^+ \qquad A_y - 12 + 6 = 0 \qquad A_y = 6k(\uparrow)$$

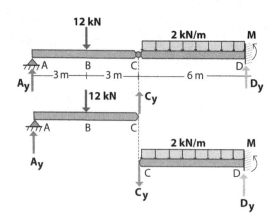

Section CD

$\Sigma M_D = 0 \circlearrowleft^+ \qquad -6*6 - 12*3 - M = 0 \qquad M = -72\,kft\,(\circlearrowleft)$

$\Sigma F_y = 0 \uparrow^+ \qquad -6 - 12 + D_y = 0 \qquad D_y = 18\,k\,(\uparrow)$

Example 10

Determine the support reactions of the frame.

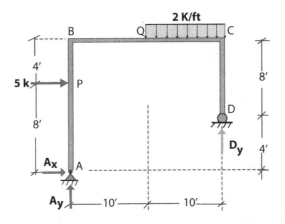

Solution

$\Sigma F_x = 0 \rightarrow \qquad A_x + 5 = 0 \qquad A_x = -5\,K\,(\leftarrow)$

$\Sigma M_A = 0 \circlearrowleft^+ \quad 5*8 + 20*15 - D_y*20 = 0 \quad D_y = +17\,k\,(\uparrow)$

$\Sigma F_y = 0 \uparrow^+ \qquad A_y - 20 + D_y = 0 \qquad A_y = +123\,k\,(\uparrow)$

2.3   CONCEPT OF STRESS AND STRAIN

2.3.1   STRESS AND STRAIN

Strain: Change of length (δ) per unit length (L).

$$\epsilon = \frac{\delta}{L}$$

Unit: in/in, mm/mm.

In mechanics, *strain* refers to the amount of deformation or change in shape that a material undergoes in response to an applied force or stress. Strain is usually represented by the symbol ε (epsilon) and is defined as the ratio of the change in length (δ) of a material to its original length (L) along a given direction.

There are several types of strain, including:

- *Tensile strain.* The elongation of a material due to a stretching force applied along its length. Tensile strain is positive.
- *Compressive strain.* The shortening of a material due to a compressive force applied along its length. Compressive strain is negative.

- *Shear strain.* The deformation of a material due to a force applied perpendicular to its length, causing the material to twist or distort.
- *Volumetric strain.* The change in volume of a material due to an applied force or stress.

Stress: Internal resistance of an elastic body to external forces, equal to the ratio of force (F) to cross-sectional area (A).

$$\sigma = \frac{F}{A}$$

Unit: lb/in² (psi), k/in² (ksi), N/mm² (MPa).
   There are several types of stress, including:

- *Tensile stress.* The stress that is exerted on a material when it is pulled or stretched along its length.
- *Compressive stress.* The stress that is exerted on a material when it is pushed or compressed along its length.
- *Shear stress.* The stress that is exerted on a material when a force is applied parallel to the surface, causing the material to twist or distort.
- *Volumetric stress.* The stress that is exerted on a material due to an applied force or stress that changes the volume of the material.

Normal Stress: Developed along normal direction of a section of an elastic body to resist normal force.

$$\sigma = \frac{P}{A}$$

Tensile Stress: Normal stress developed by tension.
Compressive Stress: normal stress developed by compression.
Shear Stress: developed along a section of an elastic body to resist shear force.

$$\tau = \frac{V}{A}$$

Shear stress is a type of stress that is exerted on a material when a force is applied parallel to its surface, causing the material to twist or distort. Shear stress is important in materials science and engineering because it can cause materials to fail, especially if the material is not able to withstand the forces applied to it. Shear stress can also cause deformation in a material, such as bending, twisting, or warping.
   Shear stress is typically expressed in units of pressure, such as pascals (Pa) or pounds per square inch (psi), and is related to the material's shear strain by its shear modulus of elasticity, also known as the modulus of rigidity. The *shear modulus of elasticity* is a measure of a material's ability to withstand deformation due to shear stress and is typically denoted by the symbol G.

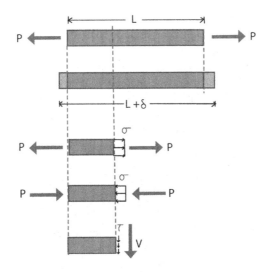

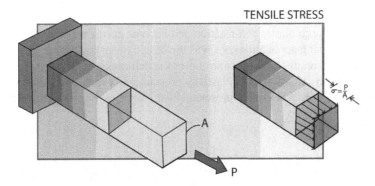

TENSILE STRESS

2.3.2 Relationship of Stress and Strain

Elasticity: Property of a material that enables it to deform in response to an applied force and to recover its original size and shape upon removal of the force.

Elasticity is the property of a material that allows it to deform when a force or stress is applied but then return to its original shape or size when the force is removed. Elasticity is a fundamental concept in materials science and engineering and is characterized by the material's elastic modulus, or Young's modulus, which is typically denoted by the symbol E.

The greater the value of E, the stiffer and more rigid the material. The elastic modulus is a measure of the material's ability to withstand deformation and return to its original shape, which is important in many engineering applications, such as the design of springs, shock absorbers, and other components that require high levels of resilience and durability. Elasticity can also be characterized by the material's yield strength, which is the maximum amount of stress that the material can withstand without undergoing permanent deformation, and its ultimate strength, which is the maximum stress that the material can withstand before it fails or breaks.

Plasticity: Ability to retain its final size and shape upon removal of the force.

Plasticity is the property of a material to undergo permanent deformation or change in shape when a stress or force is applied beyond its elastic limit. Unlike elastic deformation, which is reversible and does not result in permanent changes to the material, plastic deformation can result in permanent changes to the material's shape or size.

When a material undergoes plastic deformation, it typically results in the material being permanently stretched, compressed, or bent. The degree of plastic deformation depends on the magnitude and duration of the stress applied to the material, as well as the material's composition and structure. The point at which a material begins to undergo plastic deformation is known as the yield point or yield strength. Beyond this point, the material will undergo plastic deformation and will not return to its original shape when the stress is removed. Plasticity is an important property in materials science and engineering, as it determines the material's ability to withstand stresses and forces without failure. Materials that are highly plastic, such as many metals and polymers, are often used in applications where they will be subjected to high levels of stress or deformation, such as in the construction of bridges, aircraft, and other structures.

Hooke's Law: Stress on a body is directly proportional to the strain produced, provided the stress does not exceed the elastic limit of the material.

$$\sigma \, | \, \alpha \in$$

$$\sigma = \in E$$

Hooke's law is a fundamental principle in mechanics that describes the relationship between the deformation (stretching or compression) of a material and the applied force or stress. Hooke's law is applicable to a wide range of materials and objects, including springs, rubber bands, and other elastic materials. When a force is applied to an elastic material, it deforms or stretches, but it will return to its original shape and size when the force is removed. If the applied force exceeds the material's elastic limit, it may undergo plastic deformation and become permanently deformed. Hooke's law is used in many engineering applications, such as the design of springs, shock absorbers, and other mechanical systems that require precise control over the deformation of materials.

Modulus of Elasticity: Coefficient of elasticity of a material expressing the ratio between stress to strain derived from Hooke's law and represented by the slope of the straight-line portion of stress strain diagram.

Modulus of Elasticity,

$$E = \frac{\sigma}{\in}$$

For steel, $E = 29{,}000$ ksi, or 200 GPa.

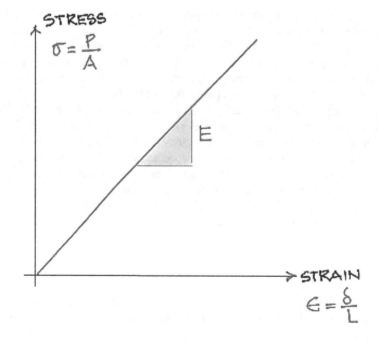

2.3.3   Tensile Test of Mild Steel

The *tensile test* of mild steel involves subjecting a sample of the material to a gradually increasing tensile load until it reaches its breaking point. The test is typically carried out using a universal testing machine which can apply the load and measure the resulting deformation.

The following are the basic steps involved in conducting a tensile test of mild steel:

- *Sample preparation.* A sample of the mild steel material is cut to a specific size and shape according to the relevant standards. The sample should be free from surface defects and deformations that could affect the test results.
- *Mounting the sample.* The sample is mounted securely in the grips of the testing machine, ensuring that it is aligned correctly and that the grips are tightened to prevent slippage.
- *Applying the load.* A tensile load is applied to the sample in a controlled manner, typically at a rate of around 1 mm/min. The load is increased gradually until the sample reaches its breaking point.
- *Measuring deformation.* During the test, the deformation of the sample is measured using an extensometer, which is attached to the sample to measure the change in length as the load is applied.
- *Recording data.* The load and deformation data is recorded throughout the test, and a stress–strain curve is generated using the data.
- *Determining mechanical properties.* From the stress–strain curve, various mechanical properties of the mild steel can be determined, including its yield strength, ultimate tensile strength, and elongation at break.

The results of a tensile test can be used to assess the suitability of a material for a particular application and to determine the design parameters for engineering structures and components.

2.3.4   Stress–Strain Diagram of Mild Steel

A stress–strain diagram is a graph that shows the relationship between the stress and strain of a material during a tensile test. The stress is plotted on the y-axis, while the strain is plotted on the x-axis. The stress–strain diagram of mild steel typically exhibits the following characteristics:

- *Elastic region.* The initial portion of the curve is linear and represents the elastic region, where the material undergoes reversible deformation under a small amount of stress. The slope of this linear region is known as the Young's modulus, which is a measure of the material's stiffness.

- *Yield point.* Beyond the elastic region, the curve becomes non-linear, and a yield point is reached where the material undergoes permanent plastic deformation. This *yield point* is defined as the stress at which the material starts to deform plastically.
- *Plastic region.* After the yield point, the curve shows a steady increase in strain with increasing stress, indicating that the material is undergoing plastic deformation. The stress required to maintain this plastic deformation is called the flow stress.
- *Ultimate tensile strength.* The highest point on the curve represents the ultimate tensile strength (UTS) of the material, which is the maximum stress that the material can withstand before it breaks.
- *Fracture point.* Beyond the UTS, the material undergoes rapid and catastrophic deformation until it reaches the point of fracture.

The stress–strain diagram of mild steel is typically characterized by a relatively high yield strength and ultimate tensile strength, combined with a high degree of ductility and toughness. This makes it a popular material for a wide range of applications, from construction and engineering to manufacturing and fabrication.

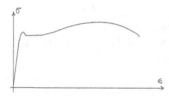

2.3.5 Salient Features of Stress–Strain Diagram

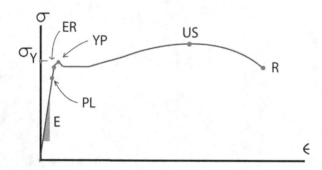

Feature	Description
Proportional Limit	Stress beyond which ratio of stress to strain for a material no longer remains constant.
Elastic Limit	Maximum stress that can be applied to a material without causing permanent deformation.
Yield Point	Stress beyond which a marked increase in strain occurs in a material without a concurrent increase in stress.
Yield Strength	Stress corresponding to yield point.
Ultimate Strength	Maximum stress a material can be expected to bear without failure.
Rupture	Breaking of a material resulting from rupture of its atomic bonds.

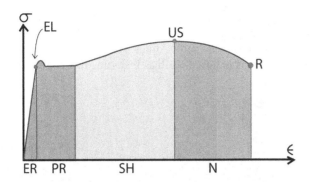

Feature	Description
Elastic Range (ER)	Material exhibits elastic deformation.
Plastic Range (PR)	Material exhibits plastic deformation.
Strain Hardening Range (SH)	Material exhibits increased strength with some loss of ductility.
Necking (N)	Marked decrease in cross section occurs.

2.4   MECHANICAL PROPERTIES OF MATERIALS

2.4.1   DUCTILITY AND BRITTLENESS

Ductility: Property of a material that enables it to undergo plastic deformation after being stressed beyond elastic limit and before rupturing.

- Desirable property of a structural material.
- Plastic behavior is an indicator of reserve strength.
- Serves as a visual warning of impending failure.

Ductility is the ability of a material to deform under tensile stress, without breaking or cracking, and to retain this deformation after the stress is removed. In other words, *ductility* is a measure of how much a material can be stretched or pulled before it breaks. Materials that are highly ductile include metals like copper, gold, and silver, as well as some nonmetals, like sulfur and phosphorus. Ductility is an important property for many engineering materials, as it can allow them to be formed into various shapes and configurations and to withstand deformation and stress without fracturing or breaking. For example, ductility is important for materials used in construction, manufacturing, and infrastructure, such as steel beams, cables, and pipes.

Brittleness: Property of a material that causes it to rupture suddenly under stress with little evident deformation.

- Undesirable property of a structural material.
- Lacks plastic behavior of ductile material, thus reserved strength.
- Gives no advance warning of impending failure.

Brittleness is the opposite of ductility; it is the property of a material that causes it to fracture or break without significant deformation when subjected to stress. A brittle material can fracture suddenly and with little warning when exposed to stress or shock, rather than deforming plastically like a ductile material. Examples of brittle materials include glass, ceramics, and some types of metals, such as cast iron. Brittle materials are generally more difficult to work with than ductile materials, as they are more prone to cracking or breaking during fabrication and processing. Brittle materials are generally not suitable for applications that require high resistance to mechanical shock or impact, such as structural components in vehicles or buildings. However, they can be useful in applications where high stiffness or hardness is required, such as in cutting tools or abrasive materials.

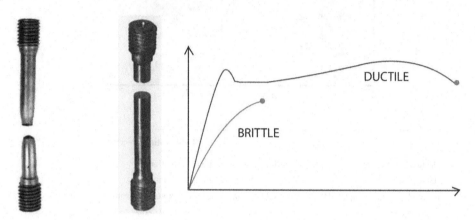

2.4.2 HIGH STRENGTH AND LOW STRENGTH

High strength and **low strength** are relative terms used to describe the strength of materials compared to each other. High-strength materials are those that have a higher resistance to deformation and failure under load or stress than other materials. Examples of high-strength materials include steel alloys, titanium alloys, and some ceramics. These materials are often used in applications that require high load-bearing capacity or where high stiffness and durability are required.

Low-strength materials, on the other hand, have a lower resistance to deformation and failure under load or stress than high-strength materials. Examples of low-strength materials include some plastics, rubber, and some low-grade metals. These materials are often used in applications where high strength is not required, such as packaging, insulation, or as seals. It is important to note that the choice of material for a particular application depends on many factors, including the specific requirements of the application, such as the level of stress, environmental conditions, and cost. Therefore, a material that is considered low-strength for one application may be high-strength for another.

- Low-strength materials usually are ductile.
- High-strength materials usually are brittle.

Toughness: Property of a material that enables it to absorb energy before rupturing.

- Represented by the area under the stress–strain curve derived from a tensile test of the material.
- Ductile materials are tougher than brittle materials.

Toughness is a mechanical property of materials that describes their ability to absorb energy without fracturing or breaking. It is the amount of energy per unit volume that a material can absorb before it fractures or breaks. A tough material can absorb a significant amount of energy without failing, even under conditions of high stress or impact. Toughness is an important property for materials that will be subjected to mechanical shock or impact, such as those used in structural components or machinery.

Materials that are tough tend to be ductile, which means that they can undergo significant plastic deformation before failure. However, toughness also depends on other factors, such as material microstructure, crystallographic defects, and the presence of impurities. Common examples of tough materials include high-strength steels, certain polymers, and some ceramics. The toughness of a material can be measured through various methods, such as the Charpy impact test or the Izod impact test.

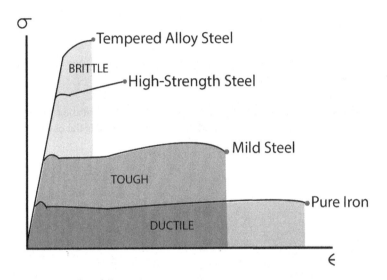

2.4.3 Loading, Unloading, and Reloading

Loading, *unloading*, and *reloading* are terms used in materials science and engineering to describe the process of applying and removing loads or stresses to a material.

- *Loading* refers to the process of applying an external force or load to a material. This force can be compressive, tensile, or shear, depending on the type of load being applied. Loading can cause deformation in the material, which can be either elastic (reversible) or plastic (irreversible).
- *Unloading* refers to the process of removing the applied load from the material. When the load is removed, the material may spring back to its original shape if the deformation was elastic, or it may retain some permanent deformation if the deformation was plastic.
- *Reloading* refers to the process of applying the load to the material again after unloading it. The behavior of the material during reloading can depend on the extent of the previous deformation and the properties of the material.

The loading, unloading, and reloading of a material can affect its mechanical properties, such as its strength, stiffness, and ductility. These processes are important to consider when designing and testing materials and structures that will be subjected to repeated loading and unloading, such as in aircraft structures, bridges, or heavy machinery.

Loading within elastic range, then unloading:

- Prior to application of loading, the stress–strain curve remains at the origin point.
- After applying a small amount of load, the material stays within the elastic range.
- If load is removed, the material goes back to the origin point, that is, the initial position.

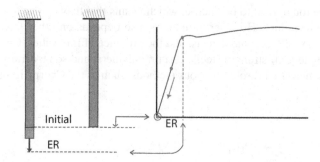

Loading beyond elastic range, then unloading:

- Prior to application of loading, the stress–strain curve remains at the origin point.
- After applying a significant amount of load, the material exceeds the elastic range and now is in the plastic range.

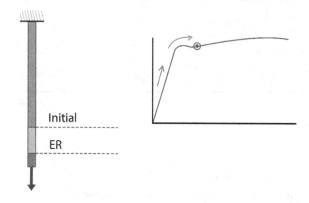

- If load is removed now, the material cannot fully go back to the origin point, that is, the initial position; thus, a permanent set (*PS*) is observed in the material.
- But the materials recover elastic deformation *ER*.

Reloading after permanent set:

- If the material is loaded again after the permanent set occurs, it starts from the new origin point.
- Also, the previous straight line (blue) is never followed, but the new one (purple) is.

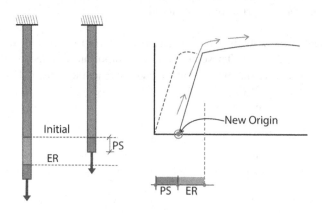

2.5   EXAMPLES ON STRESS AND STRAIN

Example 11

A short post constructed from a hollow circular tube of aluminum supports a compressive load of 26 kips. The inner and outer diameters of the tube are $d_1 = 4$ in and $d_2 = 4.5$ in, respectively, and its length is 16 *in*. The shortening of the post due to the load is measured as 0.012 in. Determine the compressive stress and strain in the post.

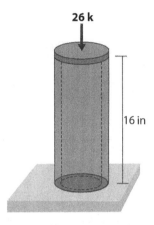

Solution

Cross-sectional area:

$$A = \frac{\pi}{4}\left(d_2^2 - d_1^2\right) = \frac{\pi}{4}\left(4.5^2 - 4^2\right) = 3.338 \ in^2$$

Compressive stress:

$$\sigma = \frac{p}{A} = \frac{26 \ kip}{3.338 \ in^2} = 7.79 \ ksi$$

Compressive strain:

$$\epsilon = \frac{\delta}{L} = \frac{0.012 \ in}{16 \ in} = 0.00075$$

Example 12

An aluminum rod shown has a circular cross section and is subjected to an axial load of 10 kN. Stress–strain diagram of the material is shown in the following. Determine the approximate elongation of the rod when the load is applied. Take $E_{al} = 70$ GPa.

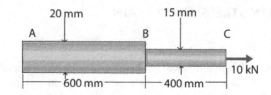

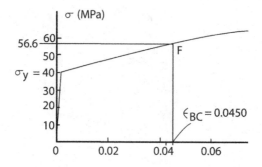

Solution

Normal stress:

$$\sigma_{AB} = \frac{P}{A} = \frac{10 \times 10^3 \ N}{\pi \times 20^2 \ / \ 4 \ mm^2} = 31.8 \ MPa$$

$$\sigma_{BC} = \frac{P}{A} = \frac{10 \times 10^3 \ N}{\pi \times 15^2 \ / \ 4 \ mm^2} = 56.6 \ MPa$$

Normal strain:

Material in segment AB is strained elastically since $\sigma_{AB} < \sigma_Y$. Using Hooke's law,

$$\epsilon_{AB} = \frac{\sigma_{AB}}{E_{al}} = \frac{31.8 \ MPa}{70 \times 10^3 \ MPa} = 0.00045$$

Material in segment BC is strained inelastically since $\sigma_{BC} < \sigma_Y$. From graph:

$$\epsilon_{BC} = 0.0450$$

Elongation of rod ABC:

$$\delta = \Sigma \epsilon L = 0.00045 \times 600 + 0.0450 \times 400 = 18.3 \ mm$$

Example 11

If the elongation of wire BC is 0.2 mm after the force P is applied, determine the magnitude of P. The wire has a diameter of 3 mm.

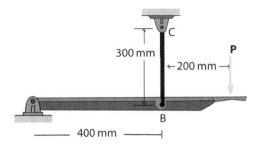

Solution

Determine the force in *BC*.

$$\epsilon = \frac{\delta}{L} = \frac{0.2\,mm}{300\,mm} = 0.000667$$

$$\sigma = \epsilon E = 0.000667 \times \left(200 \times 10^3\,MPa\right) = 133.4\,MPa$$

$$A = \frac{\pi d^2}{4} = \frac{\pi \times 3^2}{4} = 7.069\,mm^2$$

$$P_{BC} = \sigma A = 133.4\,N/mm^2 \times 7.069\,mm^2 = 943\,N$$

Determine *P*.

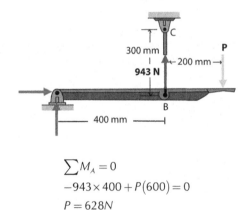

$$\sum M_A = 0$$
$$-943 \times 400 + P(600) = 0$$
$$P = 628N$$

Ans *p = 628N*

Example 13

The mast ABC is hinged at *A* and supported by a steel cable at *C*. Determine the minimum required diameter of the cable *CD* to carry the 40 kip load, given that the cable is made of A36 steel.

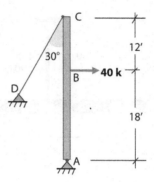

Solution

Determine the maximum tension T in the cable.

$$\Sigma M_A = 0$$
$$40 \times 18 - (T \sin 30) \times 30 = 0$$
$$T = 48\,kip$$

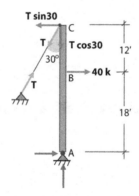

Required area to carry this tension:

$$A = \frac{T}{\sigma} = \frac{48\,kip}{36\,ksi} = 1.33\,in^2$$

Required diameter to have that area:

$$A = \frac{\pi d^2}{4}$$

$$d = \sqrt{\frac{4A}{\pi}} = \sqrt{\frac{4 \times 1.33}{\pi}} = 1.30\,in$$

Ans. $d = 1.30$ in

Example 14

The rigid bar ABC is hinged at A and supported by a steel cable at B. Determine the largest load P that can be applied at C if the stress in the cable is limited to 30 ksi and vertical movement of C must not exceed 0.10 in.

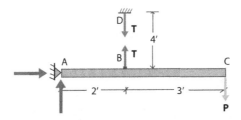

Solution

Failure criteria of cable.

$$T = \sigma A = 30 \; ksi \times 0.50 \; in^2 = 15 \; kip$$
$$\sum M_A = 0$$
$$-15 \times 2 + P \times 5 = 0$$
$$P = 6 \; kip$$

Deflection criteria of C.

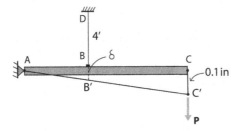

$$\delta = 0.10 \times \frac{2}{5} = 0.04 \; in$$
$$\in = \frac{\delta}{L} = \frac{0.04 \; in}{4 \times 12 \; in} = 0.000833$$
$$\sigma = \in E = 0.000833 \times 29000 \; ksi = 24.16 \; ksi$$
$$T = \sigma A = 24.16 \; ksi \times 0.50 \; in^2 = 12.08 \; kip$$

$$\sum M_A = 0$$
$$-12.08 \times 2 + P \times 5 = 0$$
$$P = 4.83 \; kip$$

Answer: P = 4.83 kip.

2.6   DEFORMATION UNDER AXIAL LOAD

Strain:

$$\epsilon = \frac{\delta}{L}$$

Stress:

$$\sigma = \frac{P}{A}$$

Hook's law:

$$\sigma = \epsilon E$$

By combining all these:

$$\delta = \epsilon L = \frac{\sigma}{E} L = \frac{P/A}{E} L$$

$$\delta = \frac{PL}{AE}$$

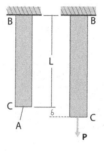

Example 15

Determine the deformation of the steel rod shown at the left under the given loads.

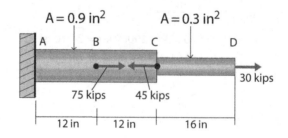

Solution

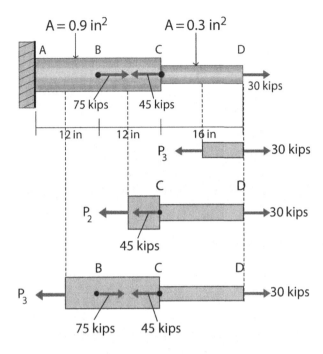

Segment CD:

$$\Sigma F_x = 0$$
$$-P_3 + 30 = 0$$
$$P_3 = 30 \text{ kip}$$

Segment BC:

$$\Sigma F_x = 0$$
$$-P_2 - 45 + 30 = 0$$
$$P_2 = -15 \text{ kip}$$

Segment BD:

$$\Sigma F_x = 0$$
$$-P_1 + 75 - 45 + 30 = 0$$
$$P_1 = 60 \text{ kip}$$

Now, applying the deformation formula of axially loaded bar:

$$\delta = \sum \frac{PL}{AE}$$

$$= \frac{1}{E}\left(\frac{P_1 L_1}{A_1} + \frac{P_2 L_2}{A_2} + \frac{P_3 L_3}{A_3}\right)$$

$$= \frac{1}{29000}\left(\frac{60 \times 12}{0.90} + \frac{-15 \times 12}{0.90} + \frac{30 \times 16}{0.30}\right)$$

$$= 0.076 \ in$$

Answer: 0.76 in

Example 16

A steel rod of $L = 24$ in, $A1 = 1:25$ in^2, and $E1 = 29,000$ ksi has been placed inside a tube made of aluminum of the same length L, but $A2 = 1:00$ in^2 and $E2 = 5,800$ ksi. The system is under compression of $P = 58$ kip.

Determine the (1) deformation of the rod and tube and the (2) load carried by the steel and the aluminum.

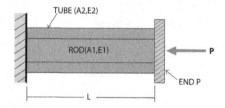

Solution

If the force carried by steel is P_1 and the force carried by aluminum is P_2, then we write:

$$P_{1+}P_2 = p = 58 \tag{1}$$

If axial deformation of steel and aluminum are _1 and _2, respectively, using the formula of deformation _, we write:

$$\delta_1 = \frac{P_1 L}{A_1 E_1}$$

$$\delta_2 = \frac{P_2 L}{A_2 E_2}$$

However, according to the geometry of the problem deformations, _1 and _2 must be equal. So we obtain:

$$\frac{P_1}{A_1 E_1} = \frac{P_2}{A_2 E_2}$$

$$P_1 = \frac{A_1 E_1 P_2}{A_2 E_2}$$

$$= \frac{1.25 \times 29000 \times P_2}{1.00 \times 5800}$$

$$P_1 = 6.25P_2$$

From Eq 1,

$$6.25p_2 + p_2 = 58$$

$$p_2 = 8\text{ kN}$$

From Eq 2,

$$p_1 = 6.25 \times 8$$

$$p_1 = 50\text{ kN}$$

$$\delta_2 = \frac{P_2 L}{A_2 E_2}$$

$$= \frac{8 \times 24}{1.00 \times 5800} = 0.033\text{ in}$$

Ans. $\delta = 0.033$ in, $p_1 = 50$ kN and $p_2 = 8$ kN

2.7 TEMPERATURE STRESS

2.7.1 DEFORMATION DUE TO TEMPERATURE CHANGE

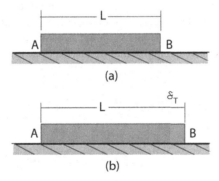

(a)

(b)

Deformation due to temperature change is a phenomenon where an object undergoes a change in its shape or size as a result of a change in temperature. This occurs because materials expand or contract when their temperature changes, and the degree of expansion or contraction depends on the material's coefficient of thermal expansion.

When an object is heated, its temperature increases, and its atoms and molecules vibrate more rapidly, causing the material to expand. Similarly, when an object is cooled, its temperature decreases, and the atoms and molecules vibrate less rapidly, causing the material to contract. The amount of deformation due to temperature change can be calculated using the material's coefficient of thermal expansion and the magnitude of the temperature change. The coefficient of thermal expansion is a measure of the material's tendency to expand or contract when its temperature changes. It is usually expressed in units of length per unit temperature change, such as mm/m/°C.

Deformation due to temperature change can have practical implications in engineering and construction, as it can cause thermal stresses in structures and equipment. These stresses can lead to deformation, cracking, or even failure of the material if they exceed the material's strength or yield strength. Therefore, it is important to consider the effects of temperature change when designing structures and equipment to ensure they can withstand the thermal stresses they will be exposed to.

$$\delta_T = \alpha\left(\Delta T\right)L$$

Symbol	Definition
δ_T	Change of length due to temperature change
α	Thermal expansion coefficient
ΔT	Change of temperature
L	Initial length of member

2.7.2   STRESS DUE TO TEMPERATURE CHANGE

Stress due to temperature change is a phenomenon where an object undergoes stress as a result of a change in temperature. This occurs because materials expand or contract when their temperature changes, and the degree of expansion or contraction depends on the material's coefficient of thermal expansion. The resulting change in size can create thermal stresses that can cause deformation, cracking, or even failure of the material.

When an object is heated, it expands, and when it is cooled, it contracts. If the object is constrained, for example, by being fastened in place, the expansion or contraction will create a stress within the material. This stress can be calculated using the material's modulus of elasticity, which is a measure of the material's stiffness, and the magnitude of the temperature change. If the stress exceeds the material's yield strength, the material may experience permanent deformation or even fracture. Therefore, it is important to consider the effects of temperature change when designing structures and equipment to ensure they can withstand the thermal stresses they will be exposed to. In some cases, it may be possible to reduce the effects of temperature change by using materials with a lower coefficient of thermal expansion, by allowing for some degree of freedom of movement in the structure or equipment, or by using thermal insulation to minimize temperature fluctuations.

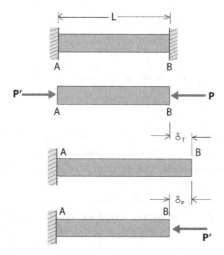

- A bar of L length is supported at both ends.
- Temperature is increased by ΔT.
- The bar tries to expand on both sides but cannot.
- Hence, compressive force P and P' generates.

To determine P:

- Release the bar at one end.
- Determine the elongation of length due to temperature, δ_T; $\delta_T = \alpha(\Delta T)L$.

- Apply force at the released end to compress the bar by

$$\delta_P = \frac{PL}{AE}$$

- By geometry:

$$\delta_B = 0$$
$$\delta_T + \delta_P = 0$$
$$\alpha(\Delta T)L + \frac{PL}{AE} = 0$$
$$P = -\alpha(\Delta T)AE$$

Example 17

Determine the support reaction at B of the steel bar shown when the temperature of the bar is +50°F, knowing that a close fit exists at both of the rigid supports when the temperature is −75°F. Use the values $E = 29,000$ ksi and $\alpha = 6.5 \times 10^{-6}/°F$ for steel. Also determine stress in both segments.

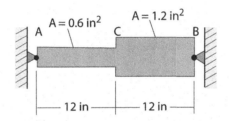

Solution

$$\Delta_T = (-75)°F - (+50)°F = -125°F$$

Deformation due to temperature change:

$$\begin{aligned}
\delta_T &= \Sigma \alpha(\Delta T)L \\
&= \alpha(\Delta T)(L_1 + L_2) \\
&= 6.5 \times 10^{-6}(-125)(12+12) \\
&= -19.5 \times 10^{-3} \text{in}
\end{aligned}$$

Deformation due to support reaction:

$$\begin{aligned}
\delta_R &= \Sigma \frac{PL}{AE} \\
&= \frac{PL}{E}\Sigma\frac{1}{A} \\
&= \frac{R_B \times 12}{29000}\left(\frac{1}{0.6} + \frac{1}{1.2}\right) \\
&= 1.0345 \times 10^{-6} R_B
\end{aligned}$$

Apply geometric condition at B:

$$\delta_B = 0$$
$$\delta_T + \delta_R = 0$$
$$-19.5 \times 10^{-3} + 1.0345 \times 10^{-6} R_B = 0$$
$$R_B = 18.85 \, \text{kip}$$

Stress at AC and BC:

$$\sigma_{AC} = \frac{F_{AC}}{A_{AC}} = \frac{18.85}{0.6} = 31.41 \, \text{ksi}$$

$$\sigma_{BC} = \frac{F_{AC}}{A_{AC}} = \frac{18.85}{1.2} = 15.71 \, \text{ksi}$$

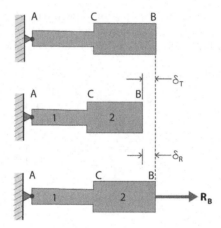

2.8   INTERNAL FORCE DIAGRAMS

Internal force diagrams are diagrams that represent the internal forces acting within a structure. These forces include axial forces, shear forces, and bending moments. The purpose of internal force diagrams is to help engineers and architects analyze the structural behavior of a building or other structure under different loads. Axial forces are forces that act along the length of a structural element, such as a column or a beam. They can be either tensile (pulling apart) or compressive (pushing together). Shear forces, on the other hand, act perpendicular to the length of a structural element and can cause it to bend or twist. *Bending moments* are forces that cause a structural element to bend or deform, and they occur when a load is applied at a distance from the center of the element.

To create an internal force diagram, engineers typically start by analyzing the external forces acting on the structure, such as the weight of the building and any loads that will be placed on it. They then use the principles of statics and mechanics to calculate the internal forces that will be present within the structure in response to those external forces. The internal force diagram will typically show the magnitude, direction, and location of the internal forces within the structure, along with any other relevant information, such as the material properties of the structural elements. This information can be used to design a structure that can withstand the expected loads and forces it will encounter over its lifetime.

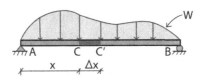

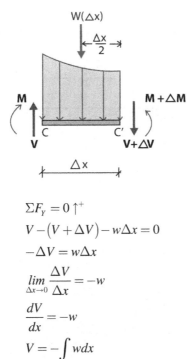

$$\Sigma F_Y = 0 \uparrow^+$$

$$V - (V + \Delta V) - w\Delta x = 0$$

$$-\Delta V = w\Delta x$$

$$\lim_{\Delta x \to 0} \frac{\Delta V}{\Delta x} = -w$$

$$\frac{dV}{dx} = -w$$

$$V = -\int w dx$$

- The slope of the shear diagram (dV/dx) is equal to the negative of a load $(-w)$.
- Shear force (V) is equal to the negative area of the load diagram $(-\int w dx)$.

$$\sum M_{C'} = 0 \curvearrowright^+$$

$$-(M + \Delta M) + M + V\Delta x - (w\Delta x)\frac{\Delta x}{2} = 0$$

$$-\Delta M = -V\Delta x + \frac{1}{2}w(\Delta x^2)$$

$$\lim_{\Delta x \to 0} \frac{\Delta M}{\Delta x} = \lim_{\Delta x \to 0}\left(V - \frac{1}{2}w\Delta x\right)$$

$$\frac{dM}{dx} = V$$

$$M = \int V dx$$

- The slope of the moment diagram (dM/dx) is equal to shear (V).
- Moment (M) is equal to the area of the shear diagram $\left(\int V dx\right)$.

2.8.1 INTERNAL FORCE DIAGRAMS FOR COMMONLY USED BEAMS

Example 18

Draw shear force and bending moment diagrams for a single point loaded beam structure.

Solution

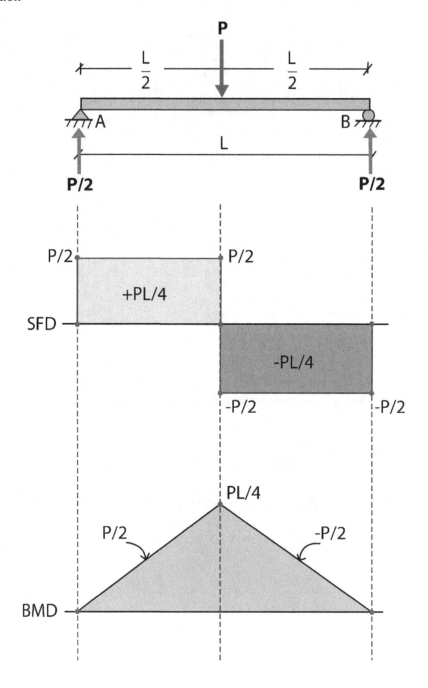

Example 19

Draw the shear force and the bending moment of the diagram for a two-point loaded beam structure.

Solution

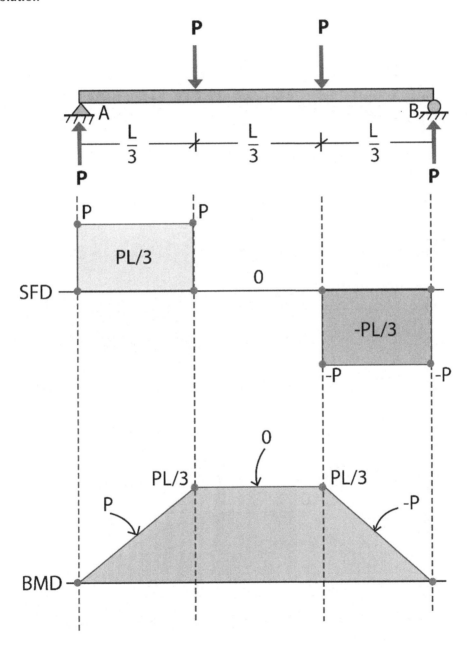

Example 20

Draw shear force and bending moment diagram for a uniformly loaded beam structure.

Solution

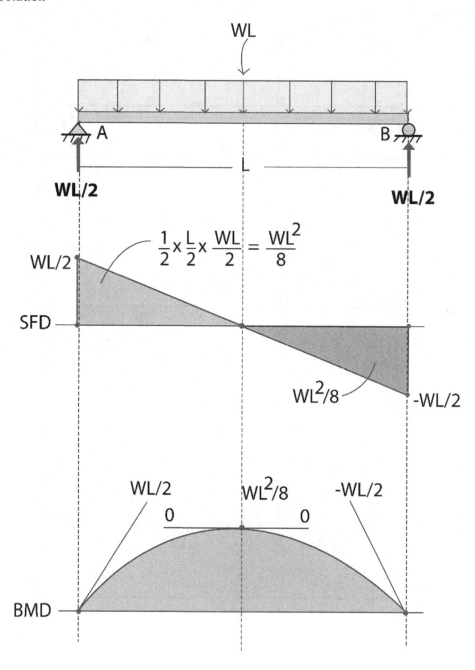

Example 21

Draw shear force and bending moment diagram for a cantilever beam.

Solution

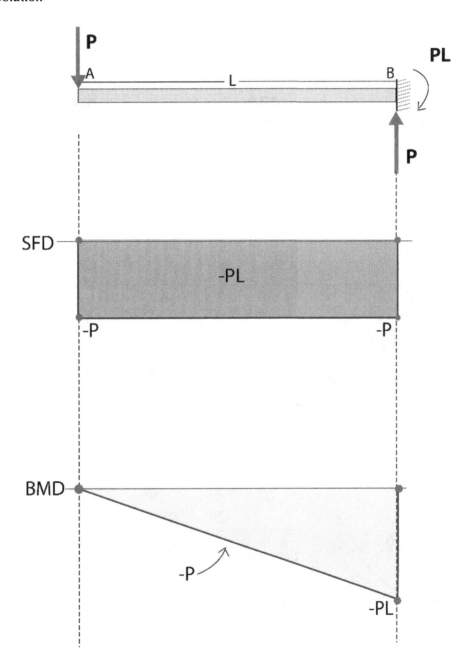

2.8.2 INTERNAL FORCE DIAGRAMS FOR BEAMS WITH MIXED TYPES OF LOADING CONDITION

Example 22

Draw shear force and bending moment diagrams for the beam structure.

Solution

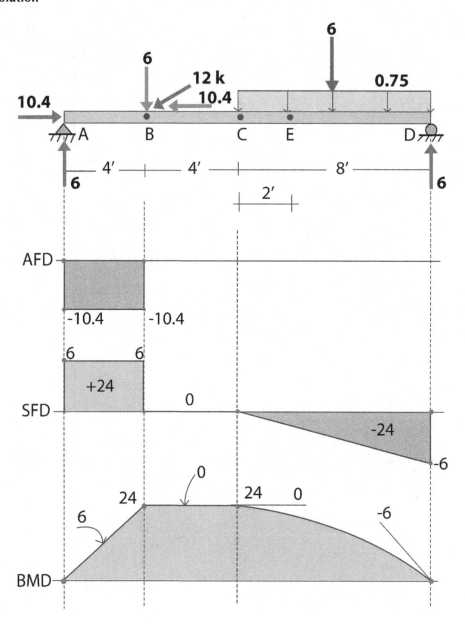

Example 23

Draw shear force and bending moment diagrams for the beam structure.

Solution

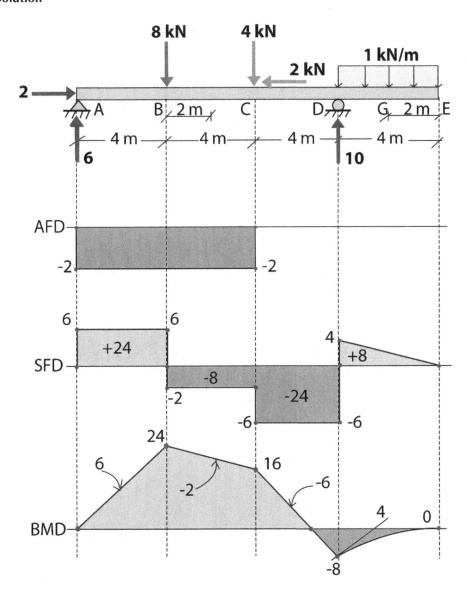

Example 24

Draw shear force and bending moment diagrams for the beam structure.

Solution

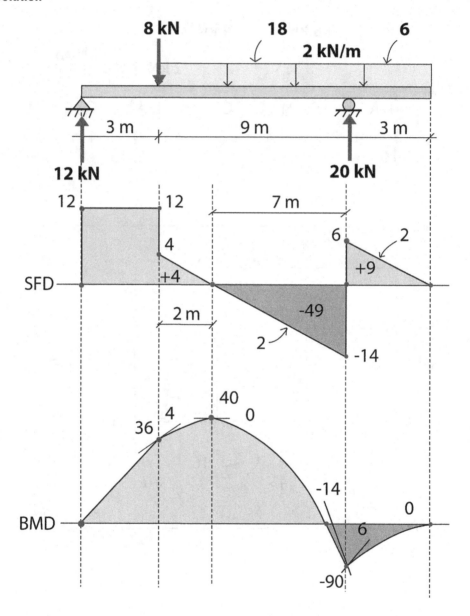

Example 25

Draw shear force and bending moment diagrams for the beam structure.

Solution

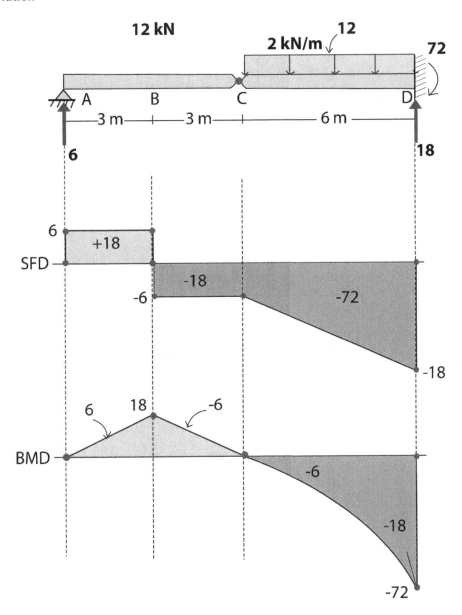

Example 26

Draw shear force and bending moment diagrams for the beam structure.

Solution

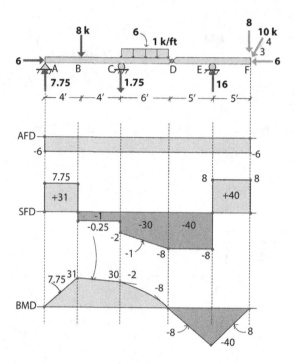

Example 27

Draw shear force and bending moment diagrams for the beam structure.

Solution

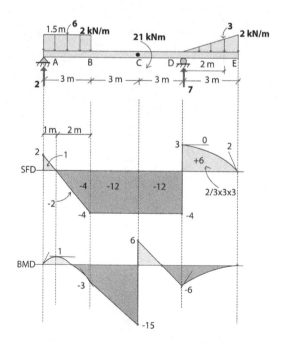

2.9   RIVET AND BOLT JOINTS

2.9.1   Bolt Joint

A *bolt joint* is a mechanical connection between two or more parts using bolts or screws. Bolt joints are widely used in various applications, including construction, automotive, aerospace, and manufacturing. The main components of a bolt joint are the bolt, nut, and parts being joined. The *bolt* is a threaded cylindrical rod that is inserted through holes in the parts to be joined. The *nut* is threaded onto the end of the bolt and tightened to create a clamping force that holds the parts together.

Bolt joints are typically designed to withstand both tensile and shear forces, depending on the application. The strength of the joint depends on the properties of the bolt, such as its diameter, material, and thread type, as well as the number and spacing of bolts used. When designing a bolt joint, it is important to consider the proper torque or tension required to achieve the desired clamping force. Over-tightening the bolt can lead to excessive stresses and potential failure of the joint, while under-tightening may result in insufficient clamping force and looseness of the joint.

To ensure the reliability and safety of bolt joints, regular inspection and maintenance are necessary, especially in high-stress applications. This includes checking for signs of corrosion, fatigue, and other types of damage that can weaken the joint over time.

2.9.2   Shearing Failure of Rivet

Shearing failure of a rivet occurs when the rivet fails due to the shearing force acting on it. A *rivet* is a mechanical fastener used to join two or more materials together permanently. It consists of a cylindrical shaft with a head at one end and a tail at the other end. The tail is inserted through holes in the materials being joined, and then it is deformed by applying pressure on the head end, forming a permanent joint.

Shearing failure of a rivet can occur when the shearing force acting on the rivet exceeds its shear strength. The *shear strength* of a rivet is the maximum load that it can withstand before it fails due to shearing. The shear strength of a rivet depends on its material, size, and shape. Shearing failure of a rivet can have serious consequences, as it can lead to the failure of the joint and the separation of the materials being joined. It is important to ensure that the rivet is properly sized and installed, and that the materials being joined are properly prepared and aligned. Proper installation involves ensuring that the rivet is correctly placed through the holes in the materials being joined, and that the rivet is properly tightened to the appropriate torque or tension.

To prevent shearing failure of a rivet, it is important to select the right type of rivet for the application, and to use the correct installation procedures. This may involve consulting with a professional engineer or a rivet manufacturer to ensure that the right type of rivet is selected and installed properly. It is also important to regularly inspect the joint to detect any signs of wear, damage, or corrosion that could weaken the rivet and lead to shearing failure.

$$P_s = Shear\ capacity\ of\ bolt\ x\ \frac{\pi d^2}{4}$$

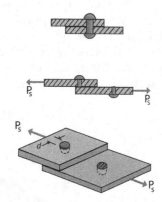

2.9.3 TENSILE FAILURE OF PLATE

Tensile failure of a plate occurs when the plate fails due to the tensile force acting on it. *Tensile force* is the force that pulls or stretches a material in opposite directions, causing it to elongate. When the tensile force acting on a plate exceeds its tensile strength, the plate will fail, resulting in a fracture or rupture. The *tensile strength* of a plate is the maximum stress that it can withstand before it fails due to tension. The tensile strength of a plate depends on its material properties, such as its yield strength, ultimate strength, and ductility. Other factors that can affect the tensile strength of a plate include its thickness, width, and length.

Tensile failure of a plate can occur due to various reasons, such as overloading, fatigue, corrosion, or improper manufacturing or installation. Overloading can cause the plate to experience a tensile force that exceeds its tensile strength, leading to fracture or rupture. Fatigue failure can occur when a plate is subjected to repeated tensile stresses over time, causing it to weaken and eventually fail. Corrosion can weaken the plate and reduce its tensile strength, making it more prone to failure. To prevent tensile failure of a plate, it is important to ensure that the plate is properly designed, manufactured, and installed. This includes selecting the right material for the application, ensuring that the plate is the correct size and thickness for the loads it will be subjected to, and ensuring that it is properly installed and maintained. It is also important to regularly inspect the plate for signs of wear, damage, or corrosion, and to replace it if necessary.

In summary, tensile failure of a plate can have serious consequences and should be avoided through proper design, material selection, manufacturing, installation, and maintenance. Regular inspection and testing of the plate can help detect any signs of potential failure before it occurs.

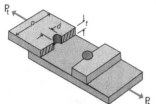

P_t = Tensile capacity of plate $(p - d_h)t$

2.9.4 BEARING FAILURE OF PLATE

Bearing failure of a plate occurs when the plate fails due to excessive pressure or load on a localized area, leading to the formation of cracks and, ultimately, fracture or collapse of the plate. Bearing failure can occur in many types of plates, including structural plates, machine parts, and bearings themselves. Bearing failure of a plate can be caused by various factors, including overloading, wear, misalignment, inadequate lubrication, and material defects. Overloading can cause excessive pressure on the plate, leading to the formation of cracks and, ultimately, bearing failure. Wear can occur due to repeated use or friction, which can cause the surface of the plate to become worn or damaged. Misalignment can cause uneven pressure on the plate, leading to bearing failure. Inadequate lubrication can cause friction, heat buildup, and wear, leading to premature failure of the plate. Material defects, such as porosity or inclusions, can also contribute to bearing failure by reducing the strength and durability of the plate.

To prevent bearing failure of a plate, it is important to ensure that the plate is properly designed and manufactured for the intended application. This includes selecting the appropriate material, size, and thickness for the loads and pressures it will be subjected to. Proper installation and maintenance of the plate, such as ensuring proper alignment and lubrication, can also help prevent bearing failure. Regular inspection of the plate for signs of wear, damage, or material defects is also important to detect and address any potential issues before they result in bearing failure. In summary, bearing failure of a plate can have serious consequences and should be avoided through proper design, material selection, manufacturing, installation, and maintenance. Regular inspection and testing of the plate can help detect any signs of potential failure before it occurs, ensuring the safe and reliable operation of the plate.

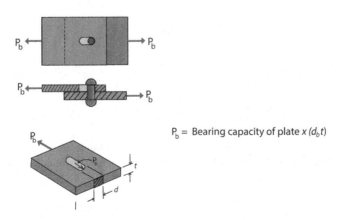

P_b = Bearing capacity of plate $x\ (d_b t)$

2.9.5   BEARING FAILURE OF RIVET

Bearing failure of a rivet occurs when the rivet fails due to excessive pressure or load on a localized area, leading to the formation of cracks and, ultimately, the separation of the materials being joined. A *rivet* is a mechanical fastener used to join two or more materials together permanently. Bearing failure of a rivet can be caused by various factors, including overloading, wear, misalignment, inadequate lubrication, and material defects. Overloading can cause excessive pressure on the rivet, leading to the formation of cracks and, ultimately, bearing failure. Wear can occur due to repeated use or friction, which can cause the surface of the rivet to become worn or damaged. Misalignment can cause uneven pressure on the rivet, leading to bearing failure. Inadequate lubrication can cause friction, heat buildup, and wear, leading to premature failure of the rivet. Material defects, such as porosity or inclusions, can also contribute to bearing failure by reducing the strength and durability of the rivet.

To prevent bearing failure of a rivet, it is important to ensure that the rivet is properly sized and installed for the intended application. This includes selecting the appropriate material, size, and shape for the loads and pressures it will be subjected to. Proper installation of the rivet, such as ensuring that it is correctly placed through the holes in the materials being joined and properly tightened to the appropriate torque or tension, can also help prevent bearing failure. Regular inspection of the rivet for signs of wear, damage, or material defects is also important to detect and address any potential issues before they result in bearing failure.

In summary, bearing failure of a rivet can have serious consequences and should be avoided through proper selection, installation, and maintenance. Regular inspection and testing of the rivet can help detect any signs of potential failure before it occurs, ensuring the safe and reliable operation of the joint.

$$P_b = Bearing\ capacity\ of\ bolt\ x\ (d_b t)$$

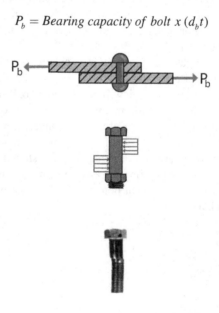

Example 28

Two steel plates, *A* (PL 8 in × 5/8 in) and *B* (PL 6 in × 3/4 in), are connected by bolts, as shown. The diameter of the rivets is 3/4 inch. Determine the tensile capacity of the steel connection.

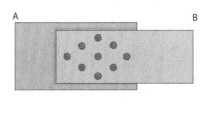

Given

Tensile capacity of plate: 25 ksi
Bearing capacity of plate: 26 ksi
Shearing capacity of bolt: 20 ksi
Bearing capacity of bolt: 24 ksi

Solution

Tensile failure of plate *B*:

$$\text{Sec } 1, P = \left[6 - 1\left(\frac{3}{4} + \frac{1}{8}\right)\right]\frac{3}{4} \times 25 = 96.1\,kip$$

$$\text{Sec } 2, P = \left[6 - 2\left(\frac{3}{4} + \frac{1}{8}\right)\right]\frac{3}{4} \times 25 + 1 \times \frac{\pi(3/4)^2}{4} \times 20 = 88.5\,kip$$

$$\text{Sec } 3, P = \left[6 - 3\left(\frac{3}{4} + \frac{1}{8}\right)\right]\frac{3}{4} \times 25 + 3 \times \frac{\pi(3/4)^2}{4} \times 20 = 89.8\,kip$$

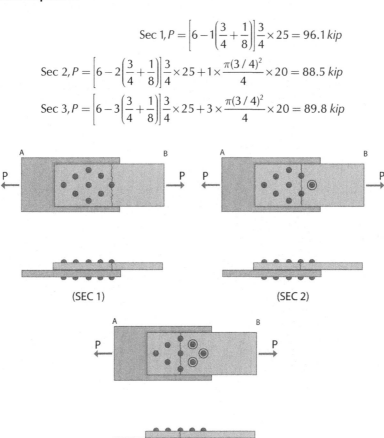

(SEC 1)   (SEC 2)

(SEC 3)

Tensile failure of plate *A*:

$$\text{Sec 4}, P = \left[8 - 1\left(\frac{3}{4} + \frac{1}{8}\right)\right]\frac{5}{8} \times 25 = 111.3 \; kip$$

$$\text{Sec 5}, P = \left[8 - 2\left(\frac{3}{4} + \frac{1}{8}\right)\right]\frac{5}{8} \times 25 + 1 \times \frac{\pi(3/4)^2}{4} \times 20 = 108.2 \; kip$$

$$\text{Sec 6}, P = \left[8 - 3\left(\frac{3}{4} + \frac{1}{8}\right)\right]\frac{5}{8} \times 25 + 3 \times \frac{\pi(3/4)^2}{4} \times 20 = 115.8 \; kip$$

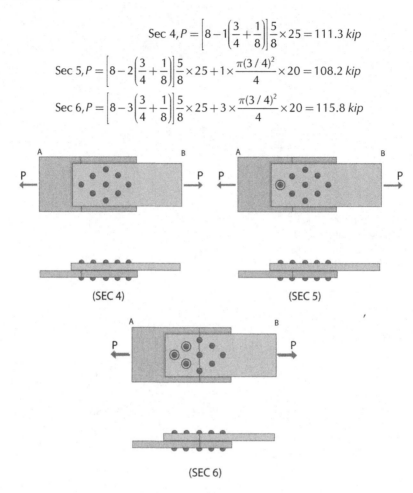

(SEC 4)   (SEC 5)

(SEC 6)

Bearing failure of plate *A*:

$$P = \left(\frac{3}{4} \times \frac{5}{8}\right) \times 26 \times 9 = 109.7 \; kip$$

Bearing failure of bolts:

$$P = \left(\frac{3}{4} \times \frac{5}{8}\right) \times 24 \times 9 = 101.3 \; kip$$

3 Flexural and Shearing Stress

Topics to be covered in this chapter are as follows:

- Bending in beam
- Shear in beam
- Stress transformation
- Deflection
- Indeterminate beam analysis
- Column

3.1 FLEXURAL STRESS

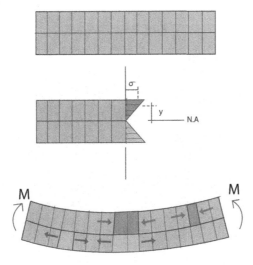

Flexural Stress: Normal stress at the section due to bending moment at that section. Flexural stress is zero at neutral axis but maximum at surface.

$$\sigma = \frac{My}{I}$$

Symbol	Description
σ	Flexural stress
M	Bending moment
y	Distance from neutral axis
I	Moment of inertia

DOI: 10.1201/9781032638072-3

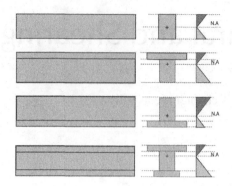

Example 1

Draw the flexural stress diagram for the following section, given that M = +40 kft.

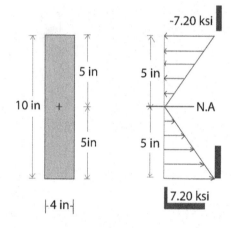

Solution

$$I = \frac{bh^3}{12} = \frac{4 \times 10^3}{12} = 333 \text{ in}^4$$

$$\sigma_{top} = \frac{My}{I} = \frac{(40 \times 12) \times 5}{333} = 7.20 \text{ ksi (comp.)}$$

$$\sigma_{bot} = \frac{My}{I} = \frac{(40 \times 12) \times 5}{333} = 7.20 \text{ ksi (tension)}$$

Example 2

Draw the flexural stress diagram for the following section, given that M = +40 kft.

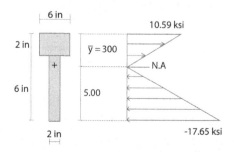

Solution

Located Centroid

$$\bar{y} = \frac{A_1Y_1 + A_2Y_2}{A_1 + A_2} = \frac{12 \times 1 + 12 \times 5}{12 + 12} = 3.00 \text{ in (from top)}$$

Moment of Inertia

$$\bar{I} = \sum \left[\frac{bh^3}{12} + Ad^2 \right] = \left[\frac{bh^3}{12} + Ad^2 \right]_{flange} + \left[\frac{bh^3}{12} + Ad^2 \right]_{web}$$

$$= \left[\frac{6 \times 2^3}{12} + 12 \times 2^2 \right] + \left[\frac{2 \times 6^3}{12} + 12 \times 2^2 \right] = 136 \text{ in}^4$$

Determine Stress

$$\sigma_{top} = \frac{My_t}{I} = \left[\frac{(40 \times 12) \times 3}{136} \right] = 10.59 \text{ ksi (tension)}$$

$$\sigma_{bot} = \frac{My_b}{I} = \left(\frac{(40 \times 12) \times 5}{136} \right) = 17.65 \text{ ksi (comp.)}$$

Example 3

Draw the flexural stress diagram at points B and D. What is the maximum tensile stress in the beam?

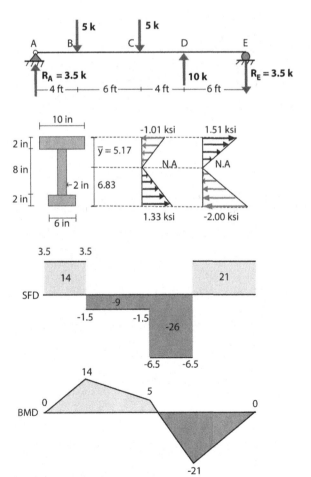

Solution

Reactions

$$\sum M_A = 00^+ (5 \times 4) + (5 \times 10) - (10 \times 14) - R_E \times 20 = 0$$

$$R_E = -3.5 \text{ kip } (\downarrow)$$

$$\sum F_y = 0 \uparrow^+ \; R_A - 5 - 5 + 10 - 3.5 = 0$$

$$R_A = +3.5 \text{ kip } (\uparrow)$$

Located Centroid and Determine the Moment of Inertia

$$\bar{y} = \frac{(20 \times 1) + (16 \times 6) + (12 \times 11)}{20 + 16 + 12} = 5.17 \text{ in (from top)}$$

$$\bar{I} = \left[\frac{10 \times 2^3}{12} + 20 \times 4.17^2 \right] + \left[\frac{2 \times 8^3}{12} + 16 \times 0.83^2 \right]$$

$$+ \left[\frac{6 \times 2^3}{12} + 12 \times 5.83^3 \right] = 862 \text{ in}^4$$

Flexural Stresses at B

$$\sigma_{top} = \frac{M_B Y_t}{I} \frac{(14 \times 12) \times 5.17}{862} = 1.01 \text{ ksi (comp.)}$$

$$\sigma_{bot} = \frac{M_B Y_b}{I} = \frac{(14 \times 12) \times 6.83}{862} = 1.33 \text{ ksi (tension)}$$

Flexural Stresses at D

$$\sigma_{top} = \frac{M_D Y_t}{I} = \frac{(21 \times 12) \times 5.17}{862} = 1.51 \text{ ksi (tension)}$$

$$\sigma_{bot} = \frac{M_D Y_b}{I} = \frac{(21 \times 12) \times 6.83}{862} = 2.00 \text{ ksi (comp.)}$$

Example 4

A timber beam *AD* carries two point loads at *B* and *C*. The cross section of the beam is rectangular of 6 in width. If the maximum normal stress of timber is limited to 2 ksi, then determine the minimum required height of the beam.

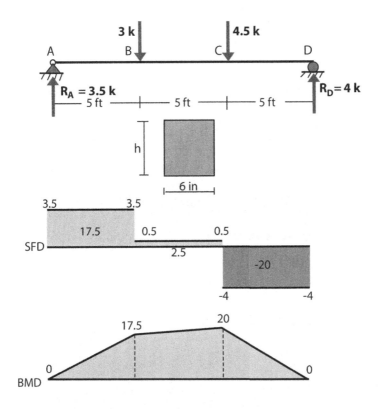

Solution

$$R_D = 4 \, \text{Kip} \, (\uparrow)$$

$$\sum F_y = 0 \uparrow^+$$

$$R_A - 3 - 4.5 + 4 = 0$$

$$R_A = +3.5 \, \text{kip} (\uparrow)$$

From the moment diagram, the maximum moment $M = 20$ kft.

Since the height of the beam section h is unknown, I can be expressed in terms of h.

$$I = \frac{bh^3}{12} = \frac{6h^3}{12} = \frac{h^3}{2}$$

Now, applying the flexural stress formula for the top- or bottommost surface, where maximum stress occurs and $y = h/2$:

$$\sigma = \frac{My}{I} = \frac{M(h/2)}{h^3/2} = \frac{M}{h^2}$$

$$h = \sqrt{\frac{M}{\sigma}} = \sqrt{\frac{20 \times 12}{2}} = 10.95 \text{ in}$$

Since cross-sectional sizes are usually provided in integer, h should be upper rounded. So provided $h = 11$ in.

Answer: $h = 11$ in.

3.2 COMPOSITE SECTION

Flexural Stress.

Example 5

Question: The following timber section is reinforced by one steel plate attached at the bottom surface. If the applied moment on the section is +50kft, then draw its flexural stress diagram, given that the moduli of elasticity of steel and timber are, respectively, 29,000ksi and 1,450 ksi.

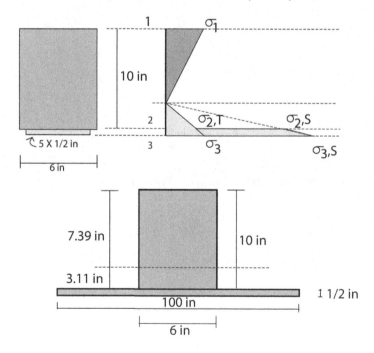

Solution

It is given that $E_s = 29,000$ ksi $E_t = 1,450$ ksi:

$$n = \frac{E_s}{E_c} = \frac{29000}{1450} = 20$$

Equivalent Homogeneous Section

The composite section should be transformed to a homogeneous one. Let us transform steel to timber by multiplying the width of steel by n. So steel width becomes $5 \times 20 = 100$ in.

Locate centroid:

$$\bar{y} = \frac{\sum Ay}{\sum A} = \frac{50 \times 0.25 + 60 \times 5.5}{50 + 60} = 3.11$$

Moment of inertia:

$$\bar{I} = \frac{100 \times 0.5^3}{12} + 50 \times (3.11 - 0.25)^2$$
$$+ \frac{6 \times 10^3}{12} + 60 \times (5.5 - 3.11)^2 = 125^4$$

Flexural stresses:

$$\sigma_1 = \frac{(50 \times 12) \times 7.39}{1253} = 3.53 \text{ ksi}$$
$$\sigma_2 = \frac{(50 \times 12) \times 2.61}{1253} = 1.25 \text{ ksi}$$
$$\sigma_3 = \frac{(50 \times 12) \times 3.11}{1253} = 1.49 \text{ ksi}$$

Corrected flexural stresses:

$$\sigma_{2,T} = \sigma_2 = 1.25 \text{ ksi}$$
$$\sigma_{2,S} = \sigma_2 \times 20 = 25.0 \text{ ksi}$$
$$\sigma_{3,S} = \sigma_3 \times 20 = 29.8 \text{ ksi}$$

Example 6

Question: The following timber is reinforced by two steel plates attached at the top and the bottom surfaces. If the applied moment on the section is 50kft, then draw its flexural stress diagram, given that the moduli of elasticity of steel and timber are, respectively, 29,000ksi and 1,450ksi.

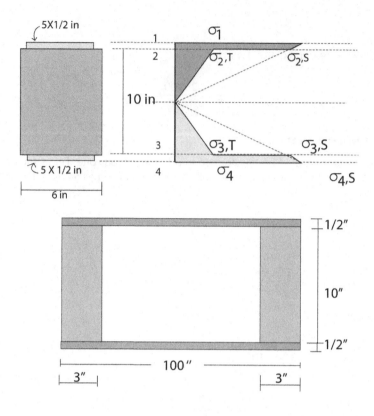

Solution

It is given that $E_s = 29,000$ksi and $E_t = 1,450$ksi. Modular ratio:

$$n = \frac{E_s}{E_c} = \frac{29000}{1450} = 20$$

Equivalent Homogeneous Section

The composite beam section should be transformed to a homogenous one. Let us transform steel to timber by multiplying the width of steel by n. So steel width becomes $5 \times 20 = 100$ in.

Moment of inertia:

$$\bar{I} = \frac{100 \times 11^3}{12} - \frac{94 \times 10^3}{12} = 3258 \text{ in}^4$$

Flexural stresses:

$$\sigma_1 = \sigma_4 = \frac{(50 \times 12) \times 5.5}{3258} = 1.01 \text{ksi}$$

$$\sigma_2 = \sigma_3 = \frac{(50 \times 12) \times 5.0}{3258} = 0.92 \text{ksi}$$

Corrected flexural stresses:

$$\sigma_{1,S} = \sigma_1 \times 20 \quad = 20.2 \text{ ksi}$$
$$\sigma_{2,S} = \sigma_2 \times 20 \quad = 18.4 \text{ ksi}$$
$$\sigma_{2,T} \qquad\qquad\quad = 0.92 \text{ ksi}$$
$$\sigma_{3,T} \qquad\qquad\quad = 0.92 \text{ ksi}$$
$$\sigma_{3,S} = \sigma_3 \times 20 \quad = 18.4 \text{ ksi}$$
$$\sigma_{4,S} = \sigma_4 \times 20 \quad = 20.2 \text{ ksi}$$

3.3 SHEAR FLOW

3.3.1 HORIZONTAL SHEAR FLOW

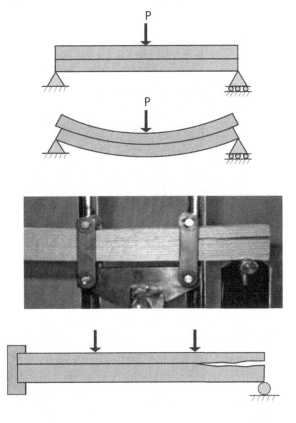

Horizontal shear stresses developed along each horizontal plane of the beam. If any such plane cannot transfer that stress, shear failure occurs along that particular plane.

Horizontal shear flow per unit length:

$$q = \frac{VQ}{I}$$

Symbol	Description
V	Shear force
Q	Static moment of segment of section above specified level
I	Moment of Inertia

Example 7

Question: Three boards, each 2 in thick, are nailed together to form a beam that is subjected to a vertical shear of 10 kip. Determine the spacing of the nail if each nail can resist 9 kip of shear force.

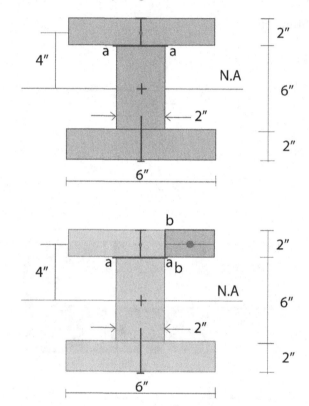

Solution

Static moment of the section up to level $a - a$:

$$Q = Ay = (6 \times 2) \times 4 = 48 \text{ in}^3$$

Moment of inertia:

$$I = 2\left[\frac{6 \times 2^3}{12} + 12 \times 4^2\right] + \frac{2 \times 6^3}{12} = 428 \text{ in}^4$$

Horizontal shear flow per unit length:

$$q = \frac{VQ}{I} = \frac{10 \text{ kip} \times 48 \text{ in}^3}{428 \text{ in}^4} = 1.12 \text{ kip/in}$$

Spacing of nail:

$$s = \frac{9\,\text{kip}}{1.12\,\text{kip}\,/\,\text{in}} = 8.03\,\text{in} \approx 8\,\text{in}$$

Static moment of the section up to level $b-b$:

$$Q = Ay = (2\times2)\times4 = 16\,\text{in}^3$$

Horizontal shear flow per unit length:

$$q = \frac{VQ}{I} = \frac{10\,\text{kip}\times16\,\text{in}^3}{428\,\text{in}^4} = 0.373\,\text{kip}\,/\,\text{in}$$

Spacing of nail:

$$s = \frac{9\,\text{kip}}{0.373\,\text{kip/in}} = 24.12\,\text{in} \approx 24\,\text{in}$$

Example 8

Question: Three boards, each 2 in thick, are nailed together to form a beam of 12 ft length that is subjected to a distributed load of $w = 2\,\text{kip}\,/\,\text{ft}$. Determine the spacing of the nail if each nail can resist 9 kip of shear force.

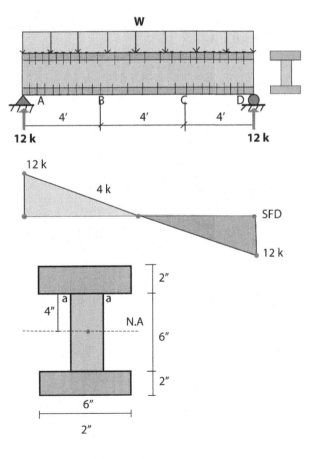

Solution

Static moment of the section up to level $a - a$:

$$Q = Ay = (6 \times 2) \times 4 = 48 \text{ in}^3$$

Moment of inertia:

$$I = 2\left[\frac{6 \times 2^3}{12} + 12 \times 4^2\right] + \frac{2 \times 6^3}{12} = 428 \text{ in}^4$$

Segment AB and CD:

Shear force should be considered at support. $V_A = V_D = 12$ kip horizontal shear flow per unit length:

$$q = \frac{VQ}{I} = \frac{12 \text{ kip} \times 48 \text{ in}^3}{428 \text{ in}^4} = 1.34 \text{ kip / in}$$

Spacing of nail:

$$s = \frac{9 \text{ kip}}{1.34 \text{ kip / in}} = 6.71 \text{ in} \approx 6 \text{ in}$$

Segment BC:

Shear force should be considered at B.

$$V_B = 12 - 2 \times 4 = 4 \text{ kip}$$

Horizontal shear flow per unit length:

$$q = \frac{VQ}{I} = \frac{4 \text{ kip} \times 48 \text{ in}^3}{428 \text{ in}^4} = 0.44 \text{ kip / in}$$

Spacing of nail:

$$s = \frac{9 \text{ kip}}{0.44 \text{ kip / in}} = 20.45 \text{ in} \approx 20 \text{ in}$$

3.4 SHEAR STRESS

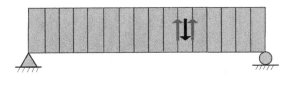

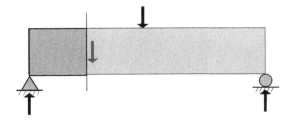

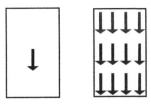

Shear stress is generated due to the action of shear force V in beam. These stresses lie in the plane of the cross section.

Shear stress $\left(\tau\right)$ at any arbitrary level in the section is:

$$\tau = \frac{VQ}{It}$$

Symbol	Description
V	*Shear force*
Q	*Static moment of segment of section above specified level*
I	*Moment of Inertia*
t	*Thickness at the specified level*

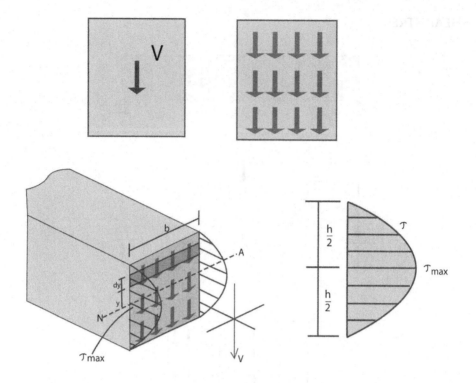

3.4.1 SHEAR STRESS FORMULA

For a rectangular section, it can be proved that:

$$\tau = \frac{VQ}{It} = \frac{6V}{bh^3}\left(\frac{h^2}{4} - y^2\right)$$

3.4.2 MAXIMUM SHEAR STRESS

The maximum shear stress occurs at neutral axis, which can be found by putting $y = 0$ in the shear stress formula.

$$\tau = \frac{6V}{bh^3}\left(\frac{h^2}{4} - 0^2\right) = \frac{3V}{2bh} = 1.5\frac{V}{A}$$

Example 9

Question: Determine the shear stress of the rectangular section at the specified levels, given that $V = 8$ kip.

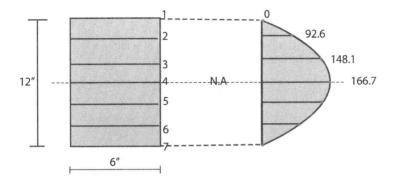

Solution

$$\tau = \frac{6V}{bh^3}\left(\frac{h^2}{4} - y^2\right)$$
$$= \frac{6 \times 8000}{6 \times 12^3}\left(\frac{12^2}{4} - y^2\right)$$
$$= 4.629\left(36 - y^2\right)$$

Shear Stress

Level	$y\,(\text{in})$	$\tau\,(\text{psi})$
1	6	0
2	4	92.6
3	2	148.1
4	0	166.7
5	−2	148.1
6	−4	92.6
7	−6	0

Sample calculation for level 3:

$$y = 2\,\text{in}$$
$$\tau = 4.626\left(36 - 2^2\right)$$
$$= 0.1481\,\text{ksi}$$
$$= 148.1\,\text{psi}$$

Example 10

Question: Determine shear stresses in the following beam at section $A-A$ at the specified levels shown in the cross section. Each level is 2 in apart.

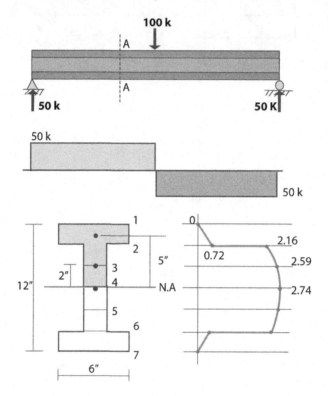

Solution

From the shear force diagram, we find $V = 50$ kip. Moment of inertia:

$$I = 2\left[\frac{6\times2^3}{12}+12\times5^2\right]+\frac{2\times8^3}{12}=693 \text{ in}^4$$

Shear stress at level 1–1:

$$Q_{1-1} = Ay = 0$$
$$\tau_{1-1} = \frac{VQ}{It} = 0$$

Shear stress at just above level 2–2.

$$Q_{2-2} = Ay = (6\times2)\times5 = 60 \text{ in}^3$$
$$\tau_{2-2} = \frac{VQ}{It} = \frac{50\times60}{693\times6} = 0.72 \text{ ksi}$$

Shear stress at just below level 2–2.

$$Q_{2-2} = Ay = (6 \times 2) \times 5 = 60 \text{ in}^3$$

$$\tau_{2-2} = \frac{VQ}{It} = \frac{50 \times 60}{693 \times 2} = 2.16 \text{ ksi}$$

Shear stress at level $3 - 3$.

$$Q_{3-3} = \sum Ay = (6 \times 2) \times 5 + (2 \times 2) \times 3 = 72 \text{ in}^3$$

$$\tau_{3-3} = \frac{VQ}{It} = \frac{50 \times 72}{693 \times 2} = 2.59 \text{ ksi}$$

Shear stress at level $4 - 4$.

$$Q_{4-4} = \sum Ay = (6 \times 2) \times 5 + (2 \times 4) \times 2 = 76 \text{ in}^3$$

$$\tau_{4-4} = \frac{VQ}{It} = \frac{50 \times 76}{693 \times 2} = 2.74 \text{ ksi}$$

3.5 STRESS TRANSFORMATION

3.5.1 ALGEBRAIC APPROACH

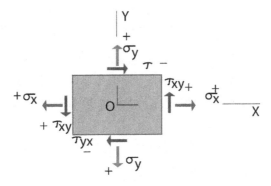

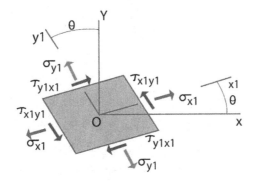

symbol	Description
σ_x	Normal stress in x direction.
σ_y	Normal stress in y direction.
τ_{xy}	xy.
τ_{yx}	Shear stress perpendicular to y and in the direction of x.
θ	Angle of rotation of xy plane.
σ_{x_1}	Normal stress in x_1 direction.
σ_{y_1}	Normal stress in y_1 direction.
$\tau_{x_1 y_1}, \tau_{y_1 x_1}$	Shear stress in $x_1 y_1$ plane.

3.5.2 STRESS FORMULAE

$$\sigma_{x_1} = \frac{\sigma_x + \sigma_y}{2} + \frac{\sigma_x - \sigma_y}{2}\cos 2\theta + \tau_{xy}\sin 2\theta$$

$$\tau_{x_1 y_1} = -\frac{\sigma_x - \sigma_y}{2}\sin 2\theta + \tau_{xy}\cos 2\theta$$

$$\sigma_{y_1} = \frac{\sigma_x + \sigma_y}{2} + \frac{\sigma_x - \sigma_y}{2}\cos 2(\theta + 90) + \tau_{xy}\sin 2(\theta + 90)$$

Example 11

For the given stress block, determine $\sigma_{x_1}, \tau_{x_1 y_1},$ and σ_{y_1} at $40°$ anticlockwise plane, then draw the corresponding stress block.

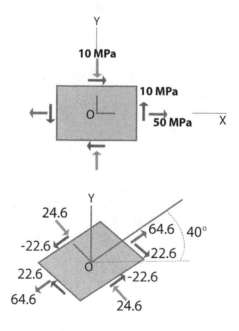

Solution

It is given that $\sigma_x = 50$ MPa, $\sigma_y = -10$ MPa, $\tau_{xy} = 40$ MPa, and $\theta = 40°$.

$$\sigma_{x_1} = \frac{\sigma_x + \sigma_y}{2} + \frac{\sigma_x - \sigma_y}{2}\cos 2\theta + \tau_{xy}\sin 2\theta$$
$$= \frac{50 + (-10)}{2} + \frac{50 - (-10)}{2}\cos 2(40) + (40)\sin 2(40)$$
$$= 64.6 \text{ MPa}$$

$$\sigma_{y_1} = \frac{\sigma_x + \sigma_y}{2} + \frac{\sigma_x - \sigma_y}{2}\cos 2(\theta + 90) + \tau_{xy}\sin 2(\theta + 90)$$
$$= \frac{50 + (-10)}{2} + \frac{50 - (-10)}{2}\cos 2(130) + (40)\sin 2(130)$$
$$= -24.6 \text{ MPa}$$

$$\tau_{x_1y_1} = -\frac{\sigma_x - \sigma_y}{2}\sin 2\theta + \tau_{xy}\cos 2\theta$$
$$= -\frac{50 - (-10)}{2}\sin 2(40) + 40\cos 2(40)$$
$$= -22.6 \text{ MPa}$$

3.6 MOHR'S CIRCLE OF STRESS

3.6.1 GRAPHICAL APPROACH TO STRESS TRANSFORMATIONS

Example 12

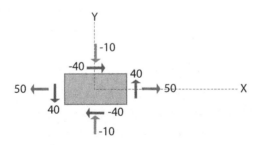

$$C_x = \frac{\sigma_x + \sigma_y}{2} = \frac{50 + (-10)}{2} = 20$$
$$C_y = 0$$

$$P_x = \sigma_x = 50$$

$$P_y = \tau_{xy} = 40$$

CP represents $\theta = 0°$

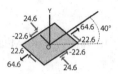

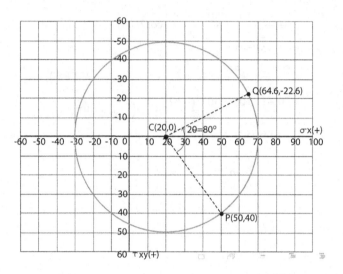

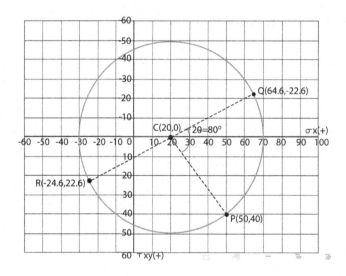

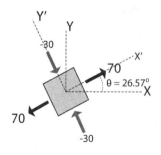

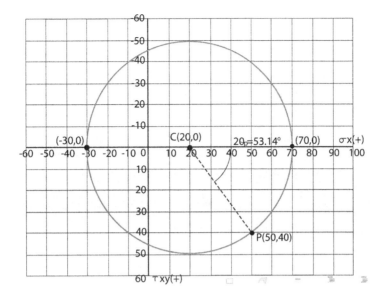

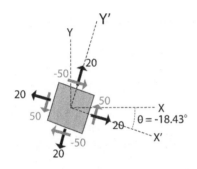

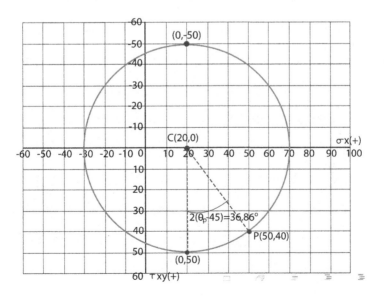

3.7 DEFLECTION OF BEAM

3.7.1 DIRECT INTEGRATION METHOD

3.7.1.1 Deflection

Deflection: The degree to which a structural element is displaced under a load. It may refer to an angle or a distance.

3.7.1.2 Significance

Deflection plays an important role in the design of high-rise buildings, as the top floors are susceptible to sway due to lateral load, such as wind or earthquake.

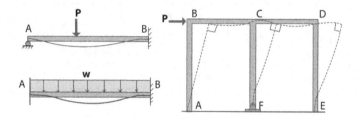

Elastic Curve: The deflected shape of the structure is known as the elastic curve.

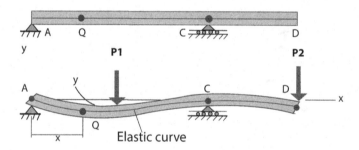

Another important consideration is serviceability. A structure is considered serviceable if the occupant feels comfortable around it. If the occupant thinks that there might be an impending failure, then the structure is not considered serviceable despite being structurally safe.

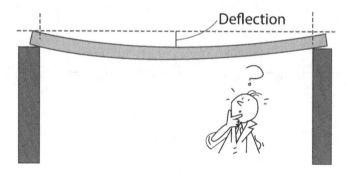

3.7.1.3 Boundary Conditions

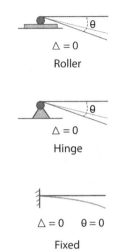

$\triangle = 0$

Roller

$\triangle = 0$

Hinge

$\triangle = 0 \quad \theta = 0$

Fixed

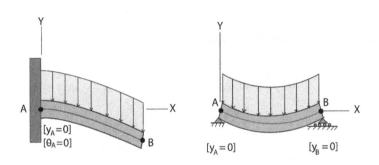

(a) Cantilever Beam

(b) Simply Supported Beam

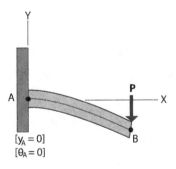

(c) Cantilever Beam

Example 13

Question: Determine the expression of deflection and slope for the cantilever beam, then evaluate that for point A.

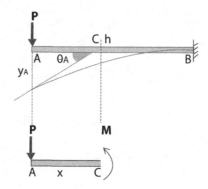

Solution

$$\sum M_C = 0 U^+ - Px - M = 0; \ M = -Px$$

3.7.1.4 Differential Equation of Deflection

$$EI\frac{d^2y}{dx^2} = M = -Px$$

$$EI\frac{dy}{dx} = -\frac{Px^2}{2} + C_1 \dots\dots\dots(1)$$

$$EIy = -\frac{Px^3}{6} + C_1 x + C_2 \dots\dots\dots(2)$$

3.7.1.5 Boundary Conditions

At $B, x = L$ and $dy/dx = 0$, substituting these in eq. (1):

$$0 = -\frac{PL^2}{2} + C_1; \ C_1 = \frac{PL^2}{2}$$

Again, at $B, x = L$ and $y = 0$, substituting these in eq. (2):

$$0 = -\frac{PL^3}{6} + \left(\frac{PL^2}{2}\right)L + C_2; \ C_2 = -\frac{PL^3}{3}$$

Substituting C_1 in eq. (1):

$$EI\frac{dy}{dx} = -\frac{Px^2}{2} + \frac{PL^2}{2}$$

$$\frac{dy}{dx} = \frac{P}{2EI}\left(-x^2 + L^2\right) \dots\dots\dots\dots(3)$$

Similarly, substituting C_1 and C_2 in eq. (2), we write:

$$Ely = -\frac{Px^3}{6} + \frac{PL^2}{2}x - \frac{PL^3}{3}$$

$$y = \frac{P}{6EI}\left(-x^3 + 3L^2x - 2L^3\right)\ldots\ldots\ldots(4)$$

Evaluation of θ_A and y_A for point A:
Putting $x = 0$ in eq. (3):

$$\theta_A = \left(\frac{dy}{dx}\right)_A = \frac{P}{2EI}\left(0 + L^2\right) = +\frac{PL^2}{2EI}(\nearrow)$$

Putting $x = 0$ in eq. (4):

$$y_A = \frac{P}{6EI}\left(0 + 0 - 2L^3\right) = -\frac{PL^3}{3EI}(\downarrow)$$

Example 14

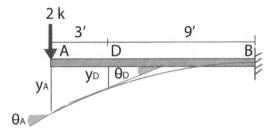

Question: The following beam AB is made of steel of $W\,8 \times 15$ section. Determine the deflection and slope both at points A and D, given that $I = 48$ in^4.

Solution

Here, $L = 12$ ft $= 144$ in, and we know that $E = 29000$ ksi for steel.
Point A:

$$y_A = -\frac{PL^3}{3EI} = -\frac{2\times(144)^3}{3\times 29000\times 48} = -1.43 \text{ in} (\downarrow)$$

Point D:

$$y_D = \frac{P}{6EI}\left(-x^3 + 3L^2x - 2L^3\right)$$

$$= \frac{2\left(-36^3 + 3\times 144^2 \times 36 - 2\times 144^3\right)}{6\times 29000\times 48}$$

$$= -0.90 \text{ in} (\downarrow)$$

$$\theta_D = \frac{P}{2EI}\left(-x^2 + L^2\right)$$

$$= \frac{2\left(-36^2 + 144^2\right)}{2\times 29000\times 48}$$

Question: For the same beam in the previous problem, determine the maximum load that can be applied at A if the deflection at A is limited to 1.0 in.

Solution

$$y_A = -\frac{PL^3}{3EI};$$

$$= -\frac{PL^3}{3EI};$$

$$P = -\frac{3EIy_A}{L^3}$$

$$= -\frac{3 \times 29000 \times 48 \times (-1.0)}{144^3} = 1.39 \text{ kip}$$

Example 15

Question: Determine the expression of deflection and the slope for the simply supported beam, then evaluate the slope at point A and deflection at midspan.

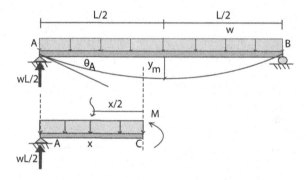

$$\sum M_C = 0 \, U^+ \left(\frac{wL}{2}\right)x - (wx)\frac{x}{2} - M = 0$$

$$M = \frac{1}{2}wLx - \frac{1}{2}wx^2$$

3.7.1.6 Differential Equation of Deflection

$$EI\frac{d^2y}{dx^2} = M = \frac{1}{2}wLx - \frac{1}{2}wx^2$$

$$EI\frac{dy}{dx} = \frac{1}{4}wLx^2 - \frac{1}{6}wx^3 + C_1 \dots \dots (1)$$

$$EIy = \frac{1}{12}wLx^3 - \frac{1}{24}wx^4 + C_1x + C_2 \dots \dots (2)$$

3.7.1.7 Boundary Conditions

At $A, x = 0$ and $y = 0$, substituting these in eq. (2):

$0 = 0 - 0 + 0 + C_2$

$C_2 = 0$

At $B, x = L$ and $y = 0$, substituting these in eq. (2):

$$0 = \frac{1}{12} wL \cdot L^3 - \frac{1}{24} wL^4 + C_1 L; \ C_1 = -\frac{wL^3}{24}$$

Expression of dy/dx and y.

Substituting C_1 in eq. (1):

$$EI \frac{dy}{dx} = \frac{1}{4} wLx^2 - \frac{1}{6} wx^3 - \frac{wL^3}{24}$$

$$\frac{dy}{dx} = \frac{w}{24EI} \left(-4x^3 + 6Lx^2 - L^3 \right) \dots\dots\dots\dots\dots (3)$$

Similarly, substituting C_1 and C_2 in eq. (2):

$$EIy = \frac{1}{12} wLx^3 - \frac{1}{24} wx^4 - \left(\frac{wL^3}{24} \right) x + 0$$

$$y = \frac{w}{24EI} \left(-x^4 + 2Lx^3 - L^3 x \right) \dots\dots\dots\dots\dots (4)$$

Evaluation of θ_A for point A:

Putting $x = 0$ in eq. (3):

$$\theta_A = \frac{w}{24EI} \left(0 + 0 - L^3 \right) = -\frac{wL^3}{24EI} (\square)$$

Evaluation of y_m for midspan. Putting $x = L/2$ in eq. (4):

$$y_m = \frac{w}{24EI} \left[-\left(\frac{L}{2} \right)^4 + 2L \left(\frac{L}{2} \right)^3 - L^3 \left(\frac{L}{2} \right) \right] = -\frac{5wL^4}{384EI} (\downarrow)$$

3.8 DEFLECTION OF BEAM

3.8.1 Moment Area Method

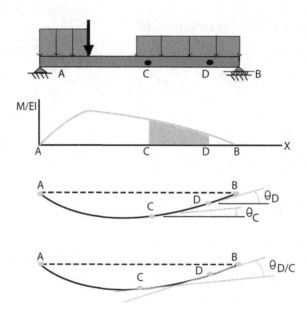

3.8.1.1 First Theorem

The change in slope between any two points on the elastic curve equals the area of the M/EI diagram between these two points.

Applying the first theorem between C and D points:

$$\theta_{C/D} = A_{CD}$$

Sym.	Description
θ_{CD}	Change in slope between C and D.
A_{CD}	Area of M/EI diagram between C and D (shaded area).

3.8.1.2 Moment Area Method

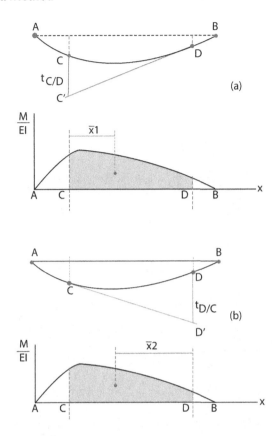

(a)

(b)

3.8.1.3 Second Theorem

The vertical deviation of the tangent at a point on the elastic curve with respect to the tangent extended from another point equals the static moment of the area under the *M/EI* diagram between the two points taken about the point where the deviation is to be determined.

For tangential deviation at C and D, respectively, we write:

$$t_{C/D} = A_{CD} \cdot \bar{x}_1 \text{ and } t_{D/C} = A_{CD} \cdot \bar{x}_2$$

Sym.	Description
$\theta_{C/D}$	Deviation of C on the tangent drawn from D.
$\theta_{D/C}$	Deviation of D on the tangent drawn from C.
A_{CD}	Area of *M/EI* diagram between C and D (shaded area).
$\bar{x}_1$	Distance from the center of gravity of A_{CD} to C.
$\bar{x}_2$	Distance from the center of gravity of A_{CD} to D.

Shape	Figure	Area	Centroidal Distance, $\bar{x}$
Right Triangle		$\dfrac{bh}{2}$	$\dfrac{b}{3}$
Parabola		$\dfrac{2bh}{3}$	$\dfrac{3b}{8}$
Parabola		$\dfrac{bh}{3}$	$\dfrac{b}{4}$

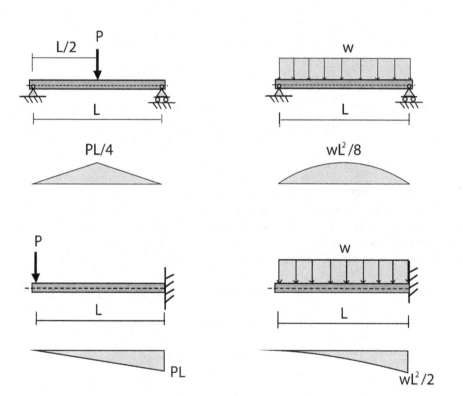

Example 16

Question: Determine the slope and deflection of the cantilever beam at point B due to point load P.

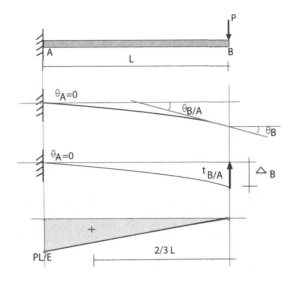

Slope at B:

Applying the first theorem of the moment–area method, the change in angle from A to B is:

$$\theta_{B/A} = \text{Area of } M/EI \text{ diagram between } A \text{ and } B$$

$$= \frac{1}{2}\left(\frac{PL}{EI}\right)L = \frac{PL^2}{2EI}$$

According to geometry of the elastic curve:

$$\theta_B = \theta_{B/A} = \frac{PL^2}{2EI}$$

Deflection at B:

Applying the second theorem of moment – area method, the deviation of B on the tangent drawn from A is:

$t_{B/A}$ = static moment of area under M/EI diagram between A and B taken about point B

$$= \left[\frac{1}{2}\left(\frac{PL}{EI}\right)L\right]\frac{2L}{3} = \frac{PL^3}{3EI}$$

According to the geometry of the elastic curve:

$$\Delta_B = t_{B/A} = \frac{PL^3}{3EI}$$

Answer: $\theta_B = \dfrac{PL^2}{2EI}$ and $\Delta_B = \dfrac{PL^3}{3EI}$

Example 17

Question: Determine the slope and deflection of the cantilever beam at point B due to uniformly distributed load w.

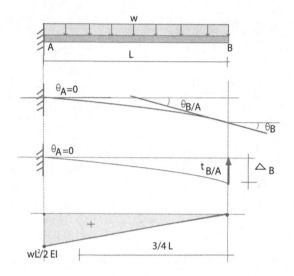

Slope at B:

Applying the first theorem of the moment–area method, the change in angle from A to B is:

$$\theta_{B/A} = \text{Area of } M/EI \text{ diagram between } A \text{ and } B$$

$$= \frac{1}{3}\left(\frac{wL^2}{2EI}\right)L = \frac{wL^3}{6EI}$$

According to geometry of the elastic curve:

$$\theta_B = \theta_{B/A} = \frac{wL^3}{6EI}$$

Deflection at B.

Applying the second theorem of the moment–area method, the deviation of B on the tangent drawn from A is:

$t_{B/A} = $ static moment of area under M/EI diagram between A and B taken about point B

$$= \left[\frac{1}{3}\left(\frac{wL^2}{2EI}\right)L\right]\frac{3L}{4} = \frac{wL^4}{8EI}$$

According to geometry of the elastic curve:

$$\Delta_B = t_{B/A} = \frac{wL^4}{8EI}$$

Answer: $\theta_B = \dfrac{PL^3}{6EI}$ and $\Delta_B = \dfrac{wL^4}{8EI}$

Example 18

Question: Determine the slope at B and deflection at C of the simply supported beam due to point load P.

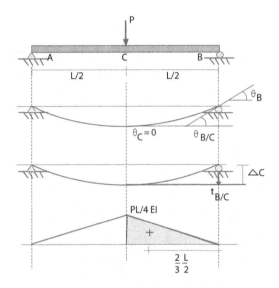

Slope at B:

Applying the first theorem of the moment–area method, the change in angle from C to B is:

$$\theta_{B/C} = \text{Area of } M/EI \text{ diagram between } C \text{ and } B$$
$$= \frac{1}{2}\left(\frac{PL}{4EI}\right)\frac{L}{2} = \frac{PL^2}{16EI}$$

According to the geometry of the elastic curve:

$$\theta_B = \theta_{B/C} = \frac{PL^2}{16EI}$$

Deflection at C:

Applying the second theorem of the moment–area method, the deviation of B on the tangent drawn from C is:

$$t_{B/C} = \text{Static moment of area under } M/EI \text{ diagram between } C \text{ and } B$$

Taken about point B:

$$= \left[\frac{1}{2} \left(\frac{PL}{4EI} \right) \frac{L}{2} \right] \left(\frac{2}{3} \cdot \frac{L}{2} \right) = \frac{PL^3}{48EI}$$

According to the geometry of the elastic curve:

$$\Delta_C = t_{B/C} = \frac{PL^3}{48EI}$$

Answer: $\theta_B = \dfrac{PL^2}{16EI}$ and $\Delta_C = \dfrac{PL^3}{48EI}$

Example 19

Question: Determine the slope at B and deflection at C of the simply supported beam due to uniformly distributed load w.

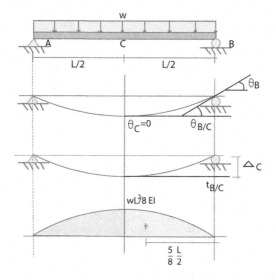

Slope at B:

Applying the first theorem of the moment–area method, the change in angle from C to B is:

$$\theta_{B/C} = \text{Area of } M/EI \text{ diagram between } C \text{ and } B$$
$$= \frac{2}{3} \left(\frac{wL^2}{8EI} \right) \frac{L}{2} = \frac{wL^3}{24EI}$$

According to the geometry of the elastic curve:

$$\theta_B = \theta_{B/C} = \frac{wL^3}{24EI}$$

Deflection at C:

Applying the second theorem of the moment–area method, the deviation of B on the tangent drawn from C is:

$$t_{B/C} = \text{Static moment of area under } M/EI \text{ diagram between } C \text{ and } B$$

Taken about point B:

$$= \left[\frac{2}{3}\left(\frac{wL^2}{8EI}\right)\frac{L}{2}\right]\left(\frac{5}{8}\cdot\frac{L}{2}\right) = \frac{5wL^4}{384EI}$$

According to the geometry of the elastic curve:

$$\Delta_C = t_{B/C} = \frac{5wL^4}{384EI}$$

Answer: $\theta_B = \dfrac{wL^3}{24EI}$ and $\Delta_C = \dfrac{5wL^4}{384EI}$

3.9 INDETERMINATE BEAM

3.9.1 DETERMINATE BEAM VS. INDETERMINATE BEAM

3.9.1.1 Equations of Statics

There are three equations of statics to solve the support reactions of a beam.

$$\sum F_x = 0; \sum F_y = 0; \sum M = 0$$

If the number of support reactions exceeds three, then all the support reactions cannot be solved by using the equation of statics.

3.9.1.2 Determinate Beam

A *beam* is classified as determinate if the number of support reactions and the number of available equations of statics are equal.

3.9.1.3 Simply Supported Beam

The number of reactions is three, which is equal to the number of equations of statics. So the beam is statically determinate.

Cantilever Beam:

The number of reactions is three, which is equal to the number of equations of statics. So the beam is statically determinate.

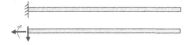

Significance: Determinate beams are easy to analyze but difficult to construct. Failure of any part of the beam results to the collapse of the whole beam.

3.9.1.4 Indeterminate Beam

A beam is classified as *indeterminate* if the number of support reactions exceeds the available equations of statics.

The number of excess reactions is known as degree of indeterminacy.

3.9.1.5 Propped Cantilever Beam

The number of unknown reactions is four, which is one greater than the equations of statics. So the beam is statically indeterminate to the first degree.

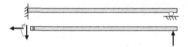

3.9.1.6 Continuous Beam

The number of unknown reactions is five, which is two greater than the equations of statics. So the beam is statically indeterminate to the second degree.

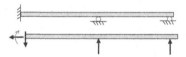

Significance: Indeterminate beams are difficult to analyze but easy to construct. Moreover, it requires the failure of several parts of the beam to result to the collapse of the whole beam.

Example 20

Question: Determine the support reactions at A and B by method of superposition of deflection.

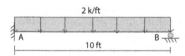

3.9.1.7 Real Beam

The deflected shape of the beam is shown here. Note that deflection at $B, y_B = 0$.

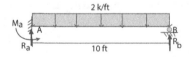

3.9.1.8 Primary Beam

Choosing R_b as the redundant reaction, by removing support B (as well as R_B), we get the primary beam. This support removal produces a deflection y_1 at B.

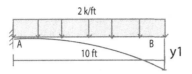

3.9.1.9 Redundant Reaction

Then, we provide R_b at B, which was removed earlier. Another deflection is produced, this time at B, which is y_2.

3.9.1.10 Determine Deflections

We can determine y_1 and y_2 by using the deflection formulae.

$$y_1 = -\frac{wL^4}{8EI} = -\frac{2 \times 10^4}{8EI} = -\frac{2500}{EI}$$

$$y_2 = +\frac{PL^3}{3EI} = \frac{R_b \times 10^3}{3EI} = \frac{1000R_b}{3EI}$$

Evaluate the redundant reaction, R_b. Now, by principle of superposition of deflection, we can write:

$$y_B = 0; \; y_1 + y_2 = 0$$

$$-\frac{2500}{EI} + \frac{1000R_b}{3EI} = 0$$

$$R_b = +7.5 \text{ kip} (\uparrow)$$

Determine M_a and R_a.

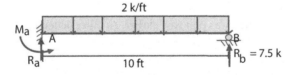

The remaining two reactions, R_a and M_a, can be easily found by equations of statics.

$$\sum F_y = 0 \uparrow^+$$

$$R_a - 2 \times 10 + 7.5 = 0;$$

$$R_a = 12.5 \text{kip} (\uparrow)$$

$$\sum M_A = 0 \cup^+$$

$$-M_a + (2 \times 10) \times 5 - 7.5 \times 10 = 0;$$

$$M_b = 25 \text{k} - \text{ft} (\cup)$$

Answer: $R_a = 12.5\text{kip}, M_a = 25\text{k} - \text{ft}$ and $R_b = 7.5\text{kip}$

Example 21

Question: Determine the support reactions at $A, B,$ and C.

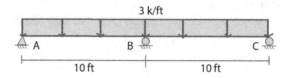

3.9.1.11 Real Beam
The deflected shape of the beam is shown here. Note that deflection at $B, y_B = 0$.

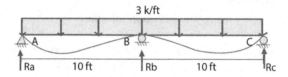

3.9.1.12 Primary Beam
Choosing R_b as the redundant reaction, by removing support B (as well as R_B), we get the primary beam. This support removal produces a deflection y_1 at B.

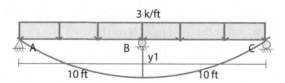

3.9.1.13 Redundant Reaction
Then we provide R_b at B, which was removed earlier. Another deflection is produced, this time at B, which is y_2 .

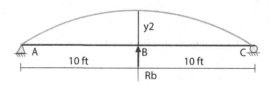

3.9.1.14 Determine Deflections

We can determine y_1 and y_2 by using the deflection formulae.

$$y_1 = -\frac{5wL^4}{384EI} = -\frac{5 \times 3 \times 20^4}{384EI} = -\frac{6250}{EI}$$

$$y_2 = +\frac{PL^3}{48EI} = \frac{R_b \times 20^3}{48EI} = \frac{500R_b}{3EI}$$

3.9.1.15 Evaluate the Redundant Reaction, R_b

Now, by principle of superposition of deflection, we can write:

$$y_B = 0; \ y_1 + y_2 = 0$$

$$-\frac{6250}{EI} + \frac{500R_b}{3EI} = 0;$$

$$R_b = +37.5\text{kip}(\uparrow)$$

Determine R_a and R_C.

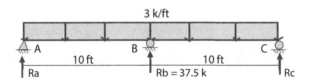

The remaining two reactions, R_a and R_C, can be easily found by equations of statics.

$$\sum M_A = 00^+$$

$$-(37.5) \times 10 + (3 \times 20) \times 10 - R_C \times 20 = 0;$$

$$R_C = 11.25\text{kip}(\uparrow)$$

$$\sum F_y = 0 \uparrow^+$$

$$R_a - 3 \times 20 + 37.5 + 11.25 = 0;$$

$$R_a = 11.25\text{kip}(\uparrow)$$

Answer: $R_a = 11.25$ kip, $R_b = 37.5$ kip and $R_C = 11.25$ kip

3.10 COLUMN ANALYSIS

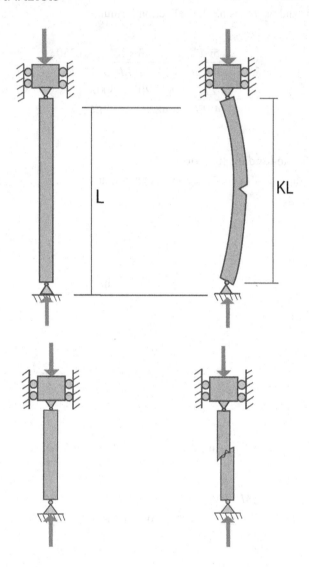

3.10.1 BUCKLING

Buckling is the sudden lateral instability of a slender structural member induced by the action of an axial load before the yield stress of the material is reached.

    L : Physical length of column
    K : Effective length factor
    KL : Effective length

3.10.2 CRUSHING

Crushing occurs when the direct stress from an axial load exceeds the compressive strength of the material available in the cross section.

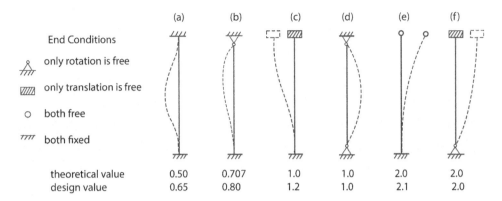

	(a)	(b)	(c)	(d)	(e)	(f)
theoretical value	0.50	0.707	1.0	1.0	2.0	2.0
design value	0.65	0.80	1.2	1.0	2.1	2.0

L	Physical length of column
K	Effective length factor
KL	Effective length
r	Radius of gyration
KL/r	Slenderness ratio
F_{cr}	Critical stress
F_y	Yield stress
E	Modulus of elasticiy
C_c	Critical coefficents

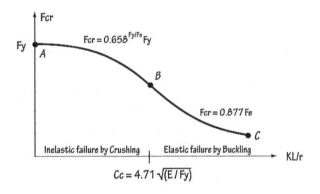

$$F_{cr} = \begin{cases} 0.658^{F_y/F_e} F_y & \text{if } KL/r < C_c \\ 0.877 F_e & \text{if } KL/r > C_c \end{cases}$$

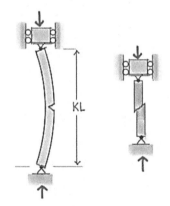

Example 22

Determine the nominal capacity of the steel column. The cross section is shown at the right, given that $F_y = 50$ ksi.

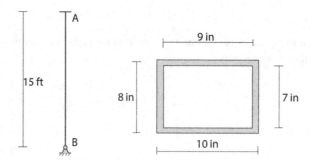

Solution

Determine the cross-sectional properties.

$$A = 10 \times 8 - 7 \times 9 = 17 \text{ in}^2$$

$$I = \frac{10 \times 8^3}{12} - \frac{9 \times 7^3}{12} = 169.41 \text{ in}^4$$

$$r = \sqrt{\frac{I}{A}} = \sqrt{\frac{241.42}{17}} = 3.157 \text{ in}$$

3.10.3 CHECK IF THE COLUMN IS SLENDER OR STOCKY

$$K = 0.80 \text{ [Since support is fixed-hinged]}$$

$$L = 15\text{ft} = 15 \times 12 = 180\text{in}$$

$$\frac{KL}{r} = \frac{0.80 \times 180}{3.157} = 45.61$$

$$C_C = 4.71\sqrt{\frac{E}{F_y}} = 4.71\sqrt{\frac{29000}{50}} = 113.43$$

Since $\left(KL / r \right) < C_C$, the column is stocky.

3.10.4 DETERMINE CAPACITY

$$F_e = \frac{\pi^2 E}{(KL / r)^2} = \frac{\pi^2 \times 29000}{45.61^2} = 137.58 \text{ ksi}$$

$$= 0.658^{\left(F_y / F_e \right)} F_y$$

$$F_{Cr} = 0.658^{(50/137.58)} \times 50$$

$$= 42.95 \text{ ksi}$$

$$P_n = F_{Cr} A = 42.95 \times 17 = 730.2 \text{ kip}$$

Answer: $P_n = 730.2\text{kip}$

Example 23

Determine the nominal capacity of the steel column AB, which is laterally braced at C in x direction, given that $F_y = 36$ ksi, $A = 11.5$ in^2, $r_x = 4.27$ in, $r_y = 1.98$ in.

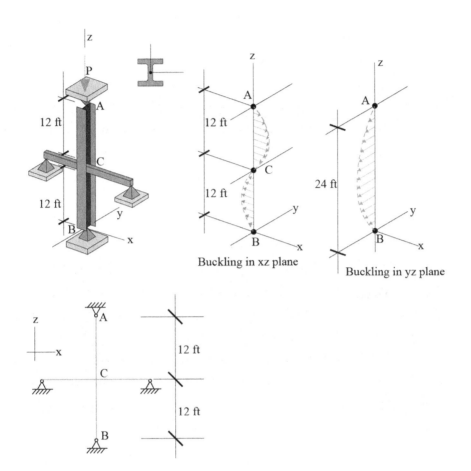

Buckling in xz plane

Buckling in yz plane

Determine the governing slenderness ratio.

$$\left(\frac{KL}{r_y}\right)_{AC} = \frac{1.0 \times 12 \times 12}{1.98} = 72.7 \; (\leftarrow)$$

$$\left(\frac{KL}{r_y}\right)_{CB} = \frac{1.0 \times 12 \times 12}{1.98} = 72.7$$

$$\left(\frac{KL}{r_x}\right)_{AB} = \frac{1.0 \times 24 \times 12}{4.27} = 67.4$$

The largest slenderness ratio $KL/r = 72.7$ governs the behavior of ABC column.

3.10.5 Check If the Column Is Slender or Stocky

$$C_C = 4.71\sqrt{\frac{E}{F_y}} = 4.71\sqrt{\frac{29000}{36}} = 133.68$$

Since $(KL/r) < C_C$, the column is stocky.

3.10.6 Determine the Capacity

$$F_e = \frac{\pi^2 E}{(KL/r)^2} = \frac{\pi^2 \times 29000}{72.7^2} = 54.1 \text{ ksi}$$

$$= 0.658^{(F_y/F_e)} F_y$$

$$= 0.658^{(36/54.1)} \times 36$$

$$F_{cr} = 27.3 \text{ ksi}$$

$$P_n = F_{cr} A = 27.3 \times 11.5 = 313.9 \text{ kip}$$

Answer: $P_n = 313.9$ kip.

4 Trusses and Space Frames

Topics to be covered in this chapter are as follows:

- Different types of trusses and space frames.
- Wind and static load analysis of trusses.
- Design of truss sections.
- Design of steel beams and columns.

4.1 TRUSS ANALYSIS

4.1.1 JOINT METHOD

Example 1

Determine the bar force in members AB, AG, GB, GF, BF.

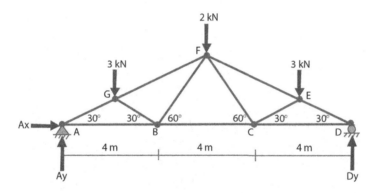

Solution

$$\sum M_D = 0 \cup^+ \quad Ay \times 12 - 3 \times 10 - 2 \times 6 - 3 \times 2 = 0; \ A_y = 4\,\text{kip}(\uparrow)$$

Joint A:

$$\sum F_y = 0\uparrow^+ \quad F_{AG}\sin 30 + 4 = 0 \qquad F_{AG} = -8\,\text{kip} \ (\text{C})$$
$$\sum F_x = 0\rightarrow^+ \quad F_{AB} + F_{AG}\cos 30 = 0 \quad F_{AB} = +6.92\,\text{kip} \ (\text{T})$$

Joint G:

$$\sum F_x = 0\rightarrow^+ \quad F_{GF}\cos 30 + F_{GB}\cos 30 - F_{AG}\cos 30 = 0$$
$$F_{GF}\cos 30 + F_{GB}\cos 30 = -6.928$$
$$\sum F_y = 0\uparrow^+ \quad F_{GF}\sin 30 - F_{GB}\sin 30 - F_{AG}\sin 30 - 3 = 0$$
$$F_{GF}\sin 30 - F_{GB}\sin 30 = -1$$

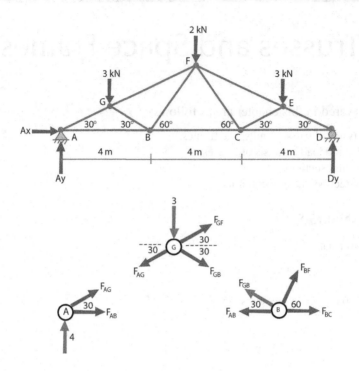

Solving, $F_{GF} = -5$ Kip (C) $F_{GB} = -3$ kip (C).

Joint B:

$$\Sigma F_y = 0 \uparrow^+ \quad F_{GB}\sin 30 + F_{BF}\sin 60 = 0$$
$$F_{BF} = +1.73 \text{ kip (T)}$$
$$\Sigma F_x = 0 \rightarrow^+ \quad -F_{AB} - F_{GB}\cos 30 + F_{BF}\cos 60 + F_{BC} = 0$$
$$F_{BC} = +3.46 \text{ kip (T)}$$

Example 2

Determine the bar force in members AB, AJ, AK, KJ, BC.

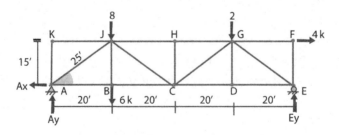

Solution

$$\Sigma M_E = 0 \cup^+$$

$$A_y \times 80 - 8 \times 60 - 6 \times 60 - 2 \times 20 + 4 \times 15 = 0$$
$$-A_x + 4 = 0$$

$$A_y = 10.25 \text{ kip}$$

$$\sum F_x = 0 \rightarrow^+$$
$$A_x = 4\,\text{kip}\,(\leftarrow)$$

Joint K:

$$\sum F_x = 0 \rightarrow^+ \quad F_{KJ} = 0;\ F_{KJ} = 0$$
$$\sum F_y = 0 \uparrow^+ \quad -F_{AK} = 0;\ F_{AK} = 0$$

Joint A: The angle $\angle JAB = \tan^{-1} 15/20 = 36.87°$.

$$\sum F_y = 0 \uparrow^+ \quad F_{AJ}\sin 36.87 + F_{AK} + 10.25 = 0$$
$$F_{AJ} = -17.08\,\text{kip}\,(C)$$
$$\sum F_x = 0 \rightarrow^+ \quad F_{AB} + F_{AJ}\cos 36.87 - 4$$
$$F_{AB} = +17.64\,\text{kip}\,(T)$$

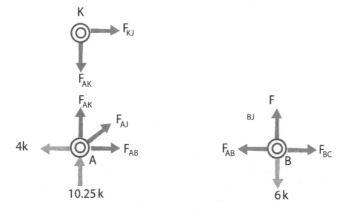

Joint B:

$$\sum F_x = 0 \rightarrow^+ \quad F_{BC} - F_{AB} = 0$$
$$F_{BC} = +17.64\,\text{kip}\,(T)$$
$$\sum F_y = 0 \uparrow^+ \quad F_{BJ} - 6 = 0$$
$$F_{BJ} = 6\,\text{kip}\,(T)$$

4.2 TRUSS ANALYSIS

4.2.1 SECTION METHOD

Example 3

Determine the bar force in members GB, GF, BF. Use the section method.

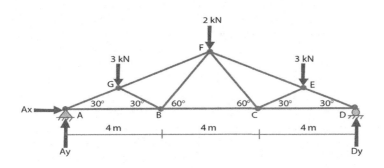

Solution

$$\sum M_D = 0 \cup^+ \; Ay \times 12 - 3 \times 10 - 2 \times 6 - 3 \times 2 = 0; \; A_y = 4 \text{ kip } (\uparrow)$$

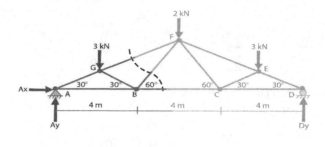

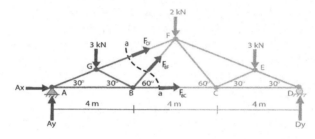

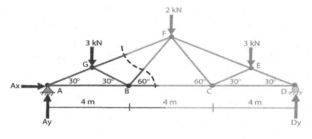

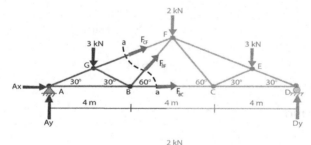

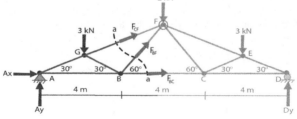

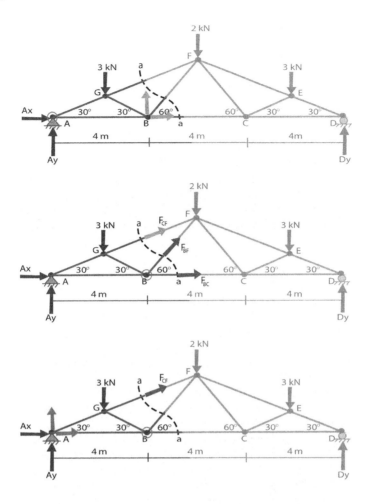

Section a − a

$$\sum M_F = 0\cup^+ \quad A_y \times 6 - 3 \times 4 - F_{BC} \times 2\tan 60 = 0 \quad F_{BC} = +3.46\, kip\,(T)$$

$$\sum M_A = 0\cup^+ \quad 3 \times 2 - (F_{BF}\sin 60) \times 4 = 0 \qquad\qquad F_{BF} = +1.73\, kip\,(T)$$

$$\sum M_B = 0\cup^+ \quad A_y \times 4 + (F_{GF}\sin 30) \times 4 - 3 \times 2 = 0 \quad F_{GF} = -5.00\, kip\,(C)$$

Example 4

Determine the bar force in members JH, JC, BC. Use the section method.

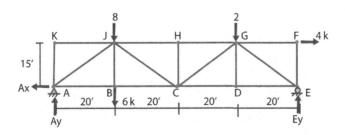

Solution

$$\sum M_E = 0 \cup^+ \quad A_y \times 80 - 8 \times 60 - 6 \times 60 - 2 \times 20 + 4 \times 15 = 0; \quad A_y = 10.25 \text{ kip} (\uparrow)$$

$$\sum F_x = 0 \rightarrow^+ \quad -A_x + 4 = 0; \qquad\qquad\qquad\qquad A_x = 4 \text{ kip} (\leftarrow)$$

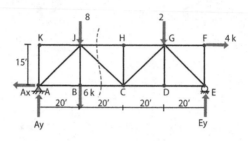

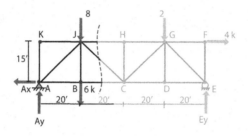

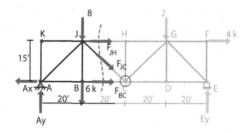

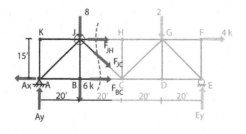

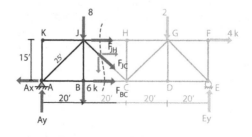

Section $a-a$:

$$\sum M_C = 0 u^+$$
$$A_y \times 40 - 8 \times 20 - 6 \times 20 + F_{JH} \times 15 = 0$$
$$F_{JH} = -8.67 \text{ kip (C)}$$

$$\sum M_J = 0 \cup^+$$
$$A_y \times 20 + A_x \times 15 - F_{BC} \times 15 = 0$$
$$F_{BC} = 17.67 \text{ kip (T)}$$

$$\sum F_y = 0 (\uparrow^+)$$
$$A_y - 8 - 6 - F_{JC} \cdot \frac{15}{25} = 0$$
$$F_{JC} = -6.25 \text{ kip (C)}$$

4.3 WIND LOADS

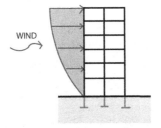

When structures block the flow of the wind, the wind's kinetic energy is converted into potential energy of pressure, which causes a wind loading.

The effect of wind on a structure depends upon:

- The density and velocity of the air
- The angle of incidence of the wind
- The shape and stiffness of the structure
- The surrounding terrain roughness

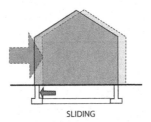

SLIDING

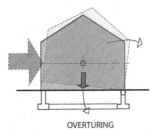

OVERTURING

4.3.1 SLIDING

Wind pressure can cause a building to translate or move laterally due to the shear forces created between the structure and its foundation. Proper anchorage is necessary to prevent this form of failure.

4.3.2 OVERTURNING

Light buildings, such as wood-framed structures, require careful detailing to prevent the effects of overturning. While heavy buildings can more easily resist overturning from wind pressure, they are susceptible to the large inertial forces they generate during an earthquake.

4.3.3 BUILDING SHAPE

Shape and form can increase or decrease the effects of wind pressure on a building.

- Aerodynamically shaped buildings, such as rounded or curved forms, generally result in lower wind resistance than rectangular buildings with flat surfaces.
- Buildings with open sides or configurations with recesses or hollows that capture the wind are subject to larger design wind pressures.
- Building projections, such as parapets, balconies, canopies, and overhangs, are subject to increased localized pressures from moving air masses.
- Wind pressure can subject very tall walls and long-spanning rafters to large bending moments and deflection.

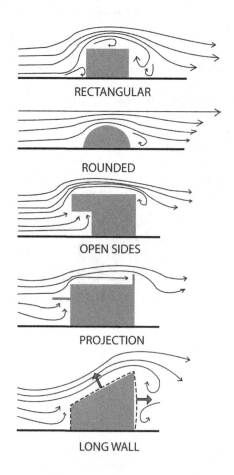

RECTANGULAR

ROUNDED

OPEN SIDES

PROJECTION

LONG WALL

4.3.4 DEFLECTION

- Wind can produce dynamic loading on tall, slender structures that exceed typical design levels. The efficient design of structural systems and cladding for tall buildings requires knowledge of how wind forces impact their slender forms.
- Tall, slender buildings with a high aspect (height-to-base width) ratio experience larger horizontal deflections at their tops and are more susceptible to overturning moments.
- Building forms that taper expose less surface area to the wind as they rise, which helps counteract the increasing wind velocities and pressures experienced higher up.

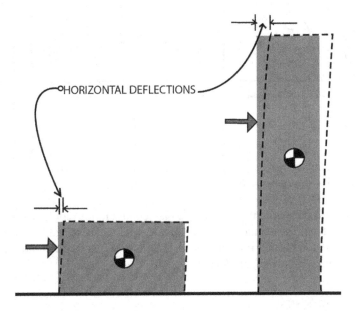

HORIZONTAL DEFLECTIONS

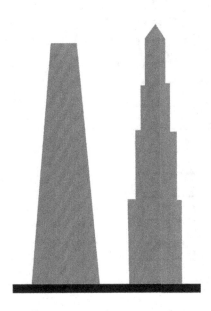

4.3.5 STABLE PLAN

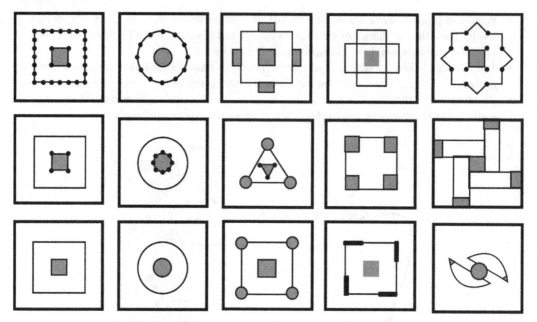

- Illustrated on this page are inherently stable plan configurations for high-rise structures.
- Open forms of bracing are inherently weak in torsional stiffness and should be avoided.
- L-, T-, and X-shaped plan arrangements are the worst in torsional resistance, while C and Z configurations are only slightly better.

4.3.6 BRACED CORE SYSTEM

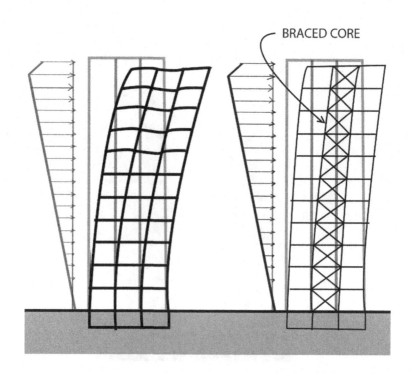

BRACED CORE

4.3.7   OUTRIGGER TRUSS SYSTEM

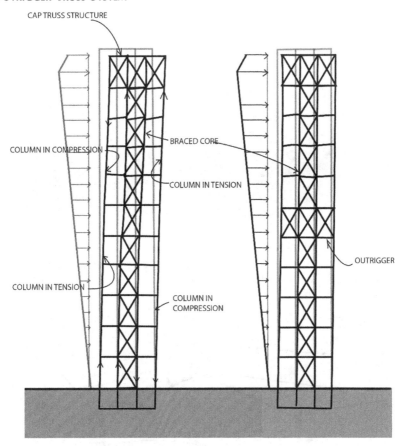

4.3.8   WIND TUNNEL TEST AND CFD ANALYSIS

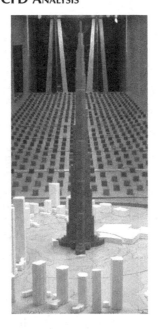

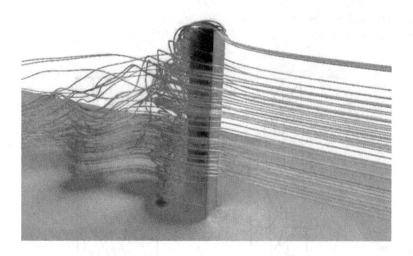

Structural designers use wind tunnel tests and computer modeling to determine the overall base shear, overturning moment, as well as floor-by-floor distribution of wind pressure on a structure, and to gather information about how the building's motion might affect occupants' comfort.

4.3.9 WIND PRESSURE

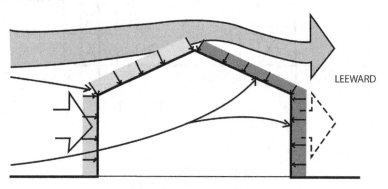

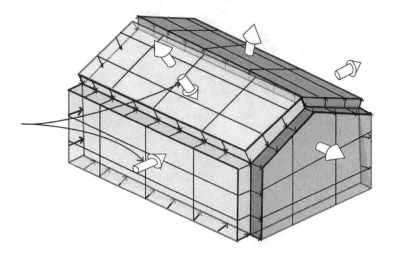

Wind exerts positive pressure horizontally on the windward vertical surfaces of a building and normal to windward roof surfaces having a slope greater than 30°.

Wind exerts negative pressure or suction on the sides and leeward surfaces and normal to windward roof surfaces having a slope less than 30°.

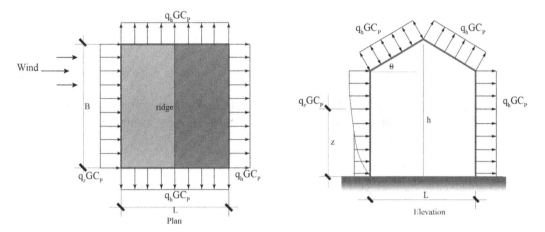

4.3.10 TERRAIN

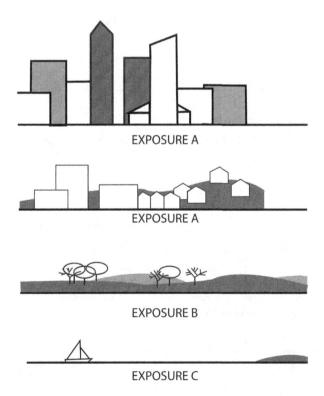

- Exposure A: Urban and suburban areas, wooded areas, other terrain with numerous closely spaced obstruction having heights of 30 ft or more.
- Exposure B: Terrain with scattered objects whose heights are generally less than 30 ft, including flat open country, grasslands.
- Exposure C: Flat, unobstructed areas and water surfaces.

4.4 WIND LOADS ANALYSIS

4.4.1 On Roof Truss

4.4.1.1 Kinetic Energy (q) and Wind Pressure (p)

4.4.1.1.1 Kinetic Energy

The fluctuating pressure caused by a constantly blowing wind is approximated by a mean velocity pressure that acts on the structure.

This pressure q is defined by its kinetic energy, $q = \frac{1}{2}\rho v^2$, where ρ is the density of the air and v is its velocity,

According to the ASCE 7–10 Standard, this equation is modified to account for the importance of the structure, its height, and the terrain in which it is located. It is represented as:

$$q = 0.00256\, K_z K_{zt} V^2$$

V is velocity in mph of a 3 sec gust of wind measured 33 ft (10 m) above the ground.

K_z is the velocity pressure exposure coefficient, which is a function of height and depends upon the ground terrain.

K_{zt} is the factor that accounts for wind speed increases due to hills and escarpments. For flat ground, $K_{zt} = 1$.

Height (ft)	K_z	Height	K_z
0–15	0.85	20	0.90
25	0.94	30	0.98
40	1.04	50	1.09

4.4.1.1.2 Wind Pressure for Enclosed Buildings

Design pressure can be determined from the following two-termed equations resulting from both external and internal pressures.

$$p = q\left(GC_p - GC_{pi}\right)$$

G	Wind-gust effect factor, which depends upon the exposure. For example, for a rigid structure, $G = 0.85$.
C_p	Roof pressure coefficient determined from the table.
GC_{pi}	Internal pressure coefficient, which depends upon the type of openings in the building. For fully enclosed buildings, $GC_{pi} = 0.18$.

Windward Angle θ

h/L	$10°$	Leeward Angle
0.25	−0.7	−0.3
0.5	−0.9	−0.5
>1.0	−1.3	−0.7

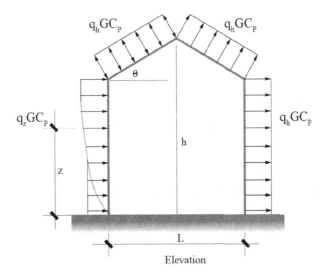

Example 5

The enclosed building shown is used for storage purposes and is located where design speed for wind is 130 mph and it is on open flat terrain. When the wind is directed as shown, determine the design wind pressure acting on the roof, design load on truss joints.

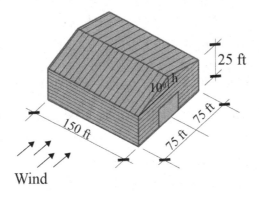

Height (ft)	K_z	Height	K_z
0–15	0.85	20	0.90
25	0.94	30	0.98
40	1.04	50	1.09

h/L	Windward C_p	Leeward C_p
≤ 0.25	−0.7	−0.3
0.50	−0.9	−0.5
≥ 1.00	−1.3	−0.7

Solution

Determination of kinetic energy:
   Wind velocity, $V = 130$mph, and for flat terrain, $K_{zt} = 1$.

$$q = 0.00256\,K_z K_{zt} V^2 = 0.0256 \times K_z \times 1 \times 130^2 = 28.22\,K_z$$

In order to determine K_z, we need the mean height of the roof.

$$h = 25 + 75\tan10 = 31.6\ \text{ft}$$

Now, from the table, we find K_z by interpolation:

$$K_z = 0.98 + \left(\frac{1.04 - 0.98}{40 - 30}\right) \times (31.6 - 30) = 0.99$$

Therefore, the kinetic energy is:

$$q = 28.22 \times 0.99 = 27.9\ \text{psf}$$

4.4.1.2   Windward Roof Pressure

For roof pressure, we need to determine C_p at first. Here:

$$h/L = 31.6/150 = 0.211 < 0.25,$$

$$\therefore C_p = -0.7$$

For rigid structure, $G = 0.85$ and $GC_{pi} = 0.18$. So, the windward pressure is:

$$p = q\left(GC_p - GC_{pi}\right) = 27.9 \times \left[0.85(-0.7) - 0.18\right] = -21.6\ \text{psf (suction)}$$

Leeward roof pressure:
   Here, $C_p = -0.3$; therefore, the leeward pressure is:
$$p = q\left(GC_p - GC_{pi}\right) = 27.9 \times \left[0.85(-0.3) - 0.18\right] = -12.2\ \text{psf (suction)}$$

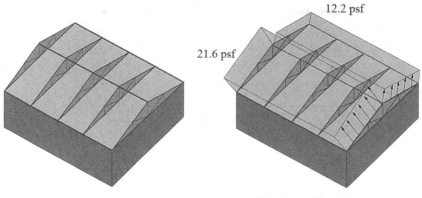

Total roof loading

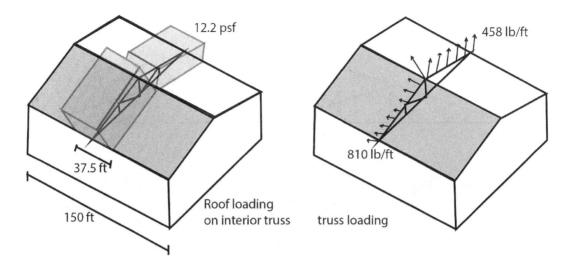

For windward side = 21:6 psf × 37:5 ft = 810 lb/ft
For leeward side = 12:2 psf × 37:5 ft = 458 lb/ft

Example (Joint Loading)

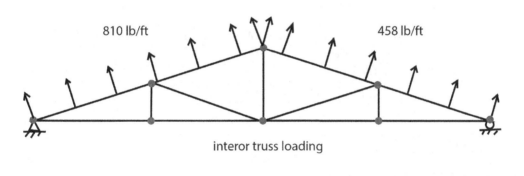

interor truss loading

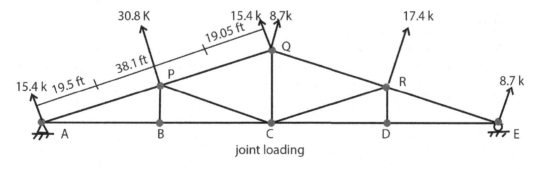

joint loading

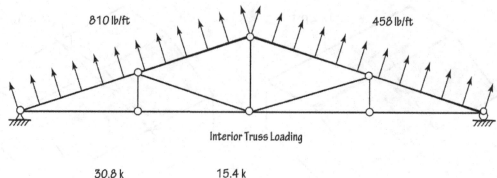

Interior Truss Loading

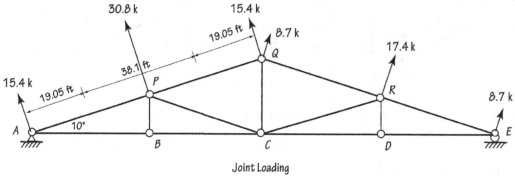

Joint Loading

Windward side and leeward side:

$$\text{Load on joint } A \text{ and } Q = 810 \text{ lb / ft} \times 19.05 \text{ ft} = 15.4 \text{ kip}$$
$$\text{Load on joint } P = 810 \text{ lb / ft} \times 38.1 \text{ ft} \quad = 30.8 \text{ kip}$$
$$\text{Load on joint } Q \text{ and } E = 458 \text{ lb / ft} \times 19.05 \text{ ft} = 8.7 \text{ kip}$$
$$\text{Load on joint } R = 458 \text{ lb / ft} \times 38.1 \text{ ft} \quad = 17.4 \text{ kip}$$

4.5 BIAXIAL BUCKLING

4.5.1 DESIGN

Example 6

Select the lightest W8 shape for the following column ABC, which is laterally braced at midsection B in x direction, given that the dead load is 200 kip and live load is 200 kip. Use A992 steel.

4.5.1.1 Design Data

$$K_x = 1.0 \ K_y = 1.0$$
$$L_x = 18 \text{ ft } L_y = 9 \text{ ft}$$
$$DL = 200 \text{ kip} ; LL = 200 \text{ kip}$$
$$F_y = 50 \text{ ksi}$$

Solution

$$P_u = 1.2DL + 1.6LL = 1.2 \times 200 + 1.6 \times 200 = 560 \text{ kip}$$

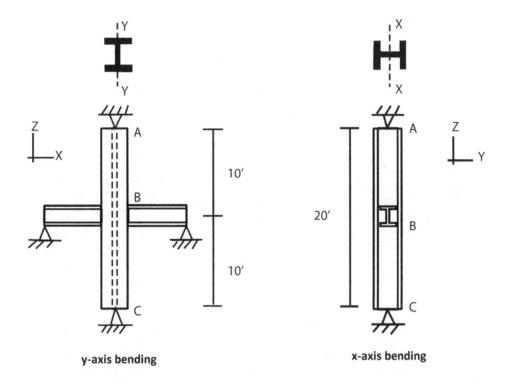

y-axis bending   **x-axis bending**

Assuming weak axis (y) controls.

$$KL = (KL)_y = 1.0 \times 9 = 9 \text{ ft}$$

From Table 4–1, for $KL = 9$, section W8 ×58 satisfies with $\phi_c P_n = 634$ kip.
    Now, check the strong axis:

$$\frac{K_x L_x}{r_x / r_y} = \frac{1 \times 18}{1.74} = 10.34 > 9 \text{ ft}$$

So the assumption was wrong; actually, strong axis (x) controls. Now, determine the strength $(\phi_c P_n)$ of W8 58 again, but using $KL = 10.34$.
    Since there is no row for $KL = 10.34$, an interpolation is required between $KL = 10$ and $KL = 11$. From Table 4–1, for W 8 × 58, the KL are listed in the following.

KL	$\phi_c P_n$
10	606
11	576

By interpolation:

$$\phi_c P_n = 606 - \frac{606 - 576}{11 - 10} * (10.34 - 10) = 596.8 \text{ kip} > P_u \, (560)$$

Note that though the assumption was wrong, it yields the right choice of section. If not, try a larger section.

Answer: Select W8 × 58

Table 4-1 (continued)
Available Strength in Axial Compression, kips
W-Shapes

$F_y = 50$ ksi

W14

Shape													
lb/ft	426[h]		398[h]		370[h]		342[h]		311[h]		283[h]	
Design	P_n/Ω_c	$\phi_c P_n$	P_n/Ω_c	$\phi_c P_n$	P_n/Ω_c	$\phi_c P_n$	P_n/Ω_c	$\phi_c P_n$	P_n/Ω_c	$\phi_c P_n$	P_n/Ω_c	$\phi_c P_n$
	ASD	LRFD	ASD	LRFD	ASD	LRFD	ASD	LRFD	ASD	LRFD	ASD	LRFD
0	3740	5620	3500	5260	3260	4900	3020	4540	2740	4110	2490	3750
11	3500	5260	3270	4920	3040	4570	2820	4230	2550	3830	2320	3480
12	3450	5190	3230	4850	3000	4510	2780	4180	2510	3770	2290	3440
13	3410	5120	3180	4780	2960	4450	2740	4120	2470	3720	2250	3380
14	3350	5040	3130	4710	2910	4380	2700	4050	2430	3660	2210	3330
15	3300	4960	3080	4630	2870	4310	2650	3980	2390	3600	2180	3270
16	3240	4870	3030	4550	2810	4230	2600	3910	2350	3530	2140	3210
17	3180	4790	2970	4470	2760	4150	2550	3840	2300	3460	2090	3150
18	3120	4690	2920	4380	2710	4070	2500	3760	2260	3390	2050	3080
19	3060	4600	2850	4290	2650	3980	2450	3680	2210	3320	2000	3010
20	2990	4500	2790	4200	2590	3890	2390	3600	2160	3240	1960	2940
22	2860	4290	2660	4000	2470	3710	2280	3420	2050	3080	1860	2800
24	2710	4080	2530	3800	2340	3520	2160	3240	1940	2920	1760	2640
26	2560	3850	2390	3590	2210	3320	2040	3060	1830	2750	1660	2490
28	2410	3630	2250	3380	2080	3120	1910	2870	1710	2580	1550	2330
30	2260	3400	2100	3160	1940	2920	1790	2680	1600	2400	1450	2170
32	2110	3170	1960	2950	1810	2720	1660	2500	1490	2230	1340	2020
34	1960	2950	1820	2730	1670	2520	1540	2310	1370	2060	1240	1860
36	1810	2730	1680	2530	1540	2320	1420	2130	1260	1900	1140	1710
38	1670	2510	1550	2320	1420	2130	1300	1950	1160	1740	1040	1560
40	1530	2300	1410	2130	1300	1950	1180	1780	1050	1580	945	1420
42	1390	2090	1290	1930	1180	1770	1070	1610	954	1430	857	1290
44	1270	1910	1170	1760	1070	1610	979	1470	869	1310	781	1170
46	1160	1750	1070	1610	980	1470	896	1350	795	1200	715	1070
48	1070	1600	985	1480	900	1350	823	1240	730	1100	656	986
50	983	1480	907	1360	830	1250	758	1140	673	1010	605	909
Properties												
P_{wo}, kips	1140	1710	1010	1520	902	1350	788	1180	672	1010	574	861
P_{wi}, kips/in.	62.7	94.0	59.0	88.5	55.3	83.0	51.3	77.0	47.0	70.5	43.0	64.5
P_{wb}, kips	10100	15100	8420	12700	6920	10400	5540	8320	4250	6390	3260	4900
P_{fb}, kips	1730	2600	1520	2280	1320	1990	1140	1720	956	1440	802	1210
L_p, ft	15.3		15.2		15.1		15.0		14.8		14.7	
L_r, ft	168		158		148		138		125		114	
A_g, in.^{2}	125		117		109		101		91.4		83.3	
I_x, in.^{4}	6600		6000		5440		4900		4330		3840	
I_y, in.^{4}	2360		2170		1990		1810		1610		1440	
r_y, in.	4.34		4.31		4.27		4.24		4.20		4.17	
r_x/r_y	1.67		1.66		1.66		1.65		1.64		1.63	
$P_{ex}(KL)^2/10^4$, k-in.^{2}	189000		172000		156000		140000		124000		110000	
$P_{ey}(KL)^2/10^4$, k-in.^{2}	67500		62100		57000		51800		46100		41200	

Effective length, KL (ft), with respect to least radius of gyration, r_y

ASD	LRFD	
$\Omega_c = 1.67$	$\phi_c = 0.90$	[h] Flange thickness is greater than 2 in. Special requirements may apply per AISC *Specification* Section A3.1c.

4.6   TRUSS DESIGN

(Tension member for dead load only.)

Example 7

Select a structural T for the bottom chord of the Warren roof truss. The trusses are spaced at 20ft. Use A992 steel and the following load data.

Purlin	M 8 × 6.5	Metal deck	2 psf
Snow	20 psf	Insulation	3 psf
Roofing	4 psf		

8 panel @ 5 ft = 40 ft

Snow
Roofing
Insulation
Metal Deck
Purlin

Solution

4.6.1   LOAD CALCULATION

Dead load intensity of roofing materials (excluding purlin) = Deck + insulation + roofing = 2 + 3 + 4.

$$= 9 \text{ psf}$$

Snow load = intensity × tributary area

$$= 20 \text{ psf} \times (40 \text{ ft} \times 20 \text{ ft}) = 16000 \text{ lb}$$

Roofing weight:

$$= \text{intensity} \times \text{tributary area}$$

$$= 9 \text{ psf} \times (40 \text{ ft x } 20 \text{ ft}) = 7200 \text{ lb}$$

Purlin weight $= Intensity \times$ length $\times$ amount $= 6.5$ lb / ft x 20 ft x 9 $= 1170$ lb

Truss weight = assuming 10% of other loads, $0.10(16000 + 7200 + 1170) = 2437$ lb

4.6.2 Joint Load

At an interior joint:

$$D = \frac{7200}{8} + \frac{2437}{8} + 6.5 \times 20 = 1335 \text{ lb}$$

$$S = \frac{16000}{8} = 2000 \ lb$$

At an exterior joint:

$$D = \frac{7200}{8 \times 2} + \frac{2437}{8 \times 2} + 6.5 \times 20 = 733 \text{ lb}$$

$$S = \frac{16000}{8 \times 2} = 1000 \ lb$$

4.6.3 Design Load

At an interior joint:

$$P_u = 1.2D + 1.6S = 1.2 \times 1335 + 1.6 \times 2000 = 4802 \text{ lb} = 4.8 \text{ kip}$$

At an exterior joint:

$$P_u = 1.2D + 1.6S = 1.2 \times 733 + 1.6 \times 1000 = 2479 \text{ lb} = 2.4 \text{ kip}$$

4.6.3.1 Support Reaction

The loads calculated so far are shown in the following figure.

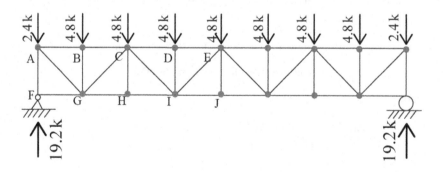

Since the truss and load are symmetric, support reaction at both ends should be half of the total downward load.

$$R_F = \frac{1}{2}(2.4 + 4.8 \times 7 + 2.4) = 19.2 \text{ kip}$$

4.6.4 Bar Force

The bottom chord is designed by determining the force in each member of the bottom chord and selecting a cross section to resist the largest force. In this example, the force in member *IJ* will control.

Since it is necessary to determine F_J, pass a section a–a through this member. Now, considering the free-body diagram at the left of section $a - a$:

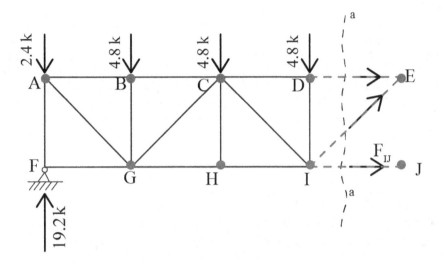

$$\sum M_E = 0$$
$$9.2 \times 20 - 2.4 \times 20 - 4.8 \times 15 - 4.8 \times 10 - 4.8 \times 5 - 4 \times F_{IJ} = 0$$
$$F_{IJ} = 48 \text{ kip (T)}$$

4.6.5 Design Section

The cross-sectional area required to resist 48 kip tensile force is:

$$A_g = \frac{F_{IJ}}{0.9 F_y} = \frac{48}{0.9 \times 50} = 1.07 \text{ in}^2$$

From Table 1–9 (AISC Design Manual 14th Edition), the section that satisfies the required area 1.07 in² is MT $5 \times 3.75 \left(A_g = 1.10 \text{ in}^2 \right)$.

Answer: Select MT5×3.75

Standard MT Shapes (Excerpted from *AISC Design Manual* 14th Edition, Part 1, p. 70)
Dimensions and Properties

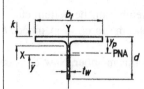

Table 1-9
MT-Shapes
Dimensions

Shape	Area, A	Depth, d	Stem Thickness, t_w		$\frac{t_w}{2}$	Area	Flange Width, b_f		Flange Thickness, t_f		k	Workable Gage
	in.²	in.	in.		in.	in.²	in.		in.		in.	in.
MT6.25×6.2[c,v]	1.82	6.27	6¼	0.155	1/8	1/16	0.971	3.75	3¾	0.228	1/4	9/16	—
×5.8[c,v]	1.70	6.25	6¼	0.155	1/8	1/16	0.969	3.50	3½	0.211	3/16	9/16	
MT6×5.9[c]	1.74	6.00	6	0.177	3/16	1/8	1.06	3.07	3⅛	0.225	1/4	9/16	—
×5.4[c,v]	1.59	5.99	6	0.160	3/16	1/8	0.958	3.07	3⅛	0.210	3/16	9/16	—
×5[c,v]	1.48	5.99	6	0.149	1/8	1/16	0.892	3.25	3¼	0.180	3/16	1/2	—
MT5×4.5[c]	1.33	5.00	5	0.157	3/16	1/8	0.785	2.69	2¾	0.206	3/16	9/16	—
×4[c]	1.19	4.98	5	0.141	1/8	1/16	0.701	2.69	2¾	0.182	3/16	9/16	—
MT5×3.75[c,v]	1.11	5.00	5	0.130	1/8	1/16	0.649	2.69	2¾	0.173	3/16	7/16	—
MT4×3.25[c,v]	0.959	4.00	4	0.135	1/8	1/16	0.540	2.28	2¼	0.189	3/16	9/16	—
×3.1[c]	0.911	4.00	4	0.129	1/8	1/16	0.516	2.28	2¼	0.177	3/16	7/16	—
MT3×2.2[c]	0.647	3.00	3	0.114	1/8	1/16	0.342	1.84	1⅞	0.171	3/16	3/8	—
×1.85[c]	0.545	2.96	3	0.0980	1/8	1/16	0.290	2.00	2	0.129	1/8	5/16	—
MT2.5×9.45[t]	2.78	2.50	2½	0.316	5/16	3/16	0.790	5.00	5	0.416	7/16	13/16	2¾[g]
MT2×3[f]	0.875	1.90	1⅞	0.130	1/8	1/16	0.247	3.80	3¾	0.160	3/16	1/2	—

[c] Shape is slender for compression with F_y = 36 ksi.
[f] Shape exceeds compact limit for flexure with F_y = 36 ksi.
[g] The actual size, combination and orientation of fastener components should be compared with the geometry of the cross section to ensure compatibility.
[t] This shape has tapered flanges while all other MT-shapes have parallel flange surfaces.
[v] Shape does not meet the h/t_w limit for shear in AISC *Specification* Section G2.1(a) with F_y = 36 ksi.
— Indicates flange is too narrow to establish a workable gage.

4.7   TRUSS DESIGN

(Compression member for dead load only.)

Example 8

Select a structural T for the top chord of the Warren roof truss. The trusses are spaced at 20 ft. Use A992 steel and the following load data.

Purlin	M8 × 6.5	Metal deck	2 psf
Snow	20 psf	Insulation	3 psf
Roofing	4 psf		

8 panel @ 5 ft = 40 ft

Solution

4.7.1   LOAD CALCULATION

Dead load intensity of roofing materials (excluding purlin) = Deck + insulation + roofing = 2 + 3 + 4 = 9 psf.

$$\text{Snow load} = \text{intensity} \times \text{tributary area} =$$

$$20 \text{ psf} \times (40 \text{ ft} \times 20 \text{ ft}) = 16000 \text{ lb}.$$

Roofing weight:

$$= \text{intensity} \times \text{tributary area}$$

$$= 9 \text{ ps} \times (40 \text{ ft x } 20 \text{ ft}) = 7200 \text{ lb}$$

Purlin weight = intensity × length × amount = 6.5 lb / ft x 20 ft x 9 = 1170 lb
Truss weight = assuming 10% of other loads = 0.10(16000 + 7200 + 1170) = 2437lb

4.7.2   JOINT LOAD

At an interior joint:

$$D = \frac{7200}{8} + \frac{2437}{8} + 6.5 \times 20 = 1335 \text{ lb}$$

$$S = \frac{16000}{8} = 2000 \text{ lb}$$

At an exterior joint:

$$D = \frac{7200}{8 \times 2} + \frac{2437}{8} + 6.5 \times 20 = 733 \text{ lb}$$

$$S = \frac{16000}{8 \times 2} = 1000 \text{ } lb$$

4.7.3 DESIGN LOAD

At an interior joint:

$$P_u = 1.2\,D + 1.6\,L = 1.2 \times 1335 + 1.6 \times 2000$$
$$= 4802 \text{ lb} = 4.8 \text{ kip}$$

At an exterior joint:

$$P_u = 1.2\,D + 1.6\,L = 1.2 \times 733 + 1.6 \times 1000$$
$$= 2479 \text{ lb} = 2.4 \text{ kip}$$

Support reaction:
   The loads calculated so far are shown in the following figure.

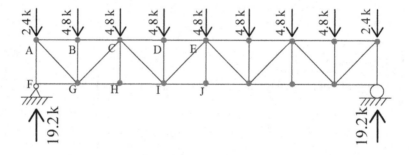

Since the truss and load are symmetric, support reaction at both ends should be half of total downward load.

$$R_F = \frac{1}{2}\left(2.4 + 4.8 \times 7 + 2.4\right) = 19.2 \text{ kip}$$

4.7.4 BAR FORCE

The top chord is designed by determining the force in each member of the top chord and selecting a cross section to resist the largest force. In this example, the force in member *DE* will control.

Since it is necessary to determine F_{DE}, pass a section a - a through this member. Now, considering the free-body diagram at left of section $a - a$:

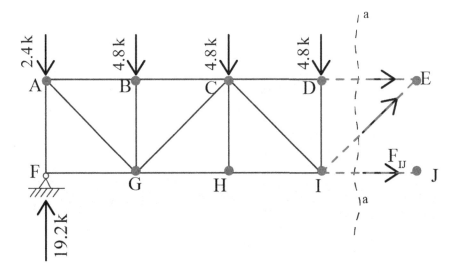

$$\sum M_I = 0$$
$$19.2 \times 15 - 2.4 \times 15 - 4.8 \times 10 - 4.8 \times 5 + 4 \times F_{DE} = 0$$
$$F_{DE} = -45 \text{ kip (C)}$$

4.7.5   DESIGN SECTION

Member DE is in compression, so a column design routine should be followed. Length of DE is 5 ft, and both D and E joints are pin connected; thus, $K = 1.0$.

$$KL = 1.0 \times 5 = 5 \text{ ft}$$

Since $KL = 5$ is missing in Table 4–7 (AISC Design Manual 14th Edition), an interpolation is required. Since weak axis $(y - y)$ will control, the section WT4 × 6.5 seems reasonable.

KL	$\phi_c P_n$
4	57.8
6	45.2

By interpolation:

$$\phi_c P_n = 57.8 - \frac{57.8 - 45.2}{6 - 4} \times (5 - 4) = 51.5 \text{ kip} > F_{DE}\,(45 \text{ kip})$$

Answer: Select WT 4 x 6.5.
Standard WT Shapes (Excerpted from *AISC Design Manual* 14th Edition, Part 4, p. 121)

Table 4-7 (continued)
Available Strength in Axial Compression, kips
WT-Shapes

F_y = 50 ksi

WT4

Shape			WT4×								
lb/ft			9		7.5		6.5		5[c]	
Design			P_n/Ω_c	$\phi_c P_n$	P_n/Ω_c	$\phi_c P_n$	P_n/Ω_c	$\phi_c P_n$	P_n/Ω_c	$\phi_c P_n$
			ASD	LRFD	ASD	LRFD	ASD	LRFD	ASD	LRFD
Effective length, KL (ft), with respect to indicated axis	X-X Axis	0	78.7	118	66.5	99.9	57.5	86.4	32.5	48.8
		4	69.2	104	59.4	89.2	51.4	77.3	29.8	44.8
		6	58.8	88.4	51.5	77.4	44.7	67.3	26.8	40.3
		8	46.9	70.5	42.3	63.5	36.8	55.3	23.1	34.7
		10	35.0	52.6	32.8	49.2	28.7	43.1	19.0	28.6
		12	24.8	37.2	24.0	36.0	21.1	31.6	15.0	22.6
		14	18.2	27.4	17.6	26.4	15.5	23.3	11.3	17.1
		16	13.9	20.9	13.5	20.2	11.8	17.8	8.69	13.1
		18	11.0	16.5	10.6	16.0	9.36	14.1	6.87	10.3
		20			8.62	13.0	7.58	11.4	5.56	8.36
	Y-Y Axis	0	78.7	118	66.5	99.9	57.5	86.4	32.5	48.8
		4	65.2	98.0	48.1	72.3	38.5	57.8	23.0	34.5
		6	57.5	86.4	37.6	56.6	30.1	45.2	19.5	29.3
		8	48.0	72.1	26.3	39.6	20.7	31.2	14.7	22.1
		10	37.9	56.9	17.3	25.9	13.6	20.5	10.1	15.2
		12	28.1	42.3	12.1	18.2	9.60	14.4	7.21	10.8
		14	20.8	31.3	8.94	13.4	7.11	10.7	5.37	8.07
		16	16.0	24.1						
		18	12.7	19.1						
		20	10.3	15.5						

Properties				
A_g, in.^{2}	2.63	2.22	1.92	1.48
r_x, in.	1.14	1.22	1.23	1.20
r_y, in.	1.23	0.876	0.843	0.840

ASD	LRFD
Ω_c = 1.67	ϕ_c = 0.90

[c] Shape is slender for compression with F_y = 50 ksi.
Note: Heavy line indicates KL/r equal to or greater than 200.

5 Stability and Determinacy of Structures

Topics to be covered in this chapter are as follows:

- Determinacy and stability of structures.
- Analysis of statically determinate trusses and arches.
- Influence lines.
- Moving loads on beams, frames, and trusses.
- Analysis of suspension bridge.
- Wind and earthquake loads.
- Approximate analysis of statically indeterminate structures.
- Deflection of beams, trusses, and frames by virtual work method.

5.1 DETERMINACY AND STABILITY OF STRUCTURES

$$R = 3$$
$$E = 3$$
$$I = R - E = 3 - 3 = 0$$

Statically determinate.

$$R = 5$$
$$E = 3$$
$$I = R - E = 5 - 3 = 2$$

DOI: 10.1201/9781032638072-5

Statically indeterminate to second degree.

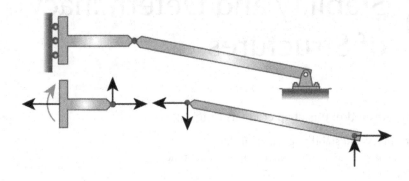

$$R = (2+2) + (2) = 6$$
$$E = 2 \times 3 = 6$$
$$I = R - E = 6 - 6 = 0$$

Statically determinate.

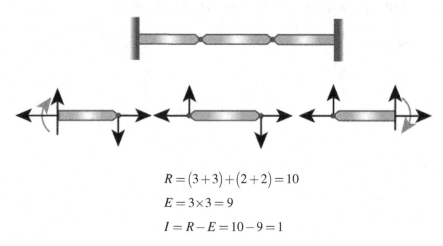

$$R = (3+3) + (2+2) = 10$$
$$E = 3 \times 3 = 9$$
$$I = R - E = 10 - 9 = 1$$

Statically indeterminate to first degree.

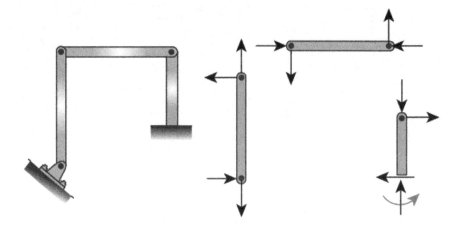

$$R = (2+3) + (2+2) = 9$$
$$E = 3 \times 3 = 9$$
$$I = R - E = 9 - 9 = 0$$

Statically determinate.

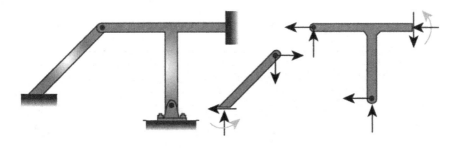

$$R = (3+2+3) + (2) = 10$$
$$E = 2 \times 3 = 6$$
$$I = R - E = 10 - 6 = 4$$

Statically indeterminate to fourth degree.

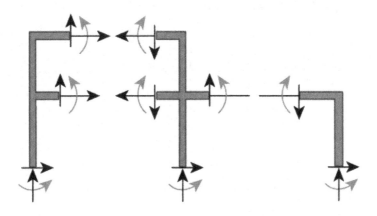

$$R = (3 \times 3) + (3 \times 3) = 18$$
$$E = 3 \times 3 = 9$$
$$I = R - E = 18 - 9 = 9$$

Statically indeterminate to ninth degree.

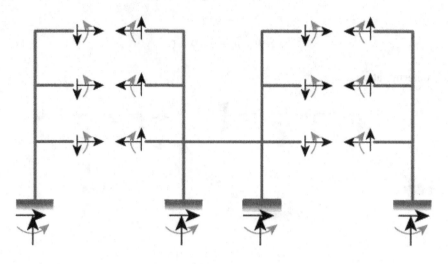

$$R = (3 \times 4) + (6 \times 3) = 30$$
$$E = 3 \times 3 = 9$$
$$I = R - E = 30 - 9 = 21$$

Statically indeterminate to twenty-first degree.

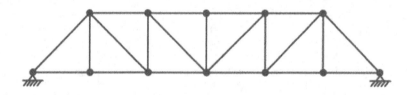

$$R = (2 + 1) + 21 = 24$$
$$E = 12 \times 2 = 24$$
$$I = R - E = 25 - 24 = 1$$

Statically determinate.

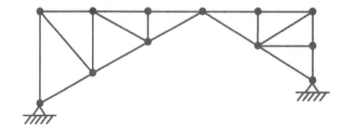

$$R = (2 \times 2) + 20 = 24$$
$$E = 12 \times 2 = 24$$
$$I = R - E = 24 - 24 = 0$$

Statically determinate.

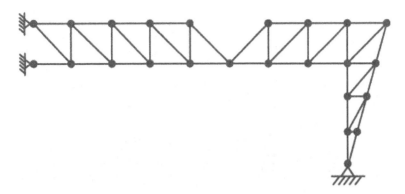

$$R = (3 \times 2) + 43 = 49$$

$$E = 24 \times 2 = 48$$

$$I = R - E = 49 - 48 = 1$$

Statically indeterminate to first degree.

5.1.1 STATICALLY UNSTABLE

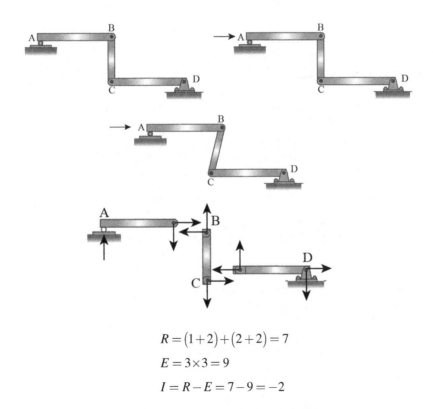

$$R = (1+2) + (2+2) = 7$$

$$E = 3 \times 3 = 9$$

$$I = R - E = 7 - 9 = -2$$

Structure is unstable.

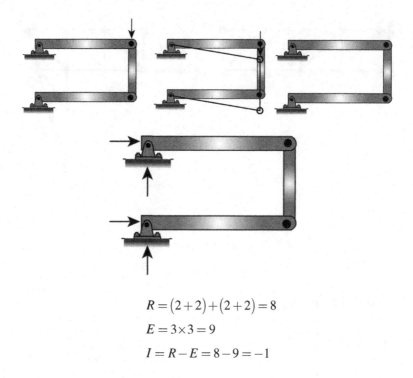

$$R = (2+2)+(2+2) = 8$$
$$E = 3 \times 3 = 9$$
$$I = R - E = 8 - 9 = -1$$

Structure is unstable.

5.1.2 GEOMETRICALLY UNSTABLE

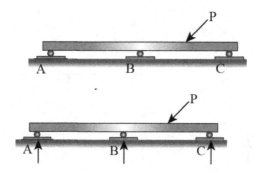

Since all reactions are parallel, structure is unstable.

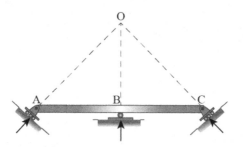

Since all reactions are concurrent, structure is unstable.

5.2   TRUSS ANALYSIS: JOINT METHOD

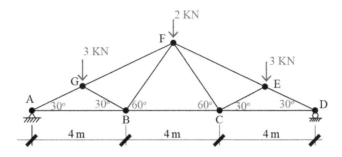

Example 1

Determine the bar force in members AB, AG, GB, GF, BF.

Solution

$$\Sigma M_D = 0 (\circlearrowleft^+)$$
$$Ay \times 12 - 3 \times 10 - 2 \times 6 - 3 \times 2 = 0; \ A_y = 4 \text{ kip}(\uparrow)$$

Joint A:

$$\uparrow + \sum F_y = 0$$
$$F_{AG} \sin 30 + 4 = 0$$
$$F_{AG} = -8 kip (C)$$

$$+ \sum F_x = 0$$
$$F_{AB} + F_{AG} \cos 30 = 0$$
$$F_{AB} = +6.92 kip (T)$$

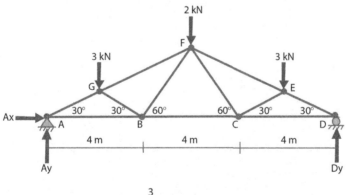

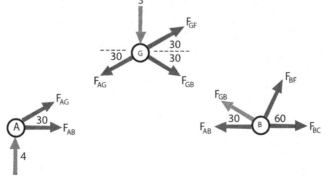

Joint G:

$$\sum F_x = 0 \rightarrow^+ \quad F_{GF}\cos 30 + F_{GB}\cos 30 - F_{AG}\cos 30 = 0$$
$$F_{GF}\cos 30 + F_{GB}\cos 30 = -6.928$$
$$\sum F_y = 0 \uparrow^+ \quad F_{GF}\sin 30 - F_{GB}\sin 30 - F_{AG}\sin 30 - 3 = 0$$
$$F_{GF}\sin 30 - F_{GB}\sin 30 = -1$$

Solving: $F_{GF} = -5\,\text{kip}\,(\text{C})$ and $F_{GB} = -3\,\text{kip}\,(\text{C})$

Joint B:

$$\sum F_y = 0 \uparrow^+ \quad F_{GB}\sin 30 + F_{BF}\sin 60 = 0$$
$$F_{BF} = +1.73\,\text{kip}\,(\text{T})$$
$$\sum F_x = 0 \rightarrow^+ \quad -F_{AB} - F_{GB}\cos 30 + F_{BF}\cos 60 + F_{BC} = 0$$
$$F_{BC} = +3.46\,\text{kip}\,(\text{T})$$

5.3   TRUSS ANALYSIS: SECTION METHOD

Example 2

Determine bar force in members GB, GF, BF. Use the section method.

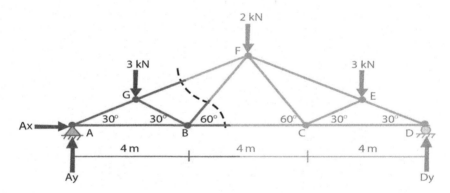

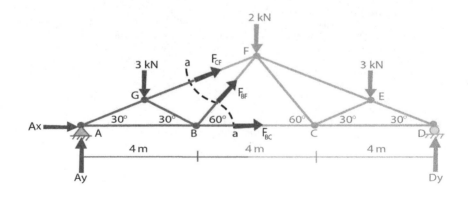

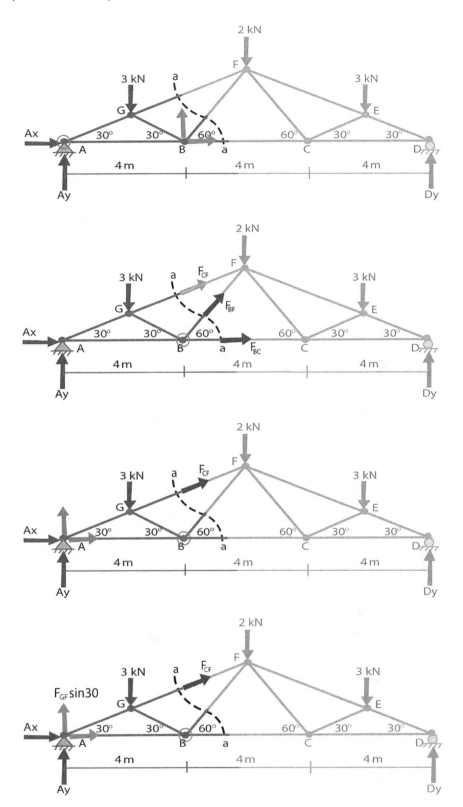

Solution

$$\Sigma M_D = 0 \circlearrowleft^+ \quad Ay \times 12 - 3 \times 10 - 2 \times 6 - 3 \times 2 = 0; \ A_y = 4 \, \text{kip}(\uparrow)$$

Section $a - a$:

$$\Sigma M_F = 0 \circlearrowleft^+ \quad A_y \times 6 - 3 \times 4 - F_{BC} \times 2 \tan 60 = 0 \quad F_{BC} = +3.46 \, \text{kip}(T)$$
$$\Sigma M_A = 0 \circlearrowleft^+ \quad 3 \times 2 - (F_{BF} \sin 60) \times 4 = 0 \quad\quad\quad F_{BF} = +1.73 \, \text{kip}(T)$$
$$\Sigma M_B = 0 \circlearrowleft^+ \quad A_y \times 4 + (F_{CF} \sin 30) \times 4 - 3 \times 2 = 0 \quad F_{CF} = -5.00 \, \text{kip}(C)$$

Example 3

Determine the bar force in members HG, BC, BC. Use the section method.

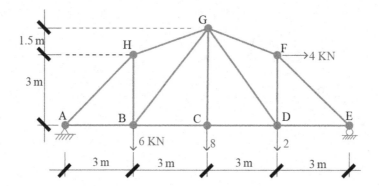

Solution

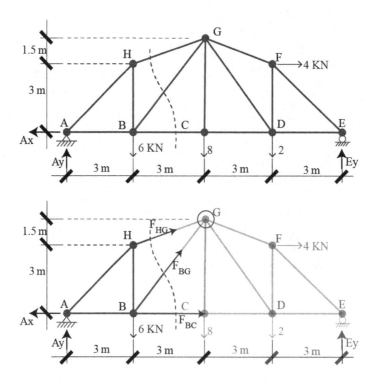

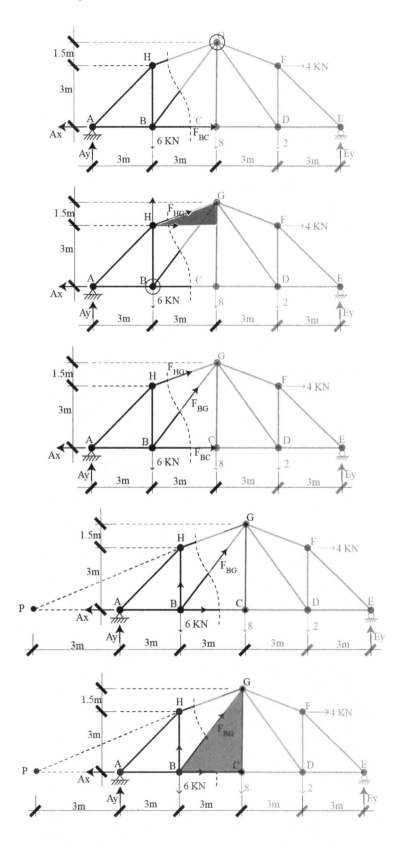

$$\sum M_E = 0 \circlearrowleft^+ \quad A_y \times 12 - 6 \times 9 - 8 \times 6 - 2 \times 3 + 4 \times 3 = 0; \quad A_y = 8 \text{ kip}(\uparrow)$$
$$\sum F_x = 0 \rightarrow^+ \quad -A_x + 4 = 0; \qquad\qquad\qquad\qquad A_x = 4 \text{ kip}(\leftarrow)$$

Section $a - a$:

$$\sum M_G = 0 \circlearrowleft^+ \quad A_y \times 6 + A_x \times 4.5 - 6 \times 3 - F_{BC} \times 4.5 = 0 \quad F_{BC} = +10.67 \text{ kip}(T)$$
$$\sum M_B = 00^+ \quad A_y \times 3 + \left[F_{GH}\cos\left(\tan^{-1}1.5/3\right)\right] \times 3 = 0 \qquad F_{GH} = -8.94 \text{ kip}(C)$$

$$\text{Similar triangle:} \quad \frac{PB}{BH} = \frac{PC}{CC} \frac{PA+3}{3} = \frac{PA+6}{4.5} \quad PA = 3 \text{ m}$$

$$\sum M_P = 0 \circlearrowleft^+ \quad -A_y \times 3 + 6 \times 6 - \left[F_{BC}\sin\left(\tan^{-1}4.5/3\right)\right] \times 6 = 0 \quad F_{GH} = +2.40 \text{ kip }(T)$$

Example 4

Determine the bar force in members JH, JC, BC. Use the section method.

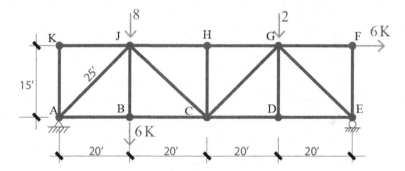

Solution

$$\sum M_E = 0 \circlearrowleft^+ \quad A_y \times 80 - 8 \times 60 - 6 \times 60 - 2 \times 20 + 4 \times 15 = 0; \quad A_y = 10.25 \text{ kip}(\uparrow)$$
$$\sum F_x = 0 \rightarrow^+ \quad -A_x + 4 = 0; \qquad\qquad\qquad\qquad\qquad A_x = 4 \text{ kip}(\leftarrow)$$

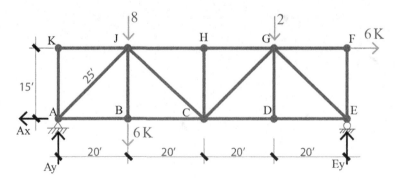

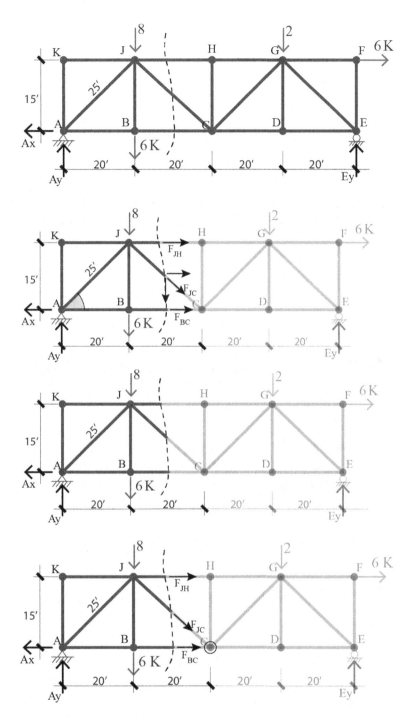

Section $a - a$:

$$\sum M_C = 0 \circlearrowleft^+ \quad A_y \times 40 - 8 \times 20 - 6 \times 20 + F_{JH} \times 15 = 0 \quad F_{JH} = -8.67 \, \text{kip(C)}$$

$$\sum M_J = 0 \circlearrowleft^+ \quad A_y \times 20 + A_x \times 15 - F_{BC} \times 15 = 0 \quad F_{BC} = 17.67 \, \text{kip(T)}$$

$$\sum F_y = 0 \uparrow^+ \quad A_y - 8 - 6 - F_{JC} \cdot \frac{15}{25} = 0 \quad F_{JC} = -6.25 \, \text{kip(C)}$$

Example 5

Determine the bar force in members a to k.

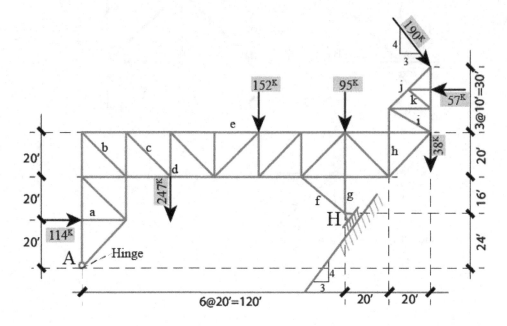

Solution

$$\Sigma M_A = 0 \circlearrowleft^+$$
$$114 \times 20 + 247 \times 40 + 152 \times 80 + 95 \times 120$$
$$+38 \times 160 - 57 \times 80 + 190 \times (3/5) \times 90 + 190 \times (4/5) \times 160$$
$$-(3/5)H \times 120 - (4/5)H \times 24 = 0 \; H = 787.5 \, \text{kip}$$
$$H_x = (4/5) \times 787.5 = 630 \, \text{kip}$$
$$H_y = (3/5) \times 787.4 = 472.5 \, \text{kip}$$

$$\Sigma F_x = 0 \rightarrow^+$$
$$A_x + 114 - 630 - 57 + 190 \times (3/5) = 0 \quad A_x = 459 \, \text{kip}$$

$$\Sigma F_y = 0 \uparrow^+$$

$$A_y - 247 - 152 - 95 + 472.5 - 38 - 190 \times (4/5) = 0 \quad A_y = 211 \, \text{kip}$$

Joint a:

$$\Sigma F_x = 0 \rightarrow^+ \; 114 + F_a = 0; \; F_a = -114 \, (\text{C})$$

Example 6

Determine the bar force in members *a* to *i*.

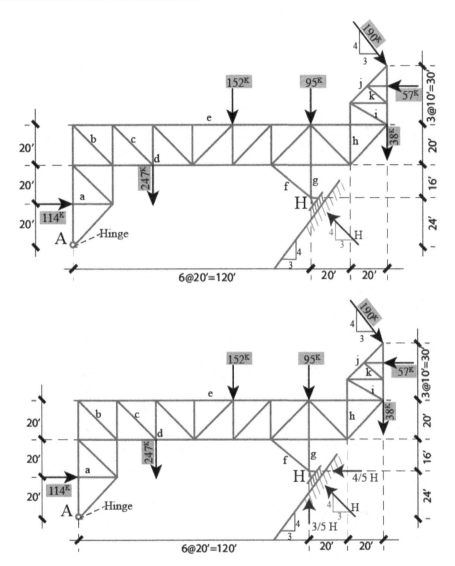

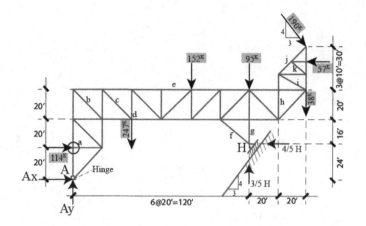

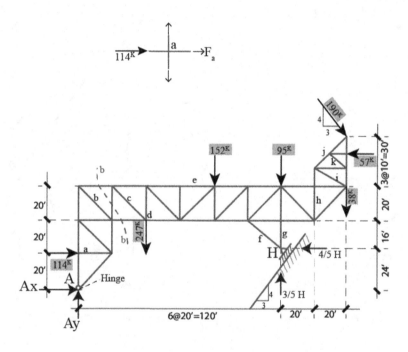

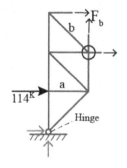

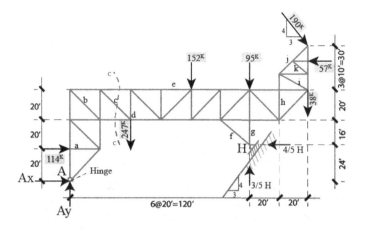

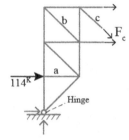

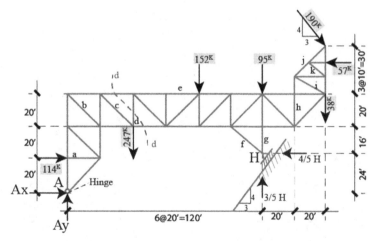

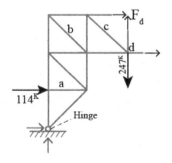

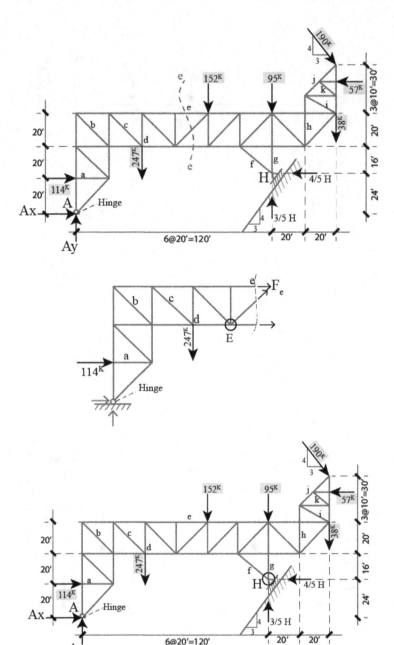

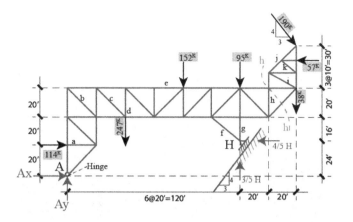

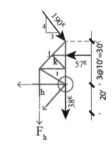

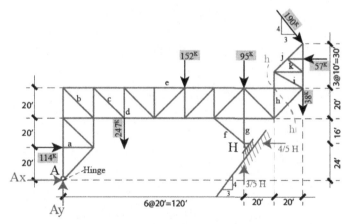

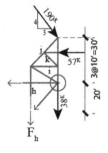

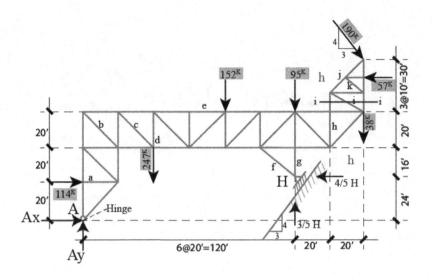

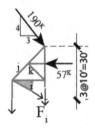

Section $b-b$:

$$\sum M_B = 0 \circlearrowleft^+ \ A_y \times 20 - A_x \times 40 - 114 \times 20 + F_1 \times 20 = 0 \ F_1 = 820.5(T)$$

Joint B:

$$\sum F_x = 0 \rightarrow^+ \ F_b \cos 45 + (820.5) = 0 \ F_b = -1161(C)$$

Section $c-c$:

$$\sum F_y = 0 \uparrow^+ \ A_y - F_c \cos 45 = 0 \ F_c = 299.1(T)$$

Section $d-d$:

$$\sum F_y = 0 \uparrow + \ A_y + F_d - 247 = 0 \ F_d = 35.5(T)$$

Section $e-e$:

$$\sum M_E = 00^+ \ A_y \times 60 - A_x \times 40 - 114 \times 20 - 247 \times 20 + F_e \times 20 = 0 \ F_e = 6$$

Joint H:

$$\sum F_x = 0 \rightarrow^+ \ -F_f(4/5) - H_x = 0 \qquad F_f = -806.1(C)$$
$$\sum F_y = 0 \uparrow^+ \ F_f(3/5) + F_g + H_y = 0 \qquad F_g = 30.4(T)$$

Section $h-h$:

$$\Sigma M = 0 \circlearrowleft^+ \; -F_h \times 20 - 57 \times 20 + 190 \times (3/5) \times 30 = 0 \; F_h = 114 \, (\text{T})$$

Section $i-i$:

$$\Sigma F_x = 0 \rightarrow^+ \; F_i \cos\left(\tan^{-1}10/20\right) - 57 + 190 \times (3/5) = 0 \; F_i = -63$$

5.4 ARCH ANALYSIS

Example 7

Determine the reactions at B and C support and axial, shear, and bending moment at D.

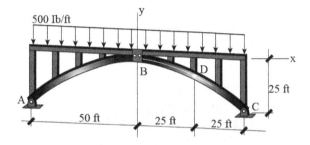

Entire arch:

$$\downarrow +\Sigma M_A = 0; \; C_y \, (100\text{ft}) - 50\text{k} \, (50\text{ft}) = 0$$
$$C_y = 25\text{k}$$

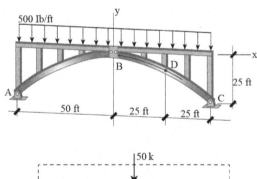

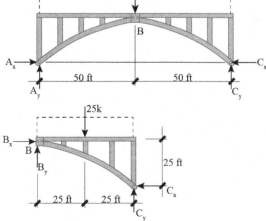

Question: Determine reactions at B and C support and axial, shear, and bending moment at D. Entire arch:

$$\downarrow +\Sigma M_A = 0; \; C_y(100\text{ft}) - 50\text{k}(50\text{ft}) = 0$$
$$C_y = 25\text{k}$$

Arch segment BC:

$$\downarrow +\Sigma M_B = 0; \quad -25\text{k}(25\text{ft}) + 25\text{k}(50\text{ft}) - C_x(25\text{ft}) = 0$$
$$C_x = 25\text{k}$$
$$\overrightarrow{\;}\Sigma F_x = 0; \quad B_x = 25\text{k}$$
$$+\uparrow \Sigma F_y = 0; \quad B_y - 25\text{k} + 25\text{k} = 0$$
$$B_y = 0$$

Example 8

Determine the reactions at B and C support and axial, shear, and bending moment at D.

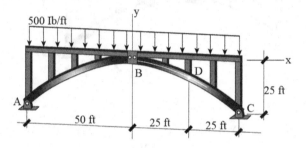

A section of the arch taken through point $D, x = 25\text{ft}$, $y = -25(25)^2 / (50)^2 = -6.25\text{ft}$, is shown in the following figure. The slope of the segment at D is:

$$\tan\theta = \frac{dy}{dx} = \frac{-50}{(50)^2}x\Big|_{x=25\text{ft}} = -0.5$$
$$\theta = -26.6°$$

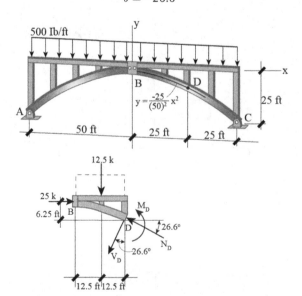

A section of the arch taken through point $D, x = 25$ ft, $y = -25(25)^2 / (50)^2 = -6.25$ ft is shown in previous figure. The slope of the segment at D is:

$$\tan \theta = \frac{dy}{dx} = \frac{-50}{(50)^2} x \bigg|_{x=25ft} = -0.5$$

$$\theta = -26.6°$$

Applying the equations of equilibrium, we have:

$$\xrightarrow{+} \Sigma F_x = 0; \quad 25k - N_D \cos 26.6° - V_D \sin 26.6° = 0$$
$$+\uparrow \Sigma F_y = 0; \quad -12.5k + N_D \sin 26.6° - V_D \cos 26.6° = 0$$
$$J + \Sigma M_D = 0; \quad M_D + 12.5k(12.5ft) - 25k(6.25ft) = 0$$

$$N_D = 28.0 \text{ k}$$
$$V_D = 0$$
$$M_D = 0$$

Example 10

Determine the reactions at A and F support and the bar force in the KC member.

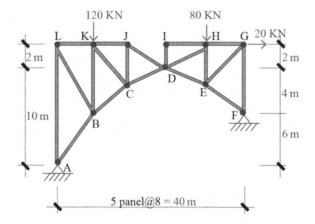

Solution

5.4.1 WHOLE FREE BODY OF ARCH

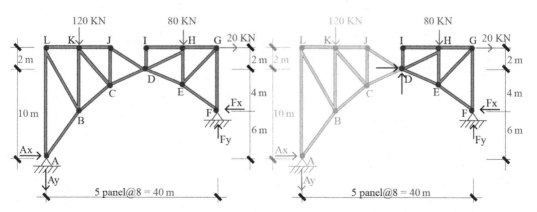

$$\sum M_A = 0 \circlearrowright^+ \quad 120 \times 8 + 80 \times 32 + 20 \times 12 - F_x \times 6 - F_y \times 40 = 0$$
$$6F_x + 40F_y = 3760$$

Free body of right side of hinge D:

$$\sum M_D = 0 \circlearrowright^+ \quad 80 \times 8 + 20 \times 2 + F_x \times 4 - F_y \times 16 = 0$$
$$4F_x - 16F_y = -680$$

Solving we find, $F_X = 128.75$kN and $F_Y = 74.68$kN.

Whole free body of arch:

$$\sum F_x = 0 \rightarrow^+ \quad A_x + 20 - 128.75 = 0 \qquad A_x = 108.75 \text{ kN}$$
$$\sum F_y = 0 \uparrow^+ \quad A_y - 120 - 80 + 74.68 = 0 \quad A_y = 125.32 \text{ kN}$$

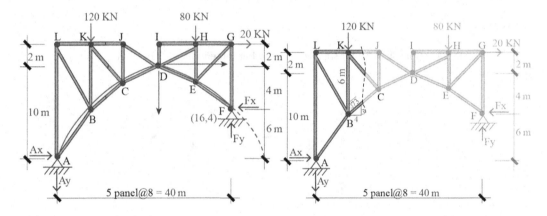

From geometry of arch:

$$\text{At } F, 4 = k \times 16^2 \quad k = 1/64$$
$$y_C = 1/64 \times 8^2 = 1 \text{ m } \quad y_B = 4 \text{ m } \quad y_{BC} = 4 - 1 = 3 \text{ m}$$

5.4.2 FROM LEFT SEGMENT

$$\sum M_I = 0 \circlearrowright^+$$
$$125.32 \times 24 - 108.75 \times 12 - 120 \times 16 - F_{KC} \times \left(3 / \sqrt{73}\right) \times 16 = 0$$
$$F_{KC} = -38.68 \text{ kN}$$

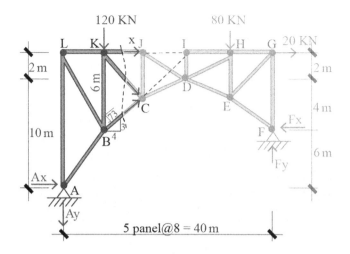

5.5   APPLICATION OF INFLUENCE LINE

Example 11

Determine the $R_A, V_P,$ and M_Q for the given load in the beam.

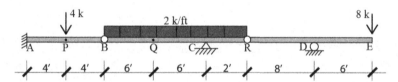

Solution

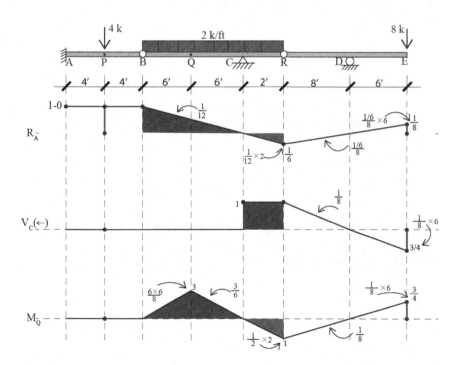

$$R_A = 4 \times 1.0 + 2 \times \left(\frac{1}{2} \times 12 \times 1 \right)$$

$$-2 \times \left(\frac{1}{2} \times 2 \times \frac{1}{6} \right) + 8 \times \frac{1}{8}$$

$$= 16.67 \text{ kip}$$

$$V_{C(\leftarrow)} = 4 \times 0 - 2 \times (1 \times 2) - 8 \times \frac{3}{4}$$

$$= -10 \text{ kip}$$

$$M_Q = 4 \times 0 + 2 \times \left(\frac{1}{2} \times 12 \times 3 \right)$$

$$-2 \times \left(\frac{1}{2} \times 2 \times 1 \right) + 8 \times \frac{3}{4}$$

$$= 40 \text{ k} - \text{ft}$$

5.6   INFLUENCE LINE FOR BEAMS

Example 12a

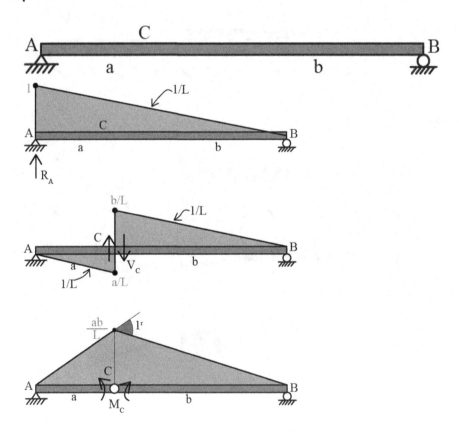

Example 12b

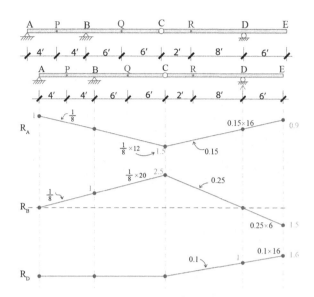

Example 12c

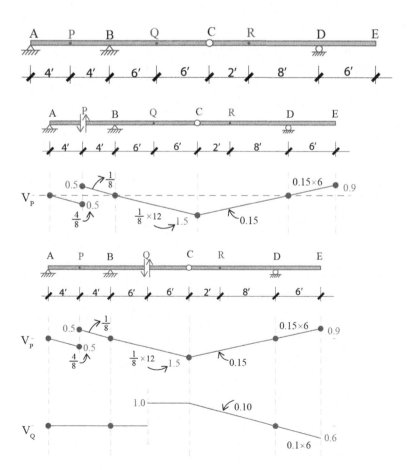

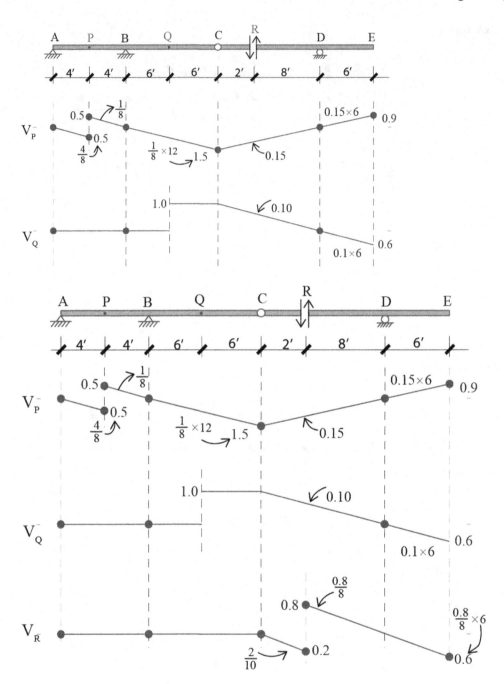

Example 12d

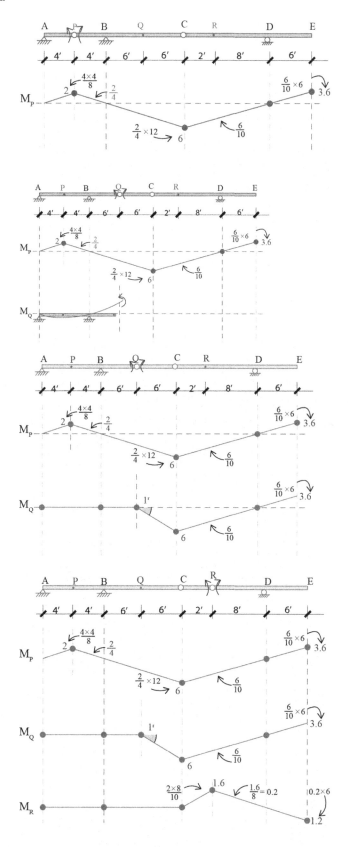

Example 13

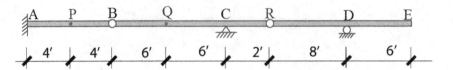

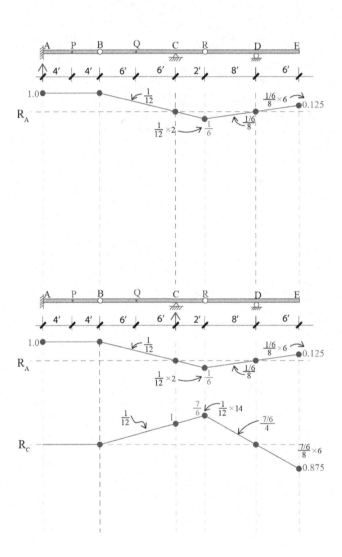

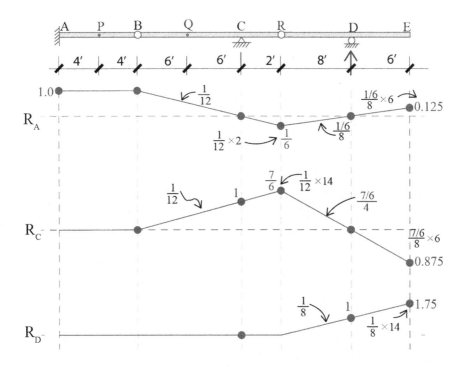

Example 14

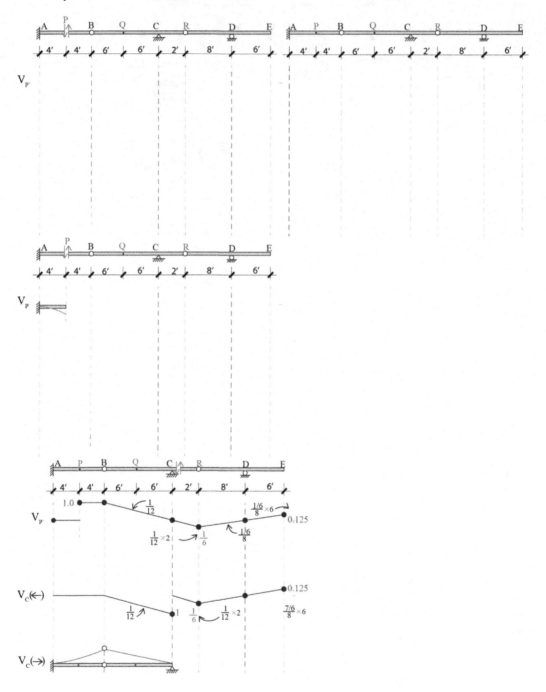

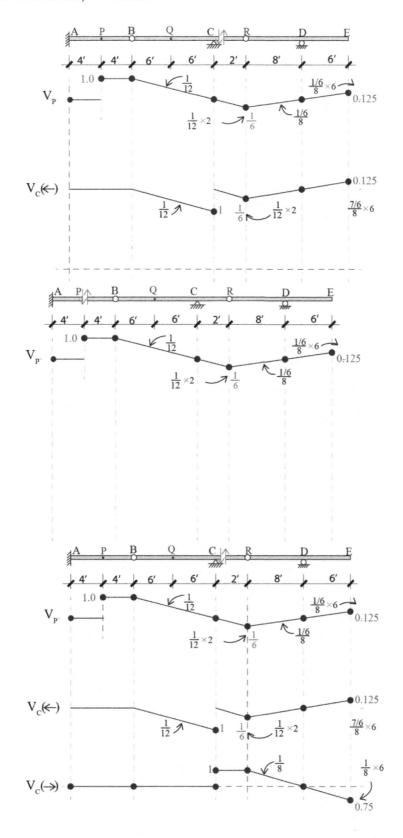

Example 15

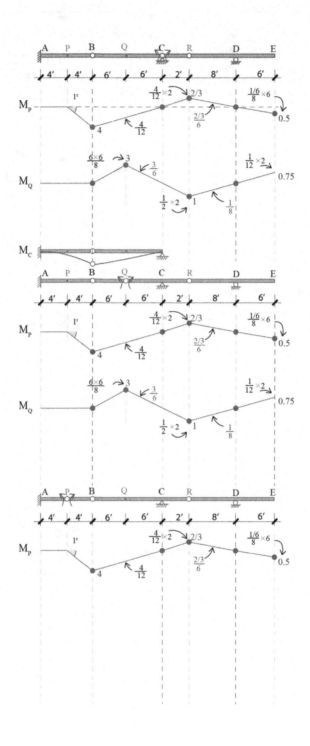

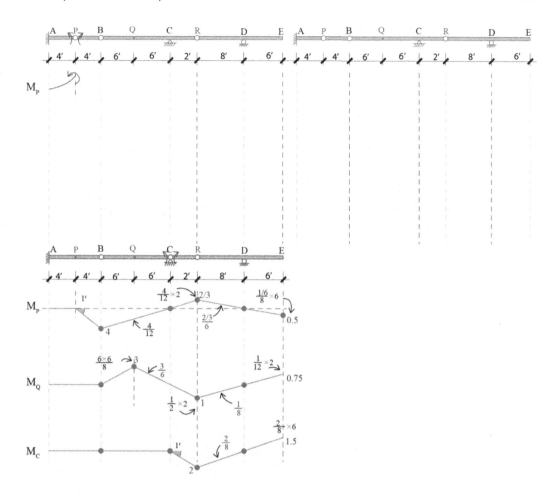

5.7   MOVING LOADS

Example 16

Determine the maximum reaction produced at support A in the beam if the wheel load moves from right to left.

Solution

$$\Delta R_{1\to 2} = -1\times 10 + \frac{1}{50}\times 5\times(10 + 4\times 40 + 2\times 10) = +9$$

$$\Delta R_{1\to 2} = -1\times 10 + \frac{1}{50}\times 5\times(10 + 4\times 40 + 2\times 10) = +9$$

$$\Delta R_{2\circledR3} = -1\times 10 + \frac{1}{50}\times 10\times(4\times 40 + 2\times 10 + 1\times 20) = +\frac{1}{50}\times 5\times 20 = +32$$

$$\Delta R_{3\circledR4} = -1\times 40 + \frac{1}{50}\times 5\times(3\times 40 + 2\times 10 + 3\times 20) = -20$$

$$R_{max} = 40\times\frac{1}{50}\times(35 + 40 + 45 + 50) + 10\times\frac{1}{50}\times(20 + 25) = +20\times\frac{1}{50}\times(0 + 5 + 10)$$

$$= 145\ kip$$

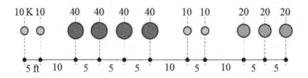

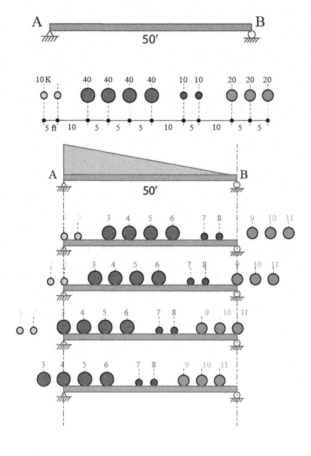

Example 17

Determine the maximum shear at C in the beam if the wheel load moves from right to left.

Solution

$$\Delta V_{1\to 2} = -1\times 10 + \frac{5}{50}\times(2\times 10 + 4\times 40) = +8$$

$$\Delta V_{2\to 3} = -1\times 10 + \frac{10}{50}\times(2\times 10 + 4\times 40) + \frac{5}{50}\times 10 = +7$$

$$\Delta V_{3\to 4} = -1\times 40 + \frac{5}{50}\times(2\times 10 + 4\times 40 + 2\times 10) = -20$$

$$V_{max} = -10\times\frac{1}{50}\times(5+10) + 40\times\frac{1}{50}\times(15 + 20 + 25 + 30)$$

$$+10\times\frac{1}{50}\times(0+5)$$

$$= 70\ kip$$

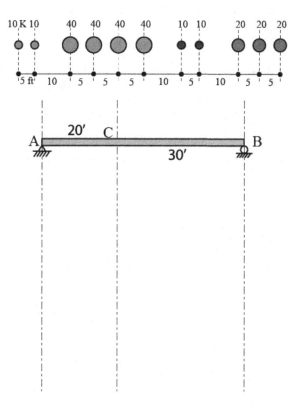

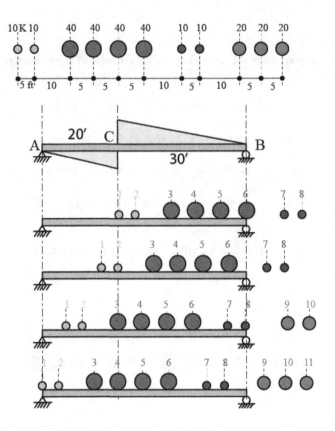

Example 18

Determine the maximum moment at C in the beam if the wheel load moves from right to left.

Solution

$$\Delta M_{1\rightarrow 2} = -\frac{3}{5}\times 5\times 10 + \frac{2}{5}\times 5\times(10 + 4\times 40) = +310$$

$$\Delta M_{2\rightarrow 3} = -\frac{3}{5}\times 10\times(2\times 10) + \frac{2}{5}\times 10\times(4\times 40)$$

$$+\frac{2}{5}\times 5\times 10$$

$$\Delta M_{3\rightarrow 4} = -\frac{3}{5}\times 5\times(2\times 10 + 40) = +540$$

$$+\frac{2}{5}\times 5\times(3\times 40 + 2\times 10)$$

$$\Delta M_{4\rightarrow 5} = -\frac{3}{5}\times 5\times(1\times 10 + 2\times 40)$$

$$+\frac{2}{5}\times 5\times(2\times 40 + 2\times 10)$$

$$M_{max} = 10\times\frac{3}{5}\times(0 + 5) + 40\times\frac{3}{5}\times(15 + 20)$$

$$+40\times\frac{2}{5}\times(25 + 20) + 10\times\frac{2}{5}\times(5 + 10)$$

$$= 1650\,k-ft$$

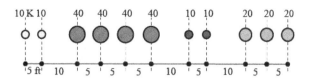

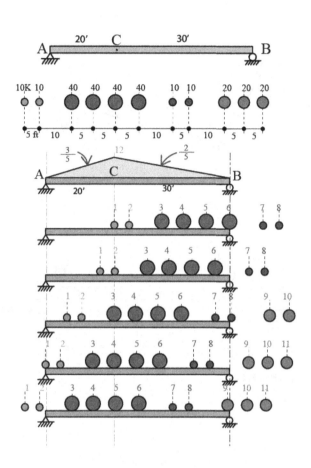

5.8 SUSPENSION BRIDGE

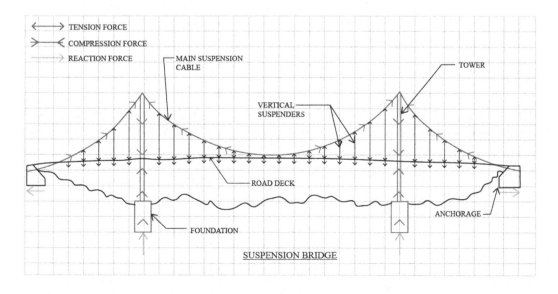

SUSPENSION BRIDGE

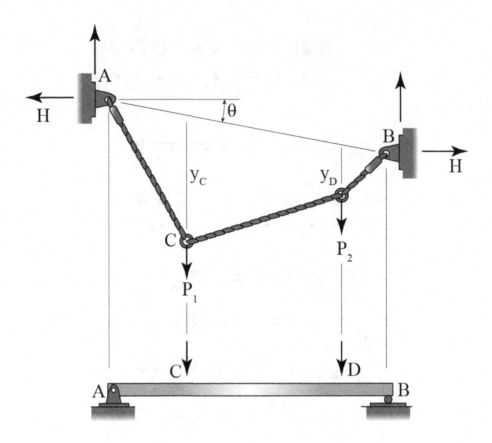

5.8.1   GENERAL CABLE THEOREM

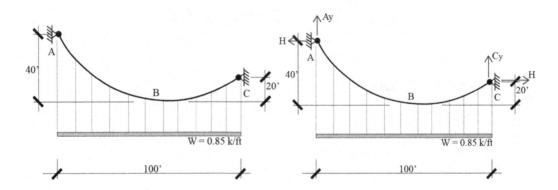

The bending moment (M) of a specific point in imaginary simply supported beam of the same span as the cable is numerically equal to the horizontal reactions (H) in cable times sag (y) from cable chord of that same point.

$$M = yH$$

For point C and D:

$$M_C = y_C H \quad \text{and} \quad M_D = y_D H$$

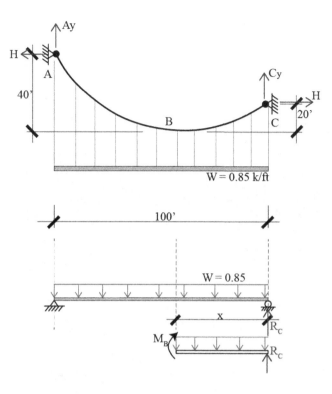

Example 19

The cable ABC carries a distributed load of $w = 0.85\text{k/ft}$ over its 100 ft span. Point B is the lowest point on the cable. Determine the tension at supports A and C in the cable. Use the cable theorem.

Solution

5.8.2 LOCATE B

Since the cable follows a parabolic profile, applying $y = kx^2$, we find:

$$\text{Point A} \quad 40 = k(100 - x)^2$$
$$\text{Point C} \quad 20 = kx^2 \qquad \text{Solving } x = 41.42\text{ft}$$

Imaginary beam AC:

$$R_C = 0.85 \times 100 / 2 = 42.5\text{kip}$$

Free body of BC of the imaginary beam:

$$\sum M_B = 0 \circlearrowleft^+$$

$$M_B + wx\left(\frac{x}{2}\right) - R_C(x) = 0 \quad M_B = 1031.2\text{k} - \text{ft}$$

From properties of similar triangles:

$$\frac{y_1}{20} = \frac{58.58}{100}; \ y_1 = 11.72\text{ft} \ \text{and} \ y_B = 40 - 11.72 = 28.28\text{ft}$$

5.8.3 Applying Cable Theorem at B

$$M_B = Hy_B; 1031.2 = H \times 28.28 \ H = 36.46 \text{ kip}$$

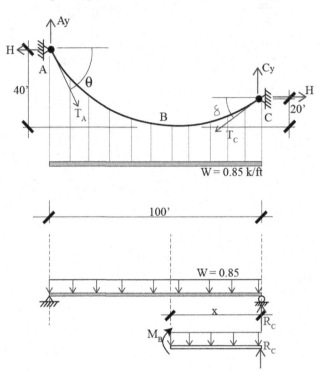

Example 20

The cable ABC carries a distributed load of $w = 0.85\text{k}/\text{ft}$ over its 100 ft span. Point B is the lowest point on the cable. Determine the maximum and minimum tensions in the cable. Use the cable theorem.

Solution

5.8.4 Whole Free Body

$$\sum M_A \quad = 0 \circlearrowright^+ \qquad 0.85 \times 100 \times 50 - 36.46 \times 20 - C_y \times 100 = 0; \quad C_y = 35.2\text{kip}$$
$$\sum F_y \quad = 0 \uparrow^+ \qquad\qquad\qquad A_y - 0.85 \times 100 + 35.2 = 0; \quad A_y = 49.8\text{kip}$$

5.8.5 JOINT A

$$\sum F_X = 0 \rightarrow^+ \quad -36.46 + T_A \cos\theta = 0; \quad T_A \cos\theta = 36.46$$
$$\sum F_y = 0 \uparrow^+ \quad 49.8 - T_A \sin\theta = 0; \quad \quad T_A \sin\theta = 49.8$$

Solving, we find:

$$T_A = \sqrt{36.46^2 + 49.8^2} = 61.72 \text{ kip}$$

Joint C:

$$\sum F_x = 0 \rightarrow^+ \quad 36.46 - T_C \cos\delta = 0; \quad T_C \cos\delta = 36.46$$
$$\sum F_y = 0 \uparrow^+ \quad 35.2 - T_C \sin\delta = 0; \quad T_C \sin\delta = 35.2$$

Solving, we find:

$$T_C = \sqrt{36.46^2 + 35.2^2} = 50.68 \text{ kip}$$

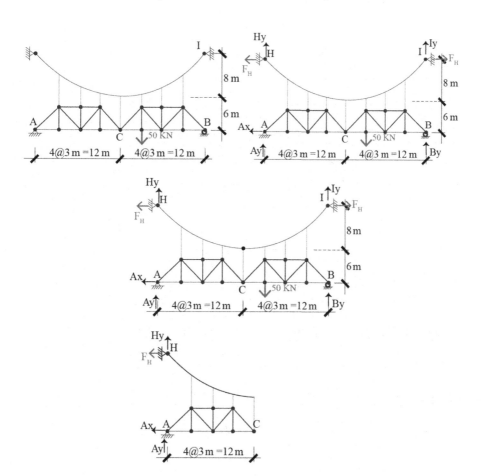

Example 21

Determine the maximum and minimum tensions in the suspension bridge cable.

Solution

5.8.6 Whole Free Body

$$\sum M_B = 0 \circlearrowright^+ \; A_y \times 24 + H_y \times 24 - 50 \times 9 = 0 \; A_y + H_y = 18.75$$

Left segment of joint C:

$$\sum M_C = 0 \circlearrowright^+ \; A_y \times 12 + H_y \times 12 - F_H \times 14 + F_H \times 6 = 0$$
$$18.75 = 0.677 F_H$$
$$F_H = 28.125 \text{ kN}$$

5.8.7 Find Equation of the Cable

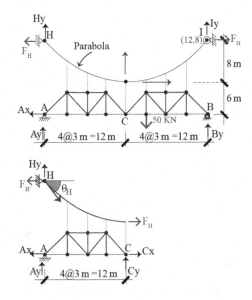

Since the cable follows a parabolic profile, applying $y = kx^2$ at $B(12,8)$:

$$8 = k(12)^2 \; k = 1/18$$

Now, find angle at H:

$$y = \left(\frac{1}{18}\right) x^2 \quad \frac{dy}{dx} = \frac{x}{9}$$

$$\tan \theta_H = \frac{dy}{dx} = \frac{x}{9} = \frac{12}{9} \quad \theta_H = \tan^{-1}\left(\frac{12}{9}\right) = 53.13°$$

Find maximum tension, joint H:

$$\Sigma F_X = 0 \rightarrow^+ T_H \cos \theta_H - F_H = 0$$
$$T_H \cos 53.13 - 28.125 = 0$$
$$T_H = 46.87 \text{ kN}$$

Answer: $T_{\text{max}} = 46.87$ kN and $T_{\text{min}} = 28.125$ kN

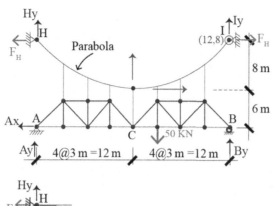

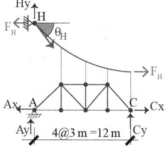

5.9   WIND LOAD ANALYSIS

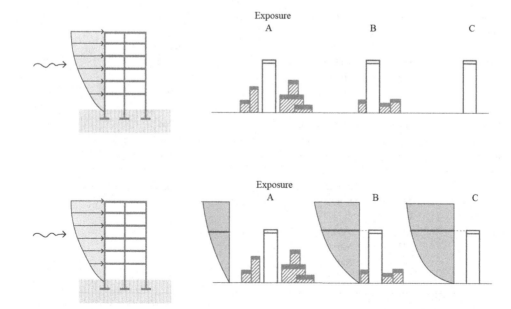

Combined Height and Exposure Coefficient, C_z

Height above ground level, z (metres)	Coefficient, c_z [1]		
	Exposure A	Exposure B	Exposure C
0–4.5	0.368	0.801	1.196
6.0	0.415	0.866	1.263
9.0	0.497	0.972	1.370
12.0	0.565	1.055	1.451
15.0	0.624	1.125	1.517
18.0	0.677	1.185	1.573
21.0	0.725	1.238	1.623
24.0	0.769	1.286	1.667
27.0	0.810	1.330	1.706
30.0	0.849	1.371	1.743
35.0	0.909	1.433	1.797
40.0	0.965	1.488	1.846

Location	Basic Wind Speed (km/h)	Location	Basic Wind Speed (km/h)	Location	Basic Wind Speed (km/h)
Angarpota	150	Joypurhat	180	Nilphamari	140
Bagerhat	252	Jamalpur	180	Noakhali	184
Bandarban	200	Jessore	205	Pabna	202
Barguna	260	Jhalakati	260	Panchagarh	130
Barisal	256	Jhenaidah	208	Patuakhali	260
Bhola	225	Khagrachhari	180	Pirojpur	260
Bogra	198	Khulna	238	Rajbari	188
Brahmanbaria	180	Kutubdia	260	Rajshahi	155
Chandpur	160	Kishoreganj	207	Rangamati	180
Chapai Nawabganj	130	Kurigram	210	Rangpur	209
Chittagong	260	Kushtia	215	Satkhira	183
Chuadanga	198	Lakshmipur	162	Shariatpur	198
Comilla	196	Lalmonirhat	204	Sherpur	200
Cox's Bazar	260	Madaripur	220	Sirajganj	160
Dahagram	150	Magura	208	Srimangal	160
Dhaka	210	Manikganj	185		
		Meherpur	185	St. Martin's Island	260
Dinajpur	130			Sunamganj	195
Faridpur	202			Sylhet	195
Feni	205	Moheshkhali	260	Sandwip	260
Gaibandha	210	Moulvibazar	168	Tangail	160
Gazipur	215	Munshiganj	184		
		Mymensingh	217		
		Naogaon	175	Teknaf	260
Gopalganj	242	Narail	222	Thakurgaon	130
Habiganj	172				
Hatiya	260	Narayanganj	195		
Ishurdi	225	Narsinghdi	190		
		Natore	198		
		Netrokona	210		

Gust Response Factors, G_h, and G_z[1]

Height above ground level (metres)	G_h[2] and G_z		
	Exposure A	Exposure B	Exposure C
0–4.5	1.654	1.321	1.154
6.0	1.592	1.294	1.140
9.0	1.511	1.258	1.121
12.0	1.457	1.233	1.107
15.0	1.418	1.215	1.097
18.0	1.388	1.201	1.089
21.0	1.363	1.189	1.082
24.0	1.342	1.178	1.077
27.0	1.324	1.170	1.072
30.0	1.309	1.162	1.067
35.0	1.287	1.151	1.061
40.0	1.268	1.141	1.055

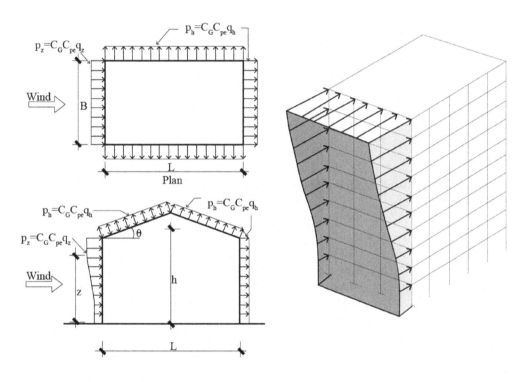

External Pressure Coefficient C_{pe} for Walls

Surface	L/B	C_{pe}	For Use With
Windward Wall	All values	0.8	$p_z = C_G C_{pe} q_z$
	≤ 0.10	−0.5	
Leeward Wall	0.65	−0.6	
	1.00	−0.5	$p_z = C_G C_{pe} q_h$
	2.00	−0.3	
	≥ 4.00	−0.2	
Sidewall	All values	−0.7	$p_z = C_G C_{pe} q_z$

Structure Importance Coefficients

Structure Importance Category (see Table 6.1.1 for occupancy)	Structure Importance Coefficient I
1. Essential Facilities	1.25
2. Hazardous Facilities	1.25
3. Special-Occupancy Structures	1.00
4. Standard-Occupancy Structures	1.00
5. Low-risk Structures	1.00

Example 22

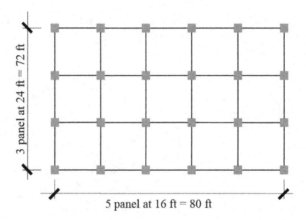

3 panel at 24 ft = 72 ft

5 panel at 16 ft = 80 ft

Determine the wind load at each story level along direction A for an internal frame. The following information is given.

Story	10
Floor Height	10 ft
Zone	Dhaka
Exposure	A
Structure	Standard-occupancy

$$p_z = C_c C_I C_{pe} C_z C_g V^2$$

Given data:

$$C_C = 47.2 \times 10^{-6}$$
$$C_I = 1.0$$
$$C_{pe} = 0.80$$
$$V = 210$$
$$p_z = \left(47.2 \times 10^{-6}\right) \times 1.0 \times 0.80 \times C_z \times C_g \times 210^2$$
$$= 1.665 C_z C_g$$

Floor	Cz	Cg	pz(kN/m²)	A(m²)	F(kN)
1	0.368	1.645	1.01	22.3	22.5
2	0.415	1.592	1.10	22.3	24.5
3	0.495	1.511	1.25	22.3	27.8
4	0.565	1.457	1.37	22.3	30.6
5	0.624	1.418	1.47	22.3	32.9
6	0.677	1.388	1.56	22.3	34.9
7	0.725	1.363	1.65	22.3	36.7
8	0.769	1.342	1.72	22.3	38.3
9	0.801	1.324	1.77	22.3	39.4
10	0.849	1.309	1.85	22.3	41.3

Sample calculation for floor 10:

$$p_z = 1.665 \times 0.849 \times 1.309 = 1.85\,\text{kN}/\text{m}^2$$
$$A = 24 \times 16 / 3.28^2 = 22.3\,\text{m}^2$$
$$F = 1.85 \times 22.3 = 41.3\,\text{kN}$$

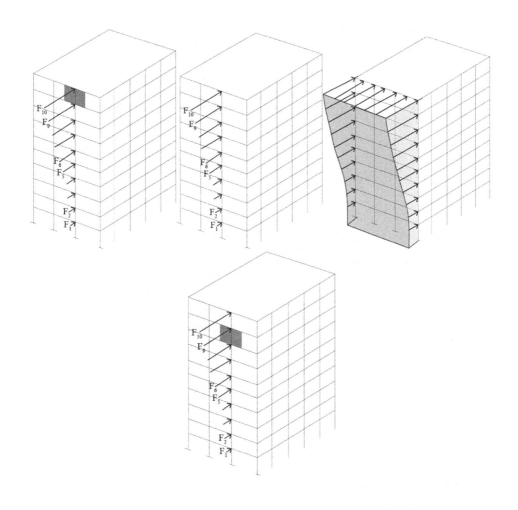

5.10   SEISMIC LOAD ANALYSIS

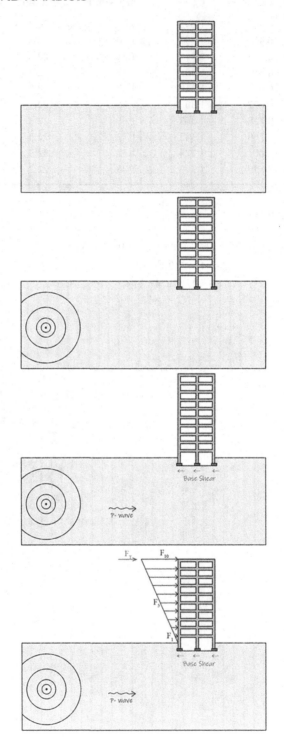

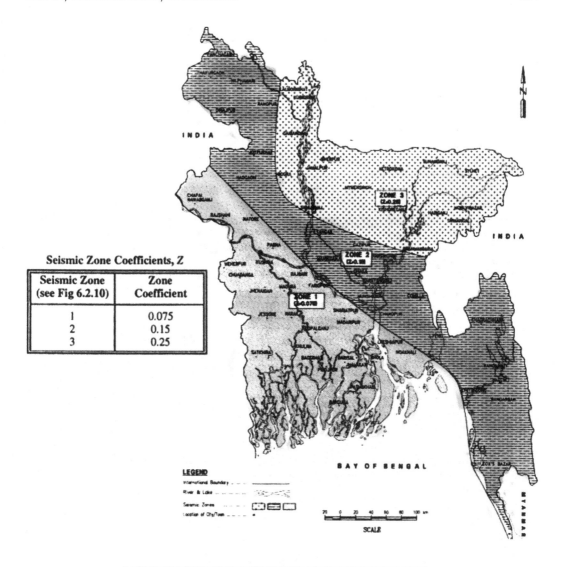

Seismic Zone Coefficients, Z

Seismic Zone (see Fig 6.2.10)	Zone Coefficient
1	0.075
2	0.15
3	0.25

Structure Importance Coefficients L

Structure Importance Category (See Table 6.1.1 for Occupancy)	Structure Importance Coefficient *I*
1. Essential Facilities	1.25
2. Hazardous Facilities	1.25
3. Special-Occupancy Structures	1.00
4. Standard-Occupancy Structures	1.00
5. Low-Risk Structures	1.00

	Site Soil Characteristics	Coefficient,
Type	Description	S
S_1	A soil profile with either:	1.0
	1. A rock-like material characterized by a shear wave velocity greater than 762 m/s or by other suitable means of classification.	
	2. Stiff or dense soil condition where the soil depth is less than 61 m.	
S_2	A soil profile with dense or stiff soil conditions, where the soil depth exceeds 61 m.	1.2
S_3	A soil profile 21 m or more in depth and containing more than 6 m of soft to medium-stiff clay but not more than 12 m of soft clay.	1.5
S_4	A soil profile containing more than 12 m of soft clay characterized by a shear wave velocity less than 152 m/s.	2.0

Basic Structural System[1]	Description of Lateral Force Resisting System	$R^{[2]}$
Moment Resisting Frame System	1. Special moment resisting frames (SMRF):	
	a. Steel	12
	b. Concrete	12
	2. Intermediate moment resisting frames (IMRF) Concrete[4]	8
	3. Ordinary moment resisting frames (OMRF)	
	a. Steel	6
	b. Concrete[5]	5

Example 23

Determine the base shear V and distribute it along each story level. The building plan is shown at the right, and information is given in the following.

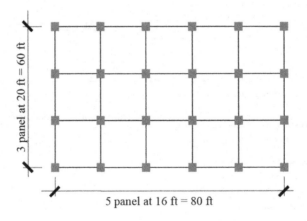

3 panel at 20 ft = 60 ft

5 panel at 16 ft = 80 ft

5.10.1 Beam

Dir. A 12×16 in

Dir. B 12×24 in

5.10.2 COLUMN

Interior 18 × 18 in
Exterior 16 × 16 in
Corner 14 × 14 in

5.10.3 SLAB

Thickness 6 in
Story: 10
Floor Height: 10 ft

5.10.4 BUILDING

Zone Dhaka
Site Stiff soil, depth less than 61 m
Structure Standard Occupancy
Frame Type Moment Resisting

	Length (*ft*)	Width (in)	Depth (in)	Quantity	Weight (kip)
Beam					
Dir. A	16	12	16	20	64
Dir. B	20	12	24	18	108
					172
Column					
Interior	10	18	18	8	27
Exterior	10	16	16	12	32
Corner	10	14	14	4	8.2
					67.2
Slab					
All	16	20	6	15	360
					360
Building					
				Per floor	599.2
				Total wt.	5,992

Beam in A Dir. $\dfrac{150}{1000}\,\text{lb}/\text{ft}^3 \times \dfrac{12\text{in}\times16\text{in}}{144}\times16\text{ft}\times20 = 64\text{ kip}$

Interior column $\dfrac{150}{1000}\,\text{lb}/\text{ft}^3 \times \dfrac{18\text{in}\times18\text{in}}{144}\times10\text{ft}\times8 = 27\text{ kip}$

Slab $\dfrac{150}{1000}\,\text{lb}/\text{ft}^3 \times16\text{ft}\times20\text{ft}\times\dfrac{6}{12}\text{in}\times15 = 360\text{ kip}$

$$h_n = (100/3.28) = 30.49 \text{ m}$$
$$C_t = 0.073(\text{for concrete frame})$$

Seismic Parameters

Z	0.15	$T = C_t h_n^{3/4} = 0.073 \times 30.49^{3/4} = 0.947$ s
I	1	
R	12	$C = \dfrac{1.25S}{T^{2/3}} = \dfrac{1.25 \times 1.0}{0.947^{2/3}} = 1.296$
S	1	

Design Base Shear

h_n	30.49 m	$V = \dfrac{ZIC}{R} W = \dfrac{0.15 \times 1.0 \times 1.296}{12} \times 5992 = 97.1$ kip
C_t	0.073	
T	0.947 s	$F_t = \begin{cases} 0, & \text{if } T < 0.7 \text{ s} \\ 0.07TV & \text{if } T \geq 0.7 \text{ s} \end{cases}$
C	1.296	
v	97.1k	
Ft	6.4	$= 0.07 \times 0.947 \times 97.1$ kip
V – Ft	90.6	$-0.07 \times 0.947 \times 37.1$ kip
		$= 6.4$ kip
		$V - F_t = 97.1 - 6.4$

Floor	Weight (wx)	Height (hx)	*wxhx*	*F* (kip)
1	599.2	10	5,992	1.6
2	599.2	20	11,983	3.3
3	599.2	30	17,975	4.9
4	599.2	40	23,967	6.6
5	599.2	50	29,958	8.2
6	599.2	60	35,950	9.9
7	599.2	70	41,942	11.5
8	599.2	80	47,933	13.2
9	599.2	90	53,925	14.8
10	599.2	100	59,917	16.5
			329,542	90.6

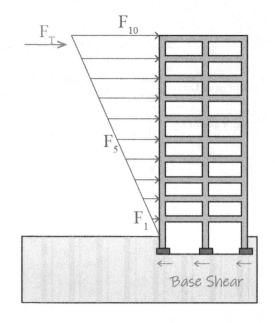

$$F_i = \frac{(V - F_t)w_x h_x}{\sum_{i=1}^{n} w_i h_i}$$

$$F_5 = \frac{90.6 \times 29958}{329542}$$

$$= 8.2 \text{ kip}$$

5.11 DEFLECTION OF BEAM

5.11.1 VIRTUAL WORK METHOD

Example 24

Determine the deflection of the beam at point C and angle at C by the unit load method.
$EI = 10,000 \text{ kft}^2$

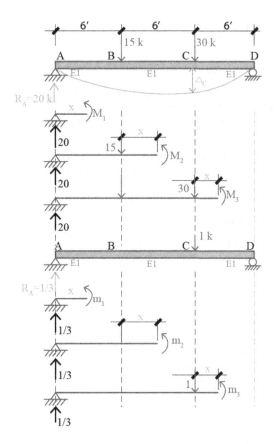

Solution

5.11.2   REAL BEAM

$$\sum M_D = 0 \circlearrowleft^+ \ R_A \times 18 - 15 \times 12 - 30 \times 6 = 0; \ R_A = 20 \text{ kip}$$
$$20x - M_1 = 0; \ M_1 = 20x$$
$$20(6+x) - 15x - M_2 = 0; M_2 = 5x + 120$$
$$20(12+x) - 15(6+x) - 30x - M_3 = 0; M_3 = -25x + 150$$

5.11.3   IMAGINARY BEAM

$$\sum M_D = 0 \circlearrowleft^+ \ r_A \times 18 - 1 \times 6 = 0; \qquad r_A = 1/3 \text{ kip}$$
$$(1/3)x - m_1 = 0 \qquad\qquad m_1 = x/3$$
$$(1/3)(6+x) - m_2 = 0 \qquad\qquad m_2 = x/3 + 2$$
$$(1/3)(12+x) - (1)x - m_3 = 0 \qquad m_3 = -2x/3 + 4$$

5.11.4   APPLYING UNIT LOAD FORMULA

$$\Delta_C = \int_A^D \frac{Mm}{EI} dx = \int_A^B \frac{M_1 m_1}{EI} dx + \int_B^C \frac{M_2 m_2}{EI} dx + \int_C^D \frac{M_3 m_3}{EI} dx$$

$$= \int_0^6 \frac{(20x)\left(\dfrac{x}{3}\right)}{EI} dx + \int_0^6 \frac{(5x+120)\left(\dfrac{x}{3}+2\right)}{EI} dx + \int_0^6 \frac{(-25x+150)\left(\dfrac{-2x}{3}+4\right)}{EI} dx$$

$$= \frac{480}{EI} + \frac{2460}{EI} + \frac{1200}{EI} = \frac{4140}{EI} = \frac{4140}{10000} \times 12 = 4.97 \text{ in} (\downarrow)$$

Real beam:

$$\sum M_D = 00^+; \ R_A \times 18 - 15 \times 12 - 30 \times 6 = 0; \ R_A = 20 \text{ kip}$$
$$20x - M_1 = 0$$
$$M_1 = 20x$$
$$20(6+x) - 15x - M_2 = 0$$
$$M_2 = 5x + 120$$
$$20(12+x) - 15(6+x) - 30x - M_3 = 0$$
$$M_3 = -25x + 150$$

Imaginary beam:

$$\sum M_D = 0^+; \ r_A \times 18 - 1 = 0; \qquad r_A = 1/18 \text{ kip}$$
$$(1/18)x - m_1 = 0 \qquad\qquad m_1 = x/18$$
$$(1/18)(6+x) - m_2 = 0 \qquad\quad m_2 = x/18 + 1/3$$
$$(1/18)(12+x) - 1 - m_3 = 0 \qquad m_3 = x/18 - 1/3$$

Applying unit load formula:

$$\theta_C = \int_0^6 \frac{(20x)\left(\dfrac{x}{18}\right)}{EI}dx + \int_0^6 \frac{(5x+120)\left(\dfrac{x}{18}+\dfrac{1}{3}\right)}{EI}dx + \int_0^6 \frac{(-25x+150)\left(\dfrac{x}{18}-\dfrac{1}{3}\right)}{EI}dx$$

$$= \frac{80}{EI} + \frac{410}{EI} - \frac{100}{EI} = \frac{390}{EI} = \frac{390}{10000} = 0.039 \text{ radian} \,(\circlearrowleft)$$

Example 25

Determine the deflection of the beam at point A by unit load method. $EI = 10,000 \text{ kft}^2$

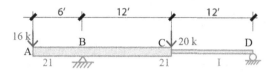

Solution

Real beam:

$$\sum M_D = 0 \circlearrowright^+ \quad -16\times30+R_B\times24-20\times12=0; \quad R_B = 30 \text{ kip}$$
$$-16x - M_1 = 0 \qquad\qquad\qquad\qquad\qquad\qquad M_1 = -16x$$
$$-16(6+x)+30x-M_2=0 \qquad\qquad\qquad M_2 = 14x-96$$
$$-16(18+x)+30(12+x)-20x-M_3=0 \qquad M_3 = -6x+72$$

Imaginary beam:

$$\sum M_D = 00^+ \quad -1\times30+r_A\times24=0; \quad r_A = 1.25 \text{ kip}$$
$$-1\times x - m_1 = 0 \qquad\qquad\qquad\qquad m_1 = -x$$
$$-1(6+x)+1.25x-m_2=0 \qquad\qquad m_2 = 0.25x-6$$
$$-1(18+x)+1.25(12+x)-m_3=0 \quad m_3 = 0.25x-3$$

Applying unit load formula:

$$\Delta_A = \int_A^D \frac{Mm}{EI}dx = \int_A^B \frac{M_1m_1}{EI}dx + \int_B^C \frac{M_2m_2}{EI}dx + \int_C^D \frac{M_3m_3}{EI}dx$$

$$= \int_0^6 \frac{(-16x)(-x)}{2EI}dx + \int_0^{12} \frac{(14x-96)(0.25x-6)}{2EI}dx$$

$$+ \int_0^{12} \frac{(-6x+72)(0.25x-3)}{EI}dx = \frac{576}{EI} + \frac{576}{EI} - \frac{864}{EI} = \frac{288}{EI}(\downarrow)$$

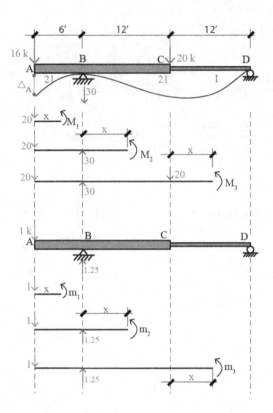

5.12   DEFLECTION OF FRAME

5.12.1   VIRTUAL WORK METHOD

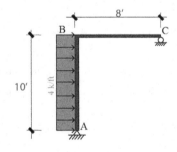

Question: Determine the deflection of the steel frame at point C and angle at B by the unit load method. $I = 600$ in^4 for both members.

Solution

5.12.2   REAL FRAME

$$\sum M_C = 0 \circlearrowleft^+ \quad 40 \times 10 - (4 \times 10) \times 5 + A_y \times 8 = 0; \quad A_y = -25 \text{ kip}$$

$$40x - (4x)(x/2) - M_1 = 0 \qquad\qquad M_1 = -2x^2 + 40x$$

$$40 \times 10 - 25x - (4 \times 10) \times 5 - M_2 = 0 \qquad M_2 = -25x + 200$$

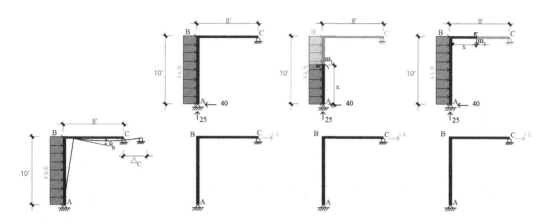

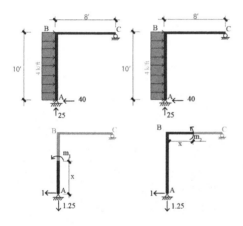

5.12.3 IMAGINARY FRAME

$$\sum M_C = 0 \circlearrowleft^+ \quad 1\times10 + a_y \times 8 = 0; \quad a_y = -1.25 \text{ kip}$$

$$1\times x - m_1 = 0 \qquad\qquad m_1 = x$$

$$1\times10 - 1.25x - m_2 = 0 \qquad m_2 = -1.25x + 10$$

5.12.4 APPLYING UNIT LOAD FORMULA

$$\Delta_C = \int_A^B \frac{M_1 m_1}{EI} dx + \int_B^C \frac{M_2 m_2}{EI} dx$$

$$= \int_0^{10} \frac{\left(-2x^2 + 40x\right)(x)}{EI} dx + \int_0^8 \frac{\left(-25x + 200\right)\left(-1.25x + 10\right)}{EI} dx$$

$$= \frac{8333}{EI} + \frac{5333}{EI} = \frac{13666}{EI} = \frac{13666}{\left(29000\times12^2\right)\left(600/12^4\right)} \times 12 - 1.36 \text{in} (\rightarrow)$$

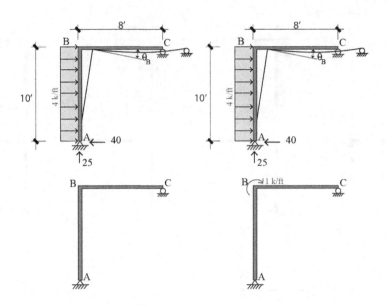

Question: Determine the deflection of the steel frame at point C and angle at B by the unit load method. $I = 600$ in^4 for both members.

Solution

5.12.5   REAL FRAME

$$\sum M_C = 0 \circlearrowleft^+ \ 40 \times 10 - (4 \times 10) \times 5 + A_y \times 8 = 0; \ A_y = -25 \text{ kip}$$
$$40x - (4x)(x/2) - M_1 = 0 \ M_1 = -2x^2 + 40x$$
$$40 \times 10 - 25x - (4 \times 10) \times 5 - M_2 = 0 \ M_2 = -25x + 200$$

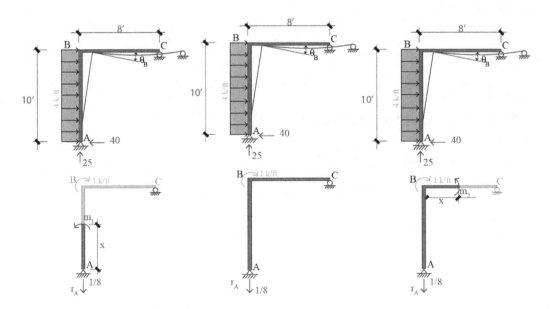

5.12.6 IMAGINARY FRAME

$$\sum M_C = 0 \circlearrowright^+ \quad r_A \times 8 + 1 = 0; \quad r_A = -1/8 \text{ kip}$$
$$-m_1 = 0 \qquad\qquad\qquad m_1 = 0$$
$$-(1/8) \times x + 1 - m_2 = 0 \qquad m_2 = -x/8 + 1$$

5.12.7 APPLYING UNIT LOAD FORMULA

$$\theta_C = \int_A^B \frac{M_1 m_1}{EI} dx + \int_B^C \frac{M_2 m_2}{EI} dx$$
$$= \int_0^{10} \frac{(-2x^2 + 40x)(0)}{EI} dx + \int_0^8 \frac{(-25x + 200)(-x/8 + 1)}{EI} dx$$
$$= 0 + \frac{533.33}{EI} = \frac{533.33}{EI} = \frac{533.33}{(29000 \times 12^2)(600 \times 12^4)} = 0.00042 \text{ radian} (\circlearrowright)$$

5.13 DEFLECTION OF TRUSS

5.13.1 VIRTUAL WORK METHOD

Question: Determine the vertical displacement of joint B of the steel truss. The cross-sectional areas are 75 mm², 50 mm², and 100 mm² for the horizontal, vertical, and diagonal members, respectively.

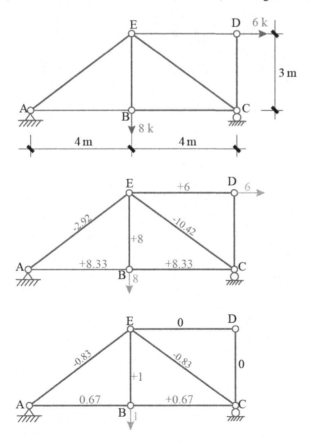

Solution

Member	N(kN)	n(kN)	L(m)	A(mm²)	$\dfrac{NnL}{A}$
AB	8.33	0.67	4	75	0.297
BC	8.33	0.67	4	75	0.297
ED	6	0	4	75	0
BE	8	1	3	50	0.480
CD	0	0	3	50	0
AE	-2.92	-0.83	5	100	0.121
EC	-10.42	-0.83	5	100	0.432

$$\Delta_B = \sum \frac{NnL}{AE} = \frac{1}{E}\sum \frac{NnL}{A} = \frac{1.627\times10^3}{200\times10^6} = 8.13 \text{ mm}(\downarrow)$$

Answer: $\Delta_B = 8.13$ mm$(\downarrow)$

Question: Determine the vertical displacement of joint C of the steel truss. The cross-sectional areas are 0.75 in^2, 0.50 in^2, and 1.00 in^2 for the horizontal, vertical, and diagonal members, respectively.

Solution

Member	N(kip)	n(kip)	L(ft)	A(in²)	$\dfrac{NnL}{A}$
AB	4	0.333	10	0.75	17.7
BC	4	0.667	10	0.75	35.5
CD	4	0.667	10	0.75	35.5
FE	-4	-0.333	10	0.75	17.7
BF	4	0.333	10	0.50	26.6
CE	4	1	10	0.50	80.0
AF	-5.66	-0.471	14.14	1.00	37.7
EB	0	-0.471	14.14	1.00	0.0
DE	-5.66	-0.943	14.14	1.00	75.4

$$\Delta_C = \sum \frac{NnL}{AE} = \frac{1}{E}\sum \frac{NnL}{A} = \frac{1}{29000}\times326.3\times12 = 0.13 \text{ in }(\downarrow)$$

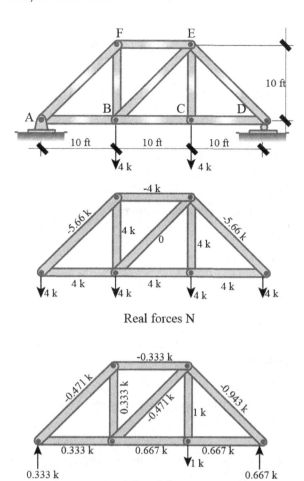

Real forces N

Virtual forces n

Answer: $\Delta_C = 0.13 \, \text{in}(\downarrow)$

6 Analysis of Statically Indeterminate Structures

Topics to be covered in this chapter are as follows:

- Moment distribution method.
- Slope deflection method.
- Influence line of statically indeterminate structure.
- Flexibility method.
- Stiffness method.
- Computer application of stiffness method.

6.1 MOMENT DISTRIBUTION FOR BEAM

6.1.1 MEMBER STIFFNESS FACTOR AND CARRY-OVER FACTOR

Consider the beam which is pinned at one end and fixed at the other. Application of the moment M causes the end A to rotate through an angle θ_A.

It can be proved that $M = (4EI / L)\theta_A$. The term in parenthesis is known as the stiffness factor.

$$K_{AB} = \frac{4EI}{L}$$

So the equation becomes, $M = K_{AB}\theta_A$. The stiffness factor at A can also be defined as the amount of moment M required to rotate the end A of the beam by 1 radian.

It also can be proved that $M' = \frac{1}{2}M$. The factor $(1/2)$ is known as the carry-over factor.

6.1.2 JOINT STIFFNESS FACTOR

If several members are fixed-connected to a joint and each of their far ends is fixed, then, by the principle of superposition, the total stiffness factor at the joint is the sum of the member stiffness factors at the joint.

$$K_A = \Sigma K = K_{AD} + K_{AB} + K_{AC}$$

DOI: 10.1201/9781032638072-6

6.1.3 DISTRIBUTION FACTOR

The *distribution factor* of a member is defined as the ratio of the member stiffness factor to the joint stiffness factor.

$$DF_{AD} = \frac{K_{AD}}{\sum K} = \frac{\left(\dfrac{4EI}{L}\right)_{AD}}{\sum \dfrac{4EI}{L}} = \frac{(I/L)_{AD}}{\sum(I/L)}$$

Similarly:

$$DF_{AB} = \frac{(I/L)_{AB}}{\sum(I/L)}; \quad DF_{AC} = \frac{(I/L)_{AC}}{\sum(I/L)}$$

It should be noted that, for fixed support, $DF = 0$, and for external hinge support, $DF = 1$.

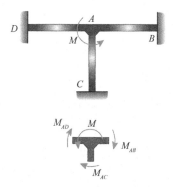

6.1.4 FIXED-END MOMENTS, *FEM*

The moments at the "walls" or fixed joints of a loaded member are called the fixed-end moments.
Clockwise moments are considered positive, whereas counterclockwise moments are negative.

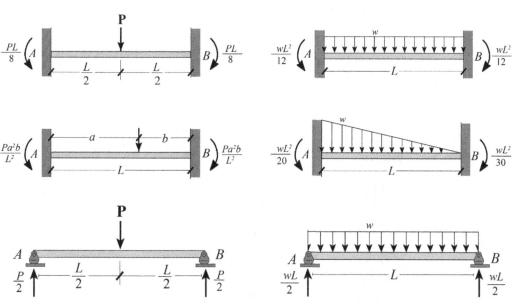

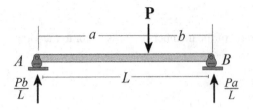

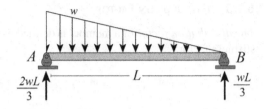

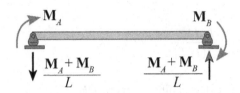

Upward reactions are considered *positive*, whereas downward reactions are *negative*.

Example 1

Determine the support reactions of the beam using the moment distribution method, given that $I_{AB} = 300\text{in}^4$ and $I_{BC} = 600\text{in}^4$.

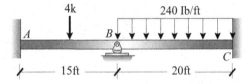

6.1.5   DISTRIBUTION FACTOR

$$DF_{AB} = 0$$

$$DF_{BA} = \frac{l/L}{\sum I/L} = \frac{300/15}{300/15 + 600/20} = 0.4$$

$$DF_{BC} = \frac{I/L}{\sum I/L} = \frac{600/20}{300/15 + 600/20} = 0.6$$

$$DF_{CB} = 0$$

6.1.6   FIXED-END MOMENTS

$$FEM_{AB} = -\frac{PL}{8} = -\frac{4 \times 15}{8} \qquad = -7.5\text{k} - \text{ft}$$

$$FEM_{BA} = +\frac{PL}{8} = +\frac{4 \times 15}{8} \qquad = +7.5\text{k} - \text{ft}$$

$$FEM_{BC} = -\frac{wL^2}{12} = -\frac{0.24 \times 20^2}{12} \qquad = -8.0\text{k} - \text{ft}$$

$$FEM_{CB} = +\frac{wL^2}{12} = +\frac{0.24 \times 20^2}{12} \qquad = +8.0\text{k} - \text{ft}$$

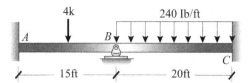

6.1.7 Moment Distribution

Joint		A		B	C
	Member	AB	BA	BC	CB
	DF	0	0.4	0.6	0
CY. 1	FEM	−7.5	+7.5	−8.0	+8.0
	Balance	0	+0.2	+0.3	0
CY. 2	C.O	+0.1	0	0	+0.15
	Balance	0	0	0	0
Sum		−7.4	+7.7	−7.7	+8.15

6.1.8 Balance at Cycle 1

$$M_{AB} = (-7.5) \times 0 \quad\quad = 0$$
$$M_{BA} = (7.5-8.0) \times 0.4 \quad = -0.2$$
$$M_{BC} = (7.5-8.0) \times 0.6 \quad = -0.3$$
$$M_{CB} = (+8.0) \times 0 \quad\quad = 0$$

6.1.9 Carry-Over at Cycle 2

$$CO_{AB} = +0.2 / 2 \quad = +0.1$$
$$CO_{BA} = 0 / 2 \quad\quad = 0$$
$$CO_{BC} = 0 / 2 \quad\quad = 0$$
$$CO_{CB} = +0.3 / 2 \quad = +0.15$$

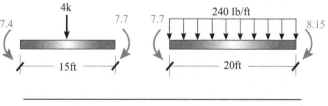

Joint	A	B		C
Member	AB	BA	BC	CB
Moment	−7.4	+7.7	−7.7	+8.15

6.1.10 REACTIONS

Load	+2	+2	+2.4	+2.4
Moment	−0.02	+0.02	−0.023	+0.023
Total	**+1.98**	**+4.397**		**+2.423**

Answer: $R_A = 1.98$ kip, $R_B = 4.397$ kip and $R_C = 2.423$ kip

6.1.11 REACTIONS FOR LOADS

$$R_{AB} = 4/2 = +2$$
$$R_{BA} = +2$$
$$R_{BC} = (0.24 \times 20)/2 = +2.4$$
$$R_{CB} = +2.4$$

6.1.12 REACTIONS FOR MOMENTS

$$R_{AB} = -(-7.4 + 7.7)/15 \quad = -0.02$$
$$R_{BA} \quad = +0.02$$
$$R_{BC} = -(-7.7 + 8.15)/20 \quad = -0.023$$
$$R_{CB} \quad = +0.023$$

Example 2

Determine the support reactions of the beam using the moment distribution method, given that I is constant all over the beam.

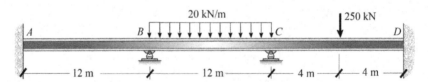

6.1.13 DISTRIBUTION FACTORS

$$DF_{AB} = DF_{DC} = 0$$

$$DF_{BA} = \frac{1/L}{\sum 1/L} = \frac{1/12}{1/12 + 1/12} = 0.5$$

$$DF_{BC} = \frac{1/L}{\sum 1/L} = \frac{1/12}{1/12 + 1/12} = 0.5$$

$$DF_{CB} = \frac{1/L}{\sum 1/L} = \frac{1/12}{1/12 + 1/8} = 0.4$$

$$DF_{CD} = \frac{1/L}{\sum 1/L} = \frac{1/8}{1/12 + 1/8} = 0.6$$

6.1.14 FIXED-END MOMENTS

$$\text{FEM}_{AB} = \text{FEM}_{BA} \qquad\qquad\qquad = 0$$

$$\text{FEM}_{BC} = -\frac{wL^2}{12} = -\frac{20 \times 12^2}{12} \quad = -240$$

$$\text{FEM}_{CB} = +\frac{wL^2}{12} = +\frac{20 \times 12^2}{12} \quad = +240$$

$$\text{FEM}_{CD} = -\frac{PL}{8} = -\frac{250 \times 8}{8} \quad = -250$$

$$\text{FEM}_{DC} = +\frac{PL}{8} = +\frac{250 \times 8}{8} \quad = +250$$

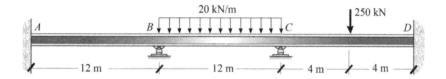

6.1.15 MOMENT DISTRIBUTION

Joint	A	B		C		D
Member	AB	BA	BC	CB	CD	DC
DF	**0**	**0.5**	**0.5**	**0.4**	**0.6**	**0.5**
FEM	0	0	−240	+240	−250	+250
Balance	0	+120	+120	+4	+6	0
CO	+60	0	+2	60	0	3
Balance	0	−1	−1	−24	−36	0
CO	−0.5	0	−12	−0.5	0	−18
Balance	0	+6	+6	+0.2	+0.3	0
CO	+3	0	+0.1	+3	0	+0.15
Balance	0	−0.05	−0.05	−1.2	−1.8	0
CO	−0.03	0	−0.6	−0.03	0	−0.9
Balance	0	+0.3	0.3	+0.01	+0.02	0
Sum	**62.48**	**125.25**	**−125.25**	**281.49**	**−281.49**	**234.25**

6.1.16 BALANCE AT CYCLE 1

$$M_{AB} = (0) \times 0 \qquad\qquad = 0$$
$$M_{BA} = (0 - 240) \times 0.5 \qquad = -120$$
$$M_{BC} = (0 - 240) \times 0.5 \qquad = -120$$
$$M_{CB} = (+240 - 250) \times 0.4 \quad = +4$$
$$M_{CD} = (+240 - 250) \times 0.6 \quad = +6$$
$$M_{DC} = (+250.0) \times 0 \qquad\quad = +0$$

Carry-Over at Cycle 2

$$C_{AB} = +120/2 \qquad\qquad = +60$$
$$C_{BA} = 0/2 \qquad\qquad = 0$$
$$C_{BC} = +4/2 \qquad\qquad = -2$$
$$C_{CB} = +120/2 \qquad\qquad = +60$$
$$C_{CD} = 0/2 \qquad\qquad = 0$$
$$C_{DC} = +6/2 \qquad\qquad = +3$$

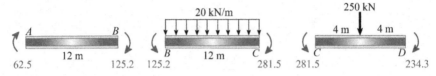

Member	AB	BA	BC	CB	CD	DC
Moment	+62.5	+125.2	−125.2	+281.5	−281.5	+234.3

Reactions	AB	BA	BC	CB	CD	DC
Load	0	+0	+120	+120	+125	+125
Moment	−15.64	+15.64	−13.02	+13.02	+5.9	−5.9
Total	−15.64	+122.62		+263.9		+119.1

6.1.17 REACTIONS FOR LOADS

6.1.17.1 Reactions for Moments

$$R_{AB} = R_{BA} \qquad\qquad = 0 \qquad R_{AB} = -(62.5 + 125.2)/12 \qquad = -15.64$$
$$R_{BC} = R_{CB} = (20 \times 12)/2 \ = +120 \quad R_{BC} = -(-125.2 + 281.5)/12 \ = -13.02$$
$$R_{CD} = R_{DC} = 250/2 \qquad = +125 \quad R_{CD} = -(-281.5 + 234.3)/8 \quad = +5.9$$

Answer: $R_A = -15.64$ kip, $R_B = 122.62$ kip, $R_C = 263.9$ kip and $R_D = 119.1$ kip

Example 3

Determine the support reactions of the beam using the moment distribution method, given that I is constant all over the beam.

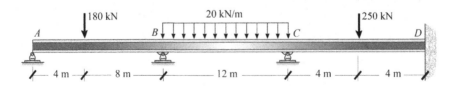

6.1.18 DISTRIBUTION FACTORS

$$DF_{AB} = 1$$

$$DF_{BA} = \frac{1/L}{\sum 1/L} = \frac{1/12}{1/12 + 1/12} = 0.5$$

$$DF_{BC} = \frac{1/L}{\sum 1/L} = \frac{1/12}{1/12 + 1/12} = 0.5$$

$$DF_{CB} = \frac{1/L}{\sum 1/L} = \frac{1/12}{1/12 + 1/8} = 0.4$$

$$DF_{CD} = \frac{1/L}{\sum 1/L} = \frac{1/8}{1/12 + 1/8} = 0.6$$

$$DF_{DC} = 0$$

6.1.19 FIXED-END MOMENTS

$$FEM_{AB} = -\frac{Pab^2}{L^2} = -\frac{180 \times 8^2 \times 4}{12^2} = -320$$

$$FEM_{BA} = +\frac{Pa^2 b}{L^2} = +\frac{180 \times 4^2 \times 8}{12^2} = +160$$

$$FEM_{BC} = -\frac{wL^2}{12} = -\frac{20 \times 12^2}{12} = -240$$

$$FEM_{CB} = +\frac{wL^2}{12} = +\frac{20 \times 12^2}{12} = +240$$

$$FEM_{CD} = -\frac{PL}{8} = -\frac{250 \times 8}{8} = -250$$

$$FEM_{DC} = +\frac{PL}{8} = +\frac{250 \times 8}{8} = +250$$

6.1.20 MOMENT DISTRIBUTION

Joint	A	B			C	
Member	AB	BA	BC	CB	CD	DC
DF	1	0.5	0.5	0.4	0.6	0
FEM	−320	160	−240	240	−250	250
Balance	+320	40	40	4	6	0
CO	+20	160	2	20	0	3
Balance	−20	−81	−81	−8	−12	0
CO	−40.5	−10	−4	−40.5	0	−6
Balance	+40.5	7	7	16.2	24.3	0
CO	+3.5	20.25	8.1	3.5	0	12.15

(Continued)

(Continued)

Joint	A	B			C	
Balance	−3.5	−14.18	−14.18	−1.4	−2.1	0
CO	−7.09	−1.75	−0.7	−7.09	0	−1.05
Balance	+7.09	1.23	1.23	2.835	4.25	0
CO	+0.61	3.54	1.42	0.61	0	2.13
Balance	−0.61	−2.48	−2.48	−0.245	−0.37	0
CO	−1.24	−0.31	−0.12	−1.24	0	0.18
Balance	+1.24	0.21	0.21	0.50	0.74	0
Sum	**0.00**	**282.52**	**−282.52**	**229.17**	**−229.17**	**2604**

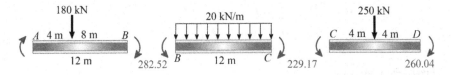

Member	BA	BC	CB	CD	DC
Moment	+282.52	−282.52	+229.17	−229.17	+260.04

6.1.21 REACTIONS

Load	+120	+60	+120	+120	+125	+125
Moment	−23.54	+23.54	+4.45	−4.45	−3.86	+3.86
Total	+96.45	+207.99		+236.69		+128.86

6.1.22 REACTIONS FOR LOADS

$$R_{AB} = (180 \times 8)/12 \quad = +120; \qquad R_{AB} = -(0 + 282.52)/12 = -23.54$$

$$R_{BA} = (180 \times 4)/12 \quad = +160; \qquad R_{BC} = -(-282.52 + 229.17)/12 = +4.45$$

$$R_{BC} = R_{CB} = (20 \times 12)/2 = +120; \qquad R_{C_D} = -(-229.17 + 260.04)/8 = -3.86$$

$$R_{CD} = R_{DC} = 250/2 \quad = +125;$$

Answer: $R_A = +96.45$ kip, $R_B = 207.99$ kip and $R_D = 128.86$ kip

6.1.23 MODIFIED STIFFNESS FACTOR

Many indeterminate beams have their far end span supported by an end pin (or roller), as in the case of joint B.

It can be proved that M = (3EI/L)*θ. Note that the constant is now 3 instead 4. So the stiffness has become:

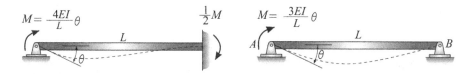

$$M = \frac{4EI}{L}\theta \qquad\qquad \frac{1}{2}M \qquad M = \frac{3EI}{L}\theta$$

Example 4 (Using Modified Stiffness)

Determine the support reactions of the beam using the moment distribution method, given that I is constant all over the beam. Use modified stiffness.

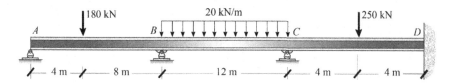

6.1.24   DISTRIBUTION FACTORS

$$DF_{AB} = 1$$

$$DF_{BA} = \frac{3EI/12}{3EI/12 + 4EI/12} = 0.429$$

$$DF_{BC} = \frac{4EI/12}{3EI/12 + 4EI/12} = 0.571$$

$$DF_{CB} = \frac{4EI/12}{4EI/12 + 4EI/8} = 0.4$$

$$DF_{CD} = \frac{4EI/8}{4EI/12 + 4EI/8} = 0.6$$

$$DF_{DC} = 0$$

6.1.25   FIXED-END MOMENTS

$$\text{FEM}_{AB} = -\frac{Pab^2}{L^2} = -\frac{180 \times 8^2 \times 4}{12^2} = -320\text{KN}$$

$$\text{FEM}_{BA} = +\frac{Pa^2b}{L^2} = -\frac{180 \times 4^2 \times 8}{12^2} = +160\text{KN}$$

$$\text{FEM}_{BC} = +\frac{WL^2}{12} = -\frac{20 \times 12^2}{12} = -240\text{KN}$$

$$\text{FEM}_{CB} = -\frac{WL^2}{12} = +\frac{20 \times 12^2}{12} = +240\text{KN}$$

$$\text{FEM}_{CD} = -\frac{PL}{8} = -\frac{250 \times 8}{8} = -250\text{KN}$$

$$\text{FEM}_{DC} = +\frac{PL}{8} = -\frac{250 \times 8}{8} = +250\text{KN}$$

6.1.26 MOMENT DISTRIBUTION

Joint	A	B		C		D
Member	AB	BA	BC	CB	CD	DC
DF	1	0.429	0.571	0.4	0.6	0
FEM	−320	+160	−240	+240	−250	250
Balance	+320	+34.32	+45.68	+4	+6	0
CO	0	+160	+2	22.84	0	+3
Balance	0	−69.50	−92.50	−9.14	−13.70	0
CO	0	0	−4.57	−46.25	0	−6.85
Balance	0	+1.96	+2.61	+18.50	+27.75	0
CO	0	0	+9.25	+1.30	0	+13.88
Balance	0	−3.97	−5.28	−0.52	−0.78	0
CO	0	0	−0.26	−2.64	0	−0.39
Balance	0	+0.11	+0.15	+1.06	+1.58	0
CO	0	0	+0.53	+0.07	0	+0.79
Balance	0	−0.23	−0.30	−0.03	−0.04	0
Sum	**0.00**	**+282.70**	**−282.70**	**+229.20**	**−229.20**	**+260.42**

Member	AB	BA	BC	CB	CD	DC
Moment	**0.00**	**282.70**	**−282.70**	**229.20**	**−229.20**	**260.42**

6.1.27 REACTIONS

Load	+120	+60	+120	+120	+125	+125
Moment	−23.56	+23.56	+4.45	−4.45	−3.90	+3.90
Total	**+96.44**	**+208.01**		**+236.65**		**+128.9**

6.1.28 REACTIONS FOR LOADS

$$R_{AB} = (180 \times 8)/12 = +120$$
$$R_{BA} = (180 \times 4)/12 = +160$$
$$R_{BC} = R_{CB} = (20 \times 12)/2 = +120$$
$$R_{CD} = R_{DC} = 250/2 = +125$$

6.1.29 REACTIONS FOR MOMENTS

$$R_{AB} = -(0 + 282.70)/12 \qquad = -23.56$$
$$R_{BC} = -(-282.70 + 229.20)/12 \quad = +4.45$$
$$R_{CD} = -(-229.20 + 260.04)/8 \quad = -3.90$$

Answer: $R_A = +96.44$ kip, $R_B = 208.01$ kip, $R_C = 236.65$ kip and $R_D = 128.9$ kip

6.2 MOMENT DISTRIBUTION FOR FRAME

Example 5

Determine the end moments of the frame using the moment distribution method. Relative stiffnesses are shown in parenthesis.

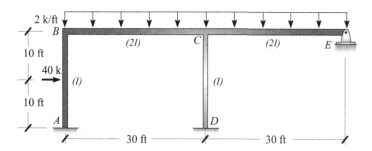

6.2.1 DISTRIBUTION FACTORS

$$\text{DF}_{AB} = 0$$

$$\text{DF}_{BA} = \frac{4I/20}{4I/20 + 4(2I)/30} = 0.429$$

$$\text{DF}_{BC} = \frac{4(2I)/30}{4I/20 + 4(2I)/30} = 0.571$$

$$\text{DF}_{CB} = \frac{4(2I)/20}{4(2I)/30 + 3(2I)/30 + 4I/20} = 0.4$$

$$\text{DF}_{CE} = \frac{3(2I)/20}{4(2I)/30 + 3(2I)/30 + 4I/20} = 0.3$$

$$\text{DF}_{CD} = \frac{4I/20}{4(2I)/30 + 3(2I)/30 + 4I/20} = 0.3$$

$$\text{DF}_{EC} = 1$$

6.2.2 FIXED-END MOMENTS

$$\text{FEM}_{AB} = -\frac{PL}{8} = -\frac{40 \times 20}{8} = -100\text{kN}$$

$$\text{FEM}_{BA} = +\frac{PL}{8} = +\frac{40 \times 20}{8} = +100\text{kN}$$

$$\text{FEM}_{BC} = -\frac{wL^2}{12} = -\frac{2 \times 30^2}{12} = -150\text{kN}$$

$$\text{FEM}_{CB} = +\frac{wL^2}{12} = +\frac{2 \times 30^2}{12} = +150\text{kN}$$

$$\text{FEM}_{CE} = -\frac{wL^2}{12} = -\frac{2 \times 30^2}{12} = -150\text{kN}$$

$$\text{FEM}_{EC} = +\frac{wL^2}{12} = +\frac{2 \times 30^2}{12} = +150\text{kN}$$

$$\text{FEM}_{DC} = \text{FEM}_{CD} = 0$$

6.2.3 Moment Distribution

Joint	A	B		C			D	E
Member	AB	BA	BC	CB	CD	CE	DC	EC
DF	0	0.429	0.571	0.4	0.3	0.3	0	1
FEM	−100	+100	−150	+150	0	−150	0	+150
Balance	0	21.45	28.55	0	0	0	0	−150
CO	10.73	0	0	14.28	0	−75	0	0
Balance	0	0	0	24.29	18.22	18.22	0	0
CO	0	0	12.15	0	0	0	9.11	0
Balance	0	−5.21	−6.93	0	0	0	0	0
CO	−2.61	0	0	−3.47	0	0	0	0
Balance	0	0	0	1.39	1.04	1.04	0	0
CO	0	0	0.69	0	0	0	0.52	0
Balance	0	−0.30	−0.40	0	0	0	0	0
Sum	−91.88	**115.94**	−115.94	186.48	19.26	−205.74	9.63	0

Member	AB	BA	BC	CB	CD	CE	DC	EC
Moment	**−91.88**	**115.94**	**−115.94**	**186.48**	**19.26**	**−205.74**	**9.63**	**0**

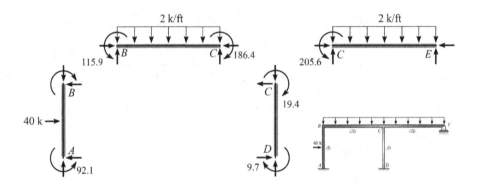

6.3 SLOPE DEFLECTION FOR BEAM

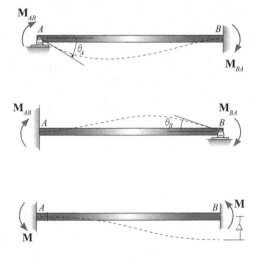

6.3.1 ANGULAR DISPLACEMENT AT A

Consider node A of the member to rotate by θ_A, while its far end node B is held fixed. It can be prove that:

$$M_{AB} = \left(\frac{4EI}{L}\right)\theta_A; \ M_{BA} = \left(\frac{2EI}{L}\right)\theta_A$$

6.3.2 ANGULAR DISPLACEMENT AT B

Similarly for node B of the member to rotate by θ_B, while its far end node A is held fixed. It can be shown that:

$$M_{AB} = \left(\frac{2EI}{L}\right)\theta_B; \ M_{BA} = \left(\frac{4EI}{L}\right)\theta_B$$

6.3.3 RELATIVE LINEAR DISPLACEMENT Δ

If the far node B of the member is displaced relative to A so that the cord of the member rotates clockwise (positive displacement) and yet both ends do not rotate.

$$M = \left(-\frac{6EI}{L^2}\right)\Delta; \ M = \left(-\frac{6EI}{L^2}\right)\Delta$$

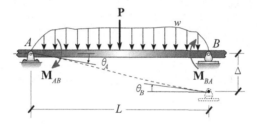

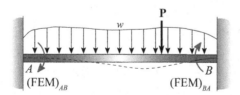

If the end moments, due to each displacement and the loading, are added together, the resultant moments at the ends can be written as:

$$M_{AB} = \left(\frac{4EI}{L}\right)\theta_A + \left(\frac{2EI}{L}\right)\theta_B - \left(\frac{6EI}{L^2}\right)\Delta + \text{FEM}_{AB}$$

$$M_{BA} = \left(\frac{2EI}{L}\right)\theta_A + \left(\frac{4EI}{L}\right)\theta_B - \left(\frac{6EI}{L^2}\right)\Delta + \text{FEM}_{BA}$$

$$\Delta = 0$$

$$M_{AB} = \left(\frac{4EI}{L}\right)\theta_A + \left(\frac{2EI}{L}\right)\theta_B + \text{FEM}_{AB}$$

$$M_{BA} = \left(\frac{2EI}{L}\right)\theta_A + \left(\frac{4EI}{L}\right)\theta_B + \text{FEM}_{BA}$$

$$M_N = \left(\frac{4EI}{L}\right)\theta_N + \left(\frac{2EI}{L}\right)\theta_F - \left(\frac{6EI}{L^2}\right)\Delta + \text{FEM}_N$$

Example 6

Determine the support reactions of the beam using the slope deflection method, given that $I_{AB} = I$ and $I_{BC} = 2I$.

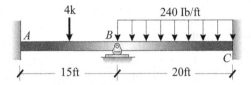

6.3.4 FIXED-END MOMENTS

$$\text{FEM}_{AB} = -\frac{PL}{8} = -\frac{4 \times 15}{8} \qquad = -7.5\text{k} - \text{ft}$$

$$\text{FEM}_{BA} = +\frac{PL}{8} = +\frac{4 \times 15}{8} \qquad = +7.5\text{k} - \text{ft}$$

$$\text{FEM}_{BC} = -\frac{wL^2}{12} = -\frac{0.24 \times 20^2}{12} \qquad = -8.0\text{k} - \text{ft}$$

$$\text{FEM}_{CB} = +\frac{wL^2}{12} = +\frac{0.24 \times 20^2}{12} \qquad = +8.0\text{k} - \text{ft}$$

$$M_N = \left(\frac{4EI}{L}\right)\theta_N + \left(\frac{2EI}{L}\right)\theta_F - \left(\frac{6EI}{L^2}\right)\Delta + \text{FEM}_N$$

Slope deflection equations for span AB:

$$M_{AB} = 0 + \left(\frac{2EI}{15}\right)\theta_B - 0 + (-7.5) \quad M_{AB} = \left(\frac{2EI}{15}\right)\theta_B - 7.5$$

$$M_{BA} = 0 + \left(\frac{4EI}{15}\right)\theta_B - 0 + 7.5 \qquad M_{BA} = \left(\frac{4EI}{15}\right)\theta_B + 7.5$$

Slope deflection equations for span BC:

$$M_{BC} = \left(\frac{4E \cdot 2I}{20}\right)\theta_B + 0 + (-8.0) \quad M_{BC} = \left(\frac{2EI}{5}\right)\theta_B - 8.0$$

$$M_{CB} = \left(\frac{2E \cdot 2I}{20}\right)\theta_B + 0 + 8.0 \qquad M_{CB} = \left(\frac{EI}{5}\right)\theta_B + 8.0$$

6.3.5   EQUILIBRIUM OF JOINT B

$$M_{BA} + M_{BC} = 0; \left[\left(\frac{4EI}{15}\right)\theta_B + 7.5\right] + \left[\left(\frac{2EI}{5}\right)\theta_B - 8.0\right] = 0;$$

$$\theta_B = 3/4EI$$

Substituting θ_B into the other equations:

$$M_{AB} = \left(\frac{2EI}{15}\right)\left(\frac{3}{4EI}\right) - 7.5 \quad = -7.4 \text{k} - \text{ft}$$

$$M_{BA} = \left(\frac{4EI}{15}\right)\left(\frac{3}{4EI}\right) + 7.5 \quad = +7.7 \text{k} - \text{ft}$$

$$M_{BC} = \left(\frac{2EI}{5}\right)\left(\frac{3}{4EI}\right) - 8.0 \quad = -7.7 \text{k} - \text{ft}$$

$$M_{CB} = \left(\frac{EI}{5}\right)\left(\frac{3}{4EI}\right) + 8.0 \quad = +8.15 \text{k} - \text{ft}$$

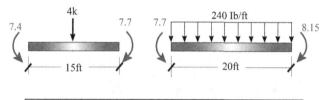

Joint	A	B		C
Member	AB	BA	BC	CB
Moment	−7.4	+7.7	−7.7	+8.15

6.3.6   REACTIONS

Load	+2	+2	+2.4	+2.4
Moment	−0.02	+0.02	−0.023	+0.023
Total	+1.98	+4.397		+2.423

Answer: $R_A = 1.98$ kip, $R_B = 4.397$ kip and $R_C = 2.423$ kip

6.3.7   REACTIONS FOR LOADS

$$R_{AB} = 4/2 = +2$$
$$R_{BA} = +2$$
$$R_{BC} = (0.24 \times 20)/2 = +2.4$$
$$R_{CB} = +2.4$$

6.3.8 REACTIONS FOR MOMENTS

$$
\begin{aligned}
R_{AB} &= -(-7.4 + 7.7)/15 &&= -0.02 \\
R_{BA} &&&= +0.02 \\
R_{BC} &= -(-7.7 + 8.15)/20 &&= -0.023 \\
R_{CB} &&&= +0.023
\end{aligned}
$$

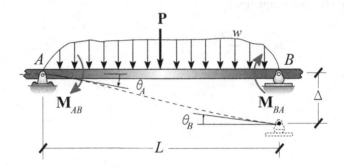

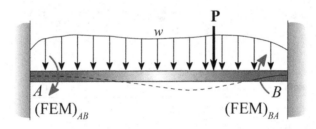

$$M_{AB} = \left(\frac{4EI}{L}\right)\theta_A + \left(\frac{2EI}{L}\right)\theta_B - \left(\frac{6EI}{L^2}\right)\Delta + \text{FEM}_{AB}$$

$$M_{BA} = \left(\frac{2EI}{L}\right)\theta_A + \left(\frac{4EI}{L}\right)\theta_B - \left(\frac{6EI}{L^2}\right)\Delta + \text{FEM}_{BA}$$

Occasionally, an end span of a beam or frame is supported by a pin or a roller at its far end, such as B. When this occurs, the moment at the roller or pin must be zero, that is, $M_{BA} = 0$. Thus, the second equation becomes:

$$0 = \left(\frac{2EI}{L}\right)\theta_A + \left(\frac{4EI}{L}\right)\theta_B - \left(\frac{6EI}{L^2}\right)\Delta + \text{FEM}_{BA}$$

Now, multiplying the first equation by 2 and subtracting the third equation, we find:

$$2M_{AB} = \left(\frac{6EI}{L}\right)\theta_A - \left(\frac{6EI}{L^2}\right)\Delta + 2\text{FEM}_{AB} - \text{FEM}_{BA}$$

$$M_N = \left(\frac{3EI}{L}\right)\theta_N - \left(\frac{3EI}{L^2}\right)\Delta + \text{FEM}_{AB} - \frac{1}{2}\text{FEM}_{BA}$$

Example 7

Determine the support reactions of the beam using the slope deflection method. *EI* is constant all over the beam.

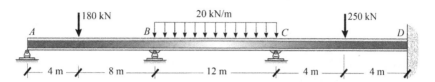

6.3.9 FIXED-END MOMENTS

$$\text{FEM}_{AB} = -\frac{Pab^2}{L^2} = -\frac{180 \times 8^2 \times 4}{12^2} = -320\text{kN}$$

$$\text{FEM}_{BA} = +\frac{Pa^2b}{L^2} = +\frac{180 \times 4^2 \times 8}{12^2} = +160\text{kN}$$

$$\text{FEM}_{BC} = -\frac{wL^2}{12} = -\frac{20 \times 12^2}{12} = -240\text{kN}$$

$$\text{FEM}_{CB} = +\frac{wL^2}{12} = +\frac{20 \times 12^2}{12} = +240\text{kN}$$

$$\text{FEM}_{CD} = -\frac{PL}{8} = -\frac{250 \times 8}{8} = -250\text{kN}$$

$$\text{FEM}_{DC} = +\frac{PL}{8} = +\frac{250 \times 8}{8} = +250\text{kN}$$

Modified slope deflection equations for span AB:

$$M_{BA} = \frac{3EI\theta_A}{12} + 160 - \frac{1}{2}(-320); \; M_{BA} = \frac{EI\theta_B}{4} + 320$$

Slope deflection equations for span BC:

$$M_{BC} = \frac{4EI\theta_B}{12} + \frac{2EI\theta_C}{12} + (-240); \quad M_{BC} = \frac{EI\theta_B}{3} + \frac{EI\theta_C}{6} - 240$$

$$M_{CB} = \frac{2EI\theta_B}{12} + \frac{4EI\theta_C}{12} + 240; \qquad M_{CB} = \frac{EI\theta_B}{6} + \frac{EI\theta_C}{3} + 240$$

Equilibrium of joint B:

$$M_{BA} + M_{BC} = 0; \left(\frac{EI\theta_B}{4} + 320\right) + \left(\frac{EI\theta_B}{3} + \frac{EI\theta_C}{6} - 240\right) = 0;$$

$$\frac{7EI\theta_B}{12} + \frac{EI\theta_C}{6} = -80$$

Slope deflection equations for span CD:

$$M_{CD} = \frac{4EI\theta_C}{8} + (-250); \quad M_{CD} = \frac{EI\theta_C}{2} - 250$$

$$M_{DC} = \frac{2EI\theta_C}{8} + 250; \quad M_{DC} = \frac{EI\theta_C}{4} + 250$$

Equilibrium of joint C:

$$M_{CB} + M_{CD} = 0; \left(\frac{EI\theta_B}{6} + \frac{EI\theta_C}{3} + 240\right) + \left(\frac{EI\theta_C}{2} - 250\right) = 0;$$

$$\frac{EI\theta_B}{6} + \frac{5EI\theta_C}{6} = 10$$

Solving, we find $\theta_B = -1640/11EI$ and $\theta_C = 460/11EI$. Substituting θ_B and θ_C into the equation of moments, we find:

$$M_{BA} = +282.73 \quad M_{BC} = -282.73$$
$$M_{CB} = +229.09 \quad M_{CD} = -229.09 \quad M_{DC} = +260.45$$

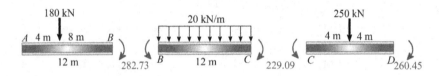

Member	AB	BA	BC	CB	CD	DC
Moment	0.00	282.73	-282.73	229.09	-229.09	260.45

6.3.10 REACTIONS

Load	+120	+60	+120	+120	+125	+125
Moment	-23.56	+23.56	+4.47	-4.47	-3.92	+3.92
Total	+96.44	+208.03		+236.61		+128.92

6.3.11 REACTIONS FOR LOADS

$$R_{AB} = (180 \times 8)/12 \quad = +120$$
$$R_{BA} = (180 \times 4)/12 \quad = +160$$
$$R_{BC} = R_{CB} \quad = (20 \times 12)/2 \quad = +120$$
$$R_{CD} = \quad R_{DC} = 250/2 \quad = +125$$

6.3.12 REACTIONS FOR MOMENTS

$$R_{AB} = -(0 + 282.73)/12 \qquad = -23.56$$
$$R_{BC} = -(-282.73 + 229.09)/12 \quad = +4.47$$
$$R_{CD} = -(-229.09 + 260.45)/8 \quad = -3.92$$

Answer: $R_A = +96.44$ kip, $R_B = 208.03$ kip, $R_C = 236.61$ kip and $R_D = 128.92$ kip

6.4 SLOPE DEFLECTION FOR FRAME

Example 8

Determine the end moments of the frame using the slope deflection method.

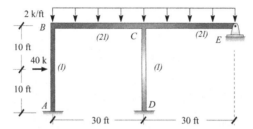

6.4.1 FIXED-END MOMENTS

$$FEM_{AB} = -\frac{PL}{8} = -\frac{40 \times 20}{8} = -100 \text{kN}$$

$$FEM_{BA} = +\frac{PL}{8} = +\frac{40 \times 20}{8} = +100 \text{kN}$$

$$FEM_{BC} = -\frac{wL^2}{12} = -\frac{2 \times 30^2}{12} = -150 \text{kN}$$

$$FEM_{CB} = +\frac{wL^2}{12} = +\frac{2 \times 30^2}{12} = +150 \text{kN}$$

$$FEM_{CE} = -\frac{wL^2}{12} = -\frac{2 \times 30^2}{12} = -150 \text{kN}$$

$$FEM_{EC} = +\frac{wL^2}{12} = +\frac{2 \times 30^2}{12} = +150 \text{kN}$$

$$FEM_{DC} = FFM_{CD} = 0$$

Slope deflection equations for span AB

$$M_{AB} = 0 + \frac{2EI\theta_B}{20} + (-100); \quad M_{AB} = \frac{EI\theta_B}{10} - 100$$

$$M_{BA} = 0 + \frac{4EI\theta_B}{20} + 100; \qquad M_{BA} = \frac{EI\theta_B}{5} + 100$$

Slope deflection equations for span BC

$$M_{BC} = \frac{4E(2I)\theta_B}{30} + \frac{2E(2I)\theta_C}{30} + (-150); \quad M_{BC} = \frac{4EI\theta_B}{15} + \frac{2EI\theta_C}{15} - 150$$

$$M_{CB} = \frac{2E(2I)\theta_B}{30} + \frac{4E(2I)\theta_C}{30} + 150; \quad M_{CB} = \frac{2EI\theta_B}{15} + \frac{4EI\theta_C}{15} + 150$$

Equilibrium of Joint B

$$M_{BA} + M_{BC} = 0; \text{ which yields, } \frac{7EI\theta_B}{15} + \frac{2EI\theta_C}{15} = 50$$

Slope deflection equations for span CD

$$M_{DC} = 0 + \frac{2EI\theta_C}{20} + 0; \quad M_{DC} = \frac{EI\theta_C}{10}$$

$$M_{CD} = 0 + \frac{4EI\theta_C}{20} + 0; \quad M_{CD} = \frac{EI\theta_C}{5}$$

Modified slope deflection equations for span CE:

$$M_{CE} = \frac{3E(2I)\theta_C}{30} - 150 - \frac{1}{2}(150); \quad M_{CE} = \frac{EI\theta_C}{5} - 225$$

Equilibrium of joint C:

$$M_{CB} + M_{CD} + M_{CE} = 0; \text{ which yields, } \frac{2EI\theta_B}{15} + \frac{2EI\theta_C}{3} = 75$$

Solving, we find $\theta_B = +875/11EI$ and $\theta_C = +2025/22EI$.
Substituting θ_B and θ_C into the equations of moment, we find:

$$\begin{aligned}
M_{AB} &= -92.1 & M_{BA} &= +115.9 \\
M_{BC} &= -115.9 & M_{CB} &= +186.4 \\
M_{CD} &= +19.4 & M_{DC} &= +9.7 \\
M_{CE} &= -205.6
\end{aligned}$$

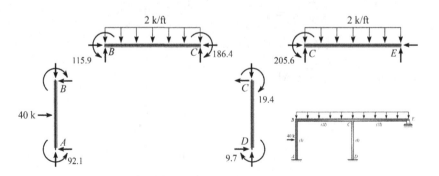

6.5   INFLUENCE LINE FOR BEAM AND FRAMES

(Qualitative.)

Example 9

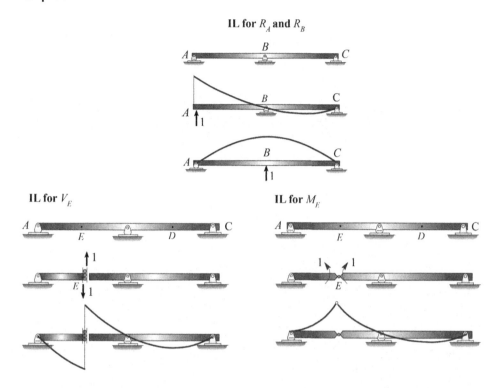

Example 10

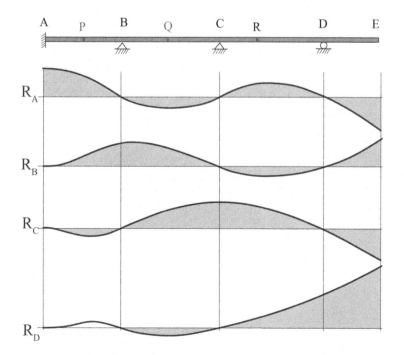

Example 11

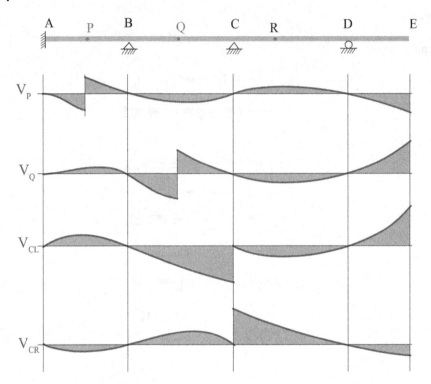

Example 12

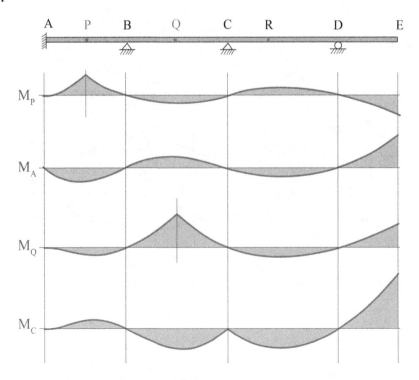

Example 13

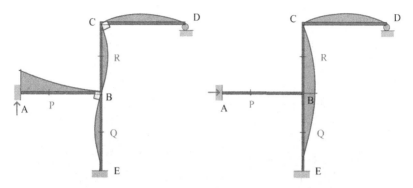

Example 14

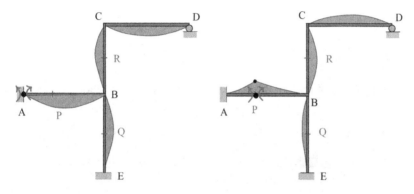

Example 15

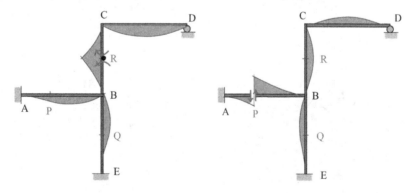

Example 16

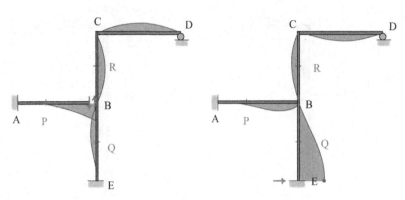

6.6   INFLUENCE LINE FOR BEAM AND FRAMES

(Quantitative.)

Example 17

Draw the influence line of reactions at B, and determine at specified locations, which are 6ft apart.

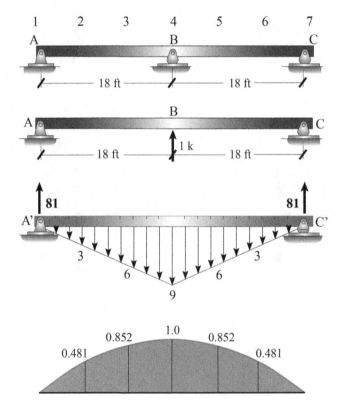

Find ordinates of the moment diagram:

$$M_B = M_4 = \frac{Pab}{L} = \frac{1 \times 18 \times 18}{36} \quad = 9$$
$$M_2 = M_6 = (1/3) \times M_4 = (1/3) \times 9 \quad = 3$$
$$M_3 = M_5 = (2/3) \times M_4 = (2/3 \times 9 \quad = 6$$

6.6.1 DETERMINE THE REACTIONS

$$R_A' = (0.5 \times 36 \times 9)/2 = 81$$

Determine the moments at specified sections:

$$M_2 = 81 \times 6 - \left[\frac{1}{2}(6 \times 3)\right] \times 2 = 468$$

$$M_3 = 81 \times 12 - \left[\frac{1}{2}(12 \times 6)\right] \times 4 = 828$$

$$M_4 = 81 \times 18 - \left[\frac{1}{2}(18 \times 9)\right] \times 6 = 972$$

$$M_5 = M_3 = 828$$
$$M_6 = M_2 = 468$$

Determine the ordinates of the influence line:
M_4 is taken as unity, and all other moments are proportioned.

Sec.	M	Ordinate of IL (M / M_4)
2	468	468/972 = 0.481
3	828	828/972 = 0.852
4	972	972/972 = 1.000
5	828	828/972 = 0.852
6	468	468/972 = 0.481

Example 18

Draw the influence line of the reactions at A, and determine at the specified locations, which are 6ft apart.

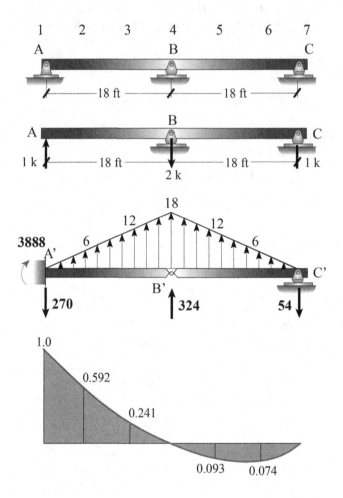

Find ordinates of moment diagram:

$$M_B = M_4 = Pab/L = (2 \times 18 \times 18) \div 36 \quad = 18$$
$$M_2 = M_6 = (1/3) \times M_4 = (1/3) \times 18 \quad\quad = 3$$
$$M_3 = M_5 = (2/3) \times M_4 = (2/3 \times 18 \quad\quad = 12$$

6.6.2 DETERMINE THE REACTIONS

$$\sum M_{B'} = 0 \circlearrowright^+ ; \quad R_C' \times 18 - (0.5 \times 18 \times 18) \times 6 = 0 ; \quad R_C' = 54$$
$$\sum F_y = 0 \uparrow^+ \quad -R_A' + 324 - 54 = 0 ; \quad R_A' = 270$$
$$\sum M_A = 0 \circlearrowright^+ ; \quad M_A' - 324 \times 18 + 54 \times 36 = 0 ; \quad M_A = 3888$$

Determine the moments at the specified sections:

$$M_2 = 3888 - 270 \times 6 - \left[\frac{1}{2}(6 \times 6)\right] \times 2 \quad = 2304$$

$$M_3 = 3888 - 270 \times 12 - \left[\frac{1}{2}(12 \times 12)\right] \times 4 = 936$$

$$M_5 = -54 \times 12 + \left[\frac{1}{2}(12 \times 12)\right] \times 4 \quad = 360$$

$$M_6 = -54 \times 6 + \left[\frac{1}{2}(6 \times 6)\right] \times 2 \quad = -288$$

Determine the ordinates of the influence line. M_A is taken as unity.

Sec.	M	Ordinate of IL $\left(M / M_4\right)$
1	3,888	3888/3888 = 1.000
2	2,304	2304/3888 = 0.592
3	936	936/3888 = 0.241
5	−360	−360/3888 = 0.093
6	−288	−288/3888 = −0.074

Example 19

Determine ordinates of line of reactions at A at specified locations which are 6ft apart. The following table represents the ordinates of the influence line of R_B.

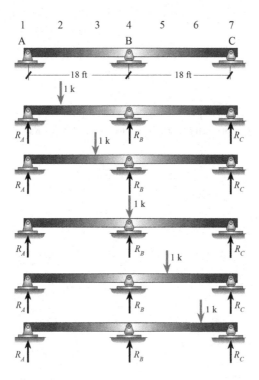

Sec.	Ordinates
1	0
2	0.481
3	0.852
4	1.000
5	0.852
6	0.481
7	0

$$\sum M_c = 0 \circlearrowleft^+$$

$$\textit{Sec. 2}: R_A \times 36 - 1 \times 30 + (0.481) \times 18 = 0 \;\; R_A = 0.592$$

$$\textit{Sec. 3}: R_A \times 36 - 1 \times 24 + (0.852) \times 18 = 0 \;\; R_A = 0.241$$

$$\textit{Sec. 4}: R_A \times 36 - 1 \times 18 + (1.000) \times 18 = 0 \;\; R_A = 0$$

$$\textit{Sec. 5}: R_A \times 36 - 1 \times 12 + (0.852) \times 18 = 0 \;\; R_A = -0.093$$

$$\textit{Sec. 6}: R_A \times 36 - 1 \times 6(0.481) \times 18 = 0 \;\; R_A = -0.074$$

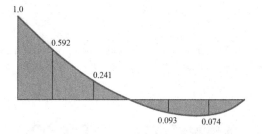

Example 20

Determine ordinates of influence line of shear at Section 6.2 for specified locations which are 6ft apart. The following table represents ordinates of influence line of R_A.

Sec.	Ordinates
1	1
2	0.592
3	0.241
4	0
5	−0.093
6	−0.074
7	0

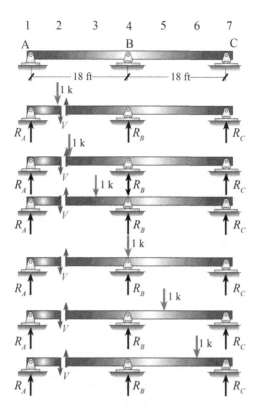

Taking $\sum F_y = 0 \uparrow^+$ for the different positions of the unit load, we can determine V.

Sec. 2	$0.592 - V - 1 = 0$	$V = -0.408$
	$0.592 - V = 0$	$V = 0.592$
Sec. 3	$0.241 - V = 0$	$V = 0.241$
Sec. 4	$0 - V = 0$	$V = 0$
Sec. 5	$-0.093 - V = 0$	$V = -0.093$
Sec. 6	$-0.074 - V = 0$	$V = -0.074$

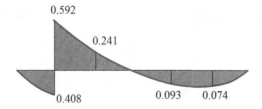

Example 21

Determine ordinates of line of reactions at A at specified locations which are 6ft apart. The following table represents the ordinates of the influence line of R_A.

Sec.	Ordinates
1	1
2	0.592
3	0.241
4	0
5	−0.093
6	−0.074
7	0

Taking $\sum M_2 = 0(\circlearrowleft^+)$ for the different positions of the unit load, we can determine M.

$$\text{Sec.2} \quad 0.592 \times 6 - M = 0; \qquad M = 2.448$$
$$\text{Sec.3} \quad 0.241 \times 6 - M = 0; \qquad M = 1.446$$
$$\text{Sec.4} \quad 0.0 \times 6 - M = 0; \qquad M = 0$$
$$\text{Sec.5} \quad -0.093 \times 6 - M = 0; \quad M = -0.558$$
$$\text{Sec.6} \quad -0.074 \times 6 - M = 0; \quad M = -0.444$$

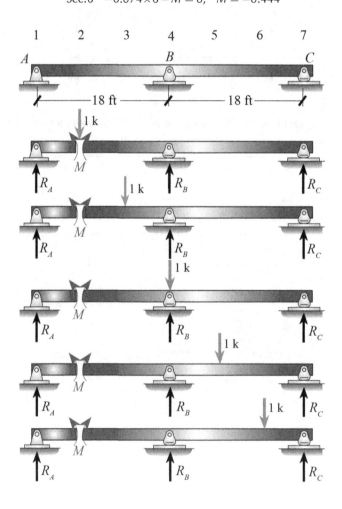

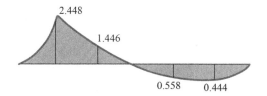

6.7   FLEXIBILITY METHOD FOR BEAM AND FRAME

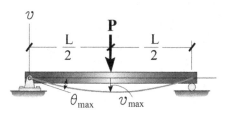

$$\theta_{max} = \frac{PL^2}{16EI}$$

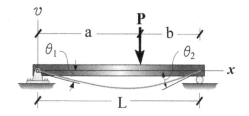

$$\theta_1 = \frac{Pab\,(L+b)}{6EIL} \qquad \theta_2 = \frac{Pab\,(L+a)}{6EIL}$$

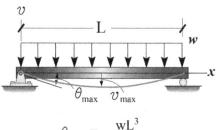

$$\theta_{max} = \frac{wL^3}{24EI}$$

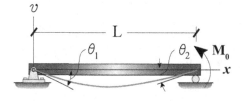

$$\theta_1 = \frac{M_0 L}{6EI} \qquad \theta_2 = \frac{M_0 L}{3EI}$$

Example 22

Determine the reactions using the flexibility method.

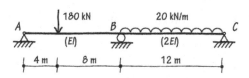

6.7.1   PRIMARY STRUCTURE

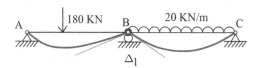

Find deformations:

$$\Delta_1 = \frac{Pab(L+a)}{6EIL} + \frac{wL^3}{24EI}$$

$$= \frac{180 \cdot 4 \cdot 8(12+4)}{6EI \cdot 12} + \frac{20 \cdot 12^3}{24E(2I)} = \frac{2000}{EI}$$

Unit moment at B:

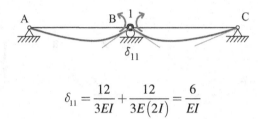

$$\delta_{11} = \frac{12}{3EI} + \frac{12}{3E(2I)} = \frac{6}{EI}$$

Flexibility equations:

$$\mathbf{FP} = -\Delta$$

$$\left[\frac{6}{EI}\right][M_B] = -\left[\frac{2000}{EI}\right]$$

$$M_B = -333.3 \, kNm$$

6.7.2   Free-Body Diagram

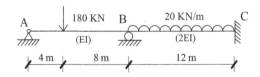

Reactions				
Load	+120	+60	+120	+120
Mom.	−27.78	+27.78	+27.78	−27.78
	+92.22	+235.56		+92.22

Answer: $R_A = 92.22 \, kN, R_B = 235.56 \, kN, R_C = 92.22 \, kN$

Example 23

Determine the reactions using the flexibility method.

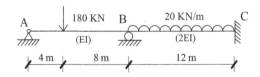

6.7.3 PRIMARY STRUCTURE

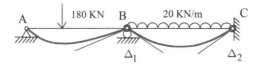

6.7.4 FIND DEFORMATIONS

$$\Delta_1 = \frac{Pab(L+a)}{6EIL} + \frac{wL^3}{24EI}$$

$$= \frac{180 \cdot 4 \cdot 8(12+4)}{6EI \cdot 12} + \frac{20 \cdot 12^3}{24E(2I)} = \frac{2000}{EI}$$

$$\Delta_2 = \frac{wL^3}{24EI} = \frac{20 \cdot 12^3}{24E(2I)} = \frac{720}{EI}$$

Unit moment at B:

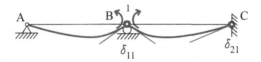

$$\delta_{11} = \frac{12}{3EI} + \frac{12}{3E(2I)} = \frac{6}{EI}; \delta_{21} = \frac{12}{6E(2I)}$$

Unit moment at C:

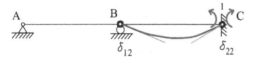

$$\delta_{12} = \frac{12}{6E(2I)}; \; \delta_{22} = \frac{12}{3E(2I)}$$

Flexibility equations:

$$\begin{bmatrix} \dfrac{6}{EI} & \dfrac{12}{6E(2I)} \\ \dfrac{12}{6E(2I)} & \dfrac{12}{3E(2I)} \end{bmatrix} \begin{bmatrix} M_B \\ M_C \end{bmatrix} = - \begin{bmatrix} \dfrac{2000}{EI} \\ \dfrac{720}{EI} \end{bmatrix}$$

Solving, $M_B = -298.2$ kNm and $M_C = -210.9$ kNm .

Free-body diagram.

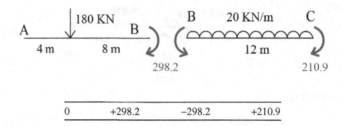

$$
\begin{array}{c|ccc}
0 & +298.2 & -298.2 & +210.9
\end{array}
$$

6.7.5 REACTIONS

Load	+120	+60	+120	+120
Mom.	−24.85	+24.85	+7.28	−7.28
	+95.15	+212.13		+112.72

Answer: $R_A = 95.15$ kN, $R_B = 212.13$ kN, $R_C = 112.72$ kN

Example 24

Determine the reactions using the flexibility method.

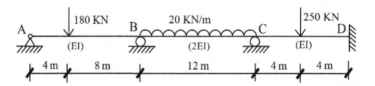

6.7.6 PRIMARY STRUCTURE

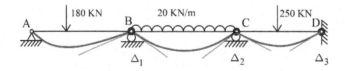

Find deformations:

$$
\begin{aligned}
\Delta_1 &= \frac{Pab(L+a)}{6EIL} + \frac{wL^3}{24EI} \\
&= \frac{180 \cdot 4 \cdot 8(12+4)}{6EI \cdot 12} + \frac{20 \cdot 12^3}{24E(2I)} = \frac{2000}{EI} \\
\Delta_2 &= \frac{wL^3}{24EI} + \frac{PL^2}{16EI} = \frac{20 \cdot 12^3}{24E(2I)} + \frac{250 \cdot 8^2}{16EI} = \frac{1720}{EI} \\
\Delta_3 &= \frac{PL^2}{16EI} = \frac{250 \cdot 8^2}{16EI} = \frac{1000}{EI}
\end{aligned}
$$

6.7.7 Unit Moment at B

$$\delta_{11} = \frac{12}{3EI} + \frac{12}{3E(2l)} = \frac{6}{EI}; \, \delta_{21} = \frac{12}{6E(2l)}; \, \delta_{31} = 0$$

6.7.8 Unit Moment at C

$$\delta_{12} = \frac{12}{6E(2I)}; \, \delta_{22} = \frac{12}{3E(2I)} + \frac{8}{3EI} = \frac{14}{3EI}; \, \delta_{32} = \frac{8}{6EI}$$

6.7.9 Unit Moment at D

Flexibility equations:

$$\delta_{13} = 0; \, \delta_{23} = \frac{8}{6EI}; \, \delta_{33} = \frac{8}{3EI}$$

$$\begin{bmatrix} \dfrac{6}{EI} & \dfrac{12}{6E(2l)} & 0 \\ \dfrac{12}{6E(2l)} & \dfrac{14}{3EI} & \dfrac{8}{6EI} \\ 0 & \dfrac{8}{6EI} & \dfrac{8}{3EI} \end{bmatrix} \begin{bmatrix} M_B \\ M_C \\ M_D \end{bmatrix} = - \begin{bmatrix} \dfrac{2000}{EI} \\ \dfrac{1720}{EI} \\ \dfrac{1000}{EI} \end{bmatrix}$$

Solving, $M_B = -294.8$ kNm, $M_C = -231.3$ kNm and $M_D = -259.3$ kNm

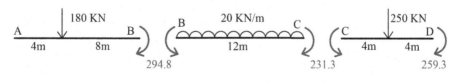

Member	AB	BA	BC	CB	CD	DC
Moment	+0	+294.8	−294.8	+231.3	−231.3	+259.3

6.7.10 REACTIONS

Load	+120	+60	+120	+120	+125	+125
Moment	−24.57	+24.57	+5.29	−5.29	−3.5	+3.5
Total	**+95.43**	**+209.86**		**+236.2**		**+128.5**

Answer: $R_A = +95.43$ kN, $R_B = 209.86$ kN, $R_C = 236.2$ kN and $R_D = 128.5$ kN

Example 25

Determine the reactions by flexibility method (2 k / ft):

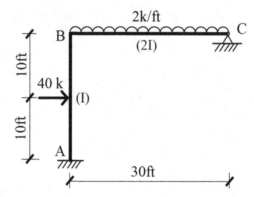

Primary structure (2 k / ft):

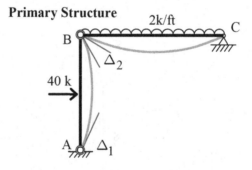

Find deformations:

$$\Delta_1 = \frac{40 \cdot 20^2}{16EI} = \frac{1000}{EI}$$

$$\Delta_2 = \frac{40 \cdot 20^2}{16EI} + \frac{2 \cdot 30^3}{24E(2I)} = \frac{2125}{EI}$$

Unit moment at A:

$$\delta_{11} = \frac{20}{3EI}; \, \delta_{21} = \frac{20}{6EI}$$

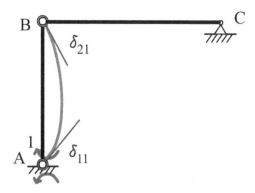

Unit moment at B:

$$\delta_{12} = \frac{20}{6EI}; \delta_{22} = \frac{20}{3EI} + \frac{30}{3E(2I)} = \frac{35}{3EI}$$

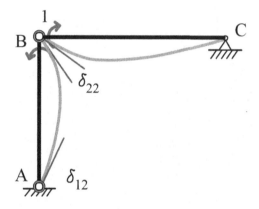

6.7.11 Flexibility Equations

$$\begin{bmatrix} \dfrac{20}{3EI} & \dfrac{20}{6EI} \\ \dfrac{20}{6EI} & \dfrac{35}{3I} \end{bmatrix} \begin{bmatrix} M_A \\ M_B \end{bmatrix} = - \begin{bmatrix} \dfrac{1000}{EI} \\ \dfrac{2125}{EI} \end{bmatrix}$$

$$M_A = -68.75 \text{ kNm}, M_B = -162.5 \text{ kNm}$$

6.7.12 FREE-BODY DIAGRAM

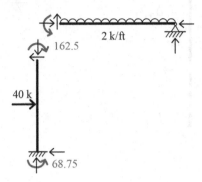

6.8 STIFFNESS METHOD FOR BEAM

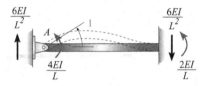

6.8.1 UNIT ROTATION APPLIED AT A

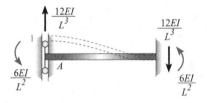

6.8.2 UNIT DISPLACEMENT APPLIED AT A

Example 26

Determine the support reactions of the beam using the stiffness method, given that $I_{AB} = I$ and $I_{BC} = 2I$.

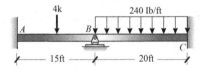

6.8.3 FIXED-END MOMENTS

$$\text{FEM}_{AB} = -\frac{PL}{8} = -\frac{4 \times 15}{8} \qquad = -7.5$$

$$\text{FEM}_{BA} = +\frac{PL}{8} = +\frac{4 \times 15}{8} \qquad = +7.5$$

$$\text{FEM}_{BC} = -\frac{wL^2}{12} = -\frac{0.24 \times 20^2}{12} \qquad = -8.0$$

$$\text{FEM}_{CB} = +\frac{wL^2}{12} = +\frac{0.24 \times 20^2}{12} \qquad = +8.0$$

6.8.4 DEGREES OF FREEDOM

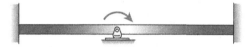

6.8.5 JOINT LOADING

$$P_1 = \text{FEM}_{BA} + \text{FEM}_{BC} = 7.5 - 8.0 = -0.5$$

6.8.6 UNIT ROTATION AT B

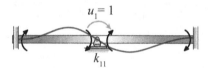

6.8.7 STIFFNESS EQUATIONS

$$k_{11} = \frac{4EI}{15} + \frac{4E(2I)}{20} = \frac{2EI}{3}$$

$$\mathbf{KU} = -\mathbf{P}; \; [k_{11}][u_1] = -[P_1]; \; \left[\frac{2EI}{3}\right][u_1] = -[-0.5]$$

Solving, we find $u_1 = 3 / 4EI$.

6.8.8 END MOMENTS USING SLOPE DEFLECTION EQUATIONS

Span AB:

$$M_{AB} = 0 + \frac{2EI}{15}\left(\frac{3}{4EI}\right) + (-7.5) \qquad = -7.4$$

$$M_{BA} = 0 + \frac{4EI}{15}\left(\frac{3}{4EI}\right) + 7.5 \qquad = 7.7$$

Span BC:

$$M_{BC} = \frac{4E \cdot 2I}{20}\left(\frac{3}{4EI}\right) + 0 + (-8) = -7.7$$

$$M_{CB} = \frac{2E \cdot 2I}{20}\left(\frac{3}{4EI}\right) + 0 + 8 = 8.15$$

Example 27

Determine the support reactions of the beam using the stiffness method, given that $I_{AB} = I$ and $I_{BC} = 2I$.

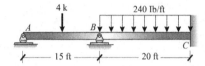

6.8.9   FIXED-END MOMENTS

$$\mathrm{FEM}_{AB} = -\frac{PL}{8} = -\frac{4 \times 15}{8} = -7.5$$

$$\mathrm{FEM}_{BA} = +\frac{PL}{8} = +\frac{4 \times 15}{8} = +7.5$$

$$\mathrm{FEM}_{BC} = -\frac{wL^2}{12} = -\frac{0.24 \times 20^2}{12} = -8.0$$

$$\mathrm{FEM}_{CB} = +\frac{wL^2}{12} = +\frac{0.24 \times 20^2}{12} = +8.0$$

6.8.10   DEGREES OF FREEDOM

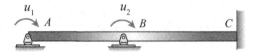

Joint loading:

$$P_1 = \mathrm{FEM}_{AB} = -7.5$$
$$P_2 = \mathrm{FEM}_{BA} + \mathrm{FEM}_{BC} = 7.5 - 8.0 = -0.5$$

$$k_{11} = \frac{4EI}{15}; \quad k_{21} = \frac{2EI}{15}$$

$u_1 = 1$

k_{11} k_{21}

Unit rotation at B:

$$k_{12} = \frac{2EI}{15}; k_{22} = \frac{4EI}{15} + \frac{4E(2I)}{20} = \frac{2EI}{3}$$

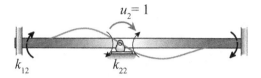

6.8.11 STIFFNESS EQUATIONS

$$\mathbf{KU} = -\mathbf{P}$$

$$\begin{bmatrix} \dfrac{4EI}{15} & \dfrac{2EI}{15} \\ \dfrac{2EI}{15} & \dfrac{2EI}{3} \end{bmatrix} \begin{bmatrix} u_1 \\ u_2 \end{bmatrix} = - \begin{bmatrix} -7.5 \\ -0.5 \end{bmatrix}$$

Solving, $u_1 = 185/6EI$ and $u_2 = -65/12EI$.
End moments using slope deflection equations:

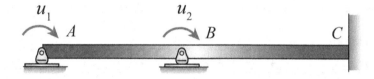

$$u_1 = 185/6EI; \ u_2 = -65/12EI$$

Span AB:

$$M_{AB} = \frac{4EI}{15} \left(\frac{185}{6EI} \right) + \frac{2EI}{15} \left(\frac{-65}{12EI} \right) + (-7.5) = 0$$

$$M_{BA} = \frac{2EI}{15} \left(\frac{185}{6EI} \right) + \frac{4EI}{15} \left(\frac{-65}{12EI} \right) + 7.5 = 10.17$$

Span BC:

$$M_{BC} = \frac{4E \cdot 2I}{20} \left(\frac{-65}{12EI} \right) + 0 + (-8) = -10.17$$

$$M_{CB} = \frac{2E \cdot 2I}{20} \left(\frac{-65}{12EI} \right) + 0 + 8 = 6.92$$

Joint	A	B		C
Member	AB	BA	BC	CB
Moment	0	+10.17	−10.17	+6.92

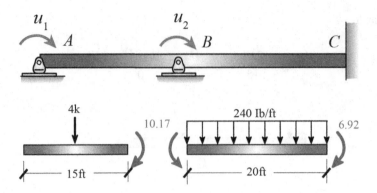

6.8.12 REACTIONS

Load	+2	+2	+2.4	+2.4
Moment	−0.68	+0.68	−0.16	+0.16
Total	**+1.32**	**+4.92**		**+2.56**

Answer: $R_A = 1.32$ kip, $R_B = 4.92$ kip, and $R_C = 2.56$ kip

Example 27

Determine the support reactions of the beam using the stiffness method. Assume EI is constant.

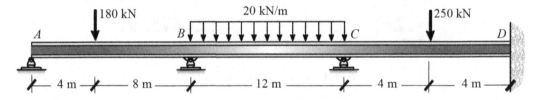

6.8.13 FIXED-END MOMENTS

$$\text{FEM}_{AB} = -Pab^2 / L^2 = -\left(180 \times 4 \times 8^2\right)/12^2 \quad = -320$$
$$\text{FEM}_{BA} = +Pa^2b / L^2 = +\left(180 \times 4^2 \times 8\right)/12^2 \quad = +160$$
$$\text{FEM}_{BC} = -wL^2 / 12 = -\left(20 \times 12^2\right)/12 \quad = -240$$
$$\text{FEM}_{CB} = +wL^2 / 12 = +\left(20 \times 12^2\right)/12 \quad = +240$$
$$\text{FEM}_{CD} = -PL / 8 = -\left(250 \times 8\right)/8 \quad = -250$$
$$\text{FEM}_{DC} = +PL / 8 = +\left(250 \times 8\right)/8 \quad = +250$$

6.8.14 DEGREES OF FREEDOM

6.8.15 Joint Loading

$$
\begin{aligned}
P_1 &= \mathrm{FEM}_{AB} & &= -320 \\
P_2 &= \mathrm{FEM}_{BA} + \mathrm{FEM}_{BC} = 160 - 240 & &= -80 \\
P_3 &= \mathrm{FEM}_{CB} + \mathrm{FEM}_{CD} = 240 - 250 & &= -10
\end{aligned}
$$

6.8.16 Unit Rotation at A $u_1 = 1$

Unit rotation at A

$u_1 = 1$

$$
k_{11} = \frac{4EI}{12}; \; k_{21} = \frac{2EI}{12}; \; k_{31} = 0
$$

6.8.17 Unit Rotation at B $u_2 = 1$

$$
k_{12} = \frac{2EI}{12}; \; k_{22} = \frac{4EI}{12} + \frac{4E}{12} = \frac{2EI}{3}; \; k_{32} = \frac{2EI}{12}
$$

Unit rotation at B

$u_2 = 1$

6.8.18 Unit Rotation at C

Unit rotation at C

$u_3 = 1$

$$
k_{13} = 0; \; k_{23} = \frac{2EI}{12}; \; k_{33} = \frac{4EI}{12} + \frac{4EI}{8} = \frac{5EI}{6}
$$

6.8.19 STIFFNESS EQUATIONS

$$
\begin{bmatrix}
\dfrac{4EI}{12} & \dfrac{2EI}{12} & 0 \\[2mm]
\dfrac{2EI}{12} & \dfrac{2EI}{3} & \dfrac{2EI}{12} \\[2mm]
0 & \dfrac{2EI}{12} & \dfrac{5EI}{6}
\end{bmatrix}
\begin{bmatrix} u_1 \\ u_2 \\ u_3 \end{bmatrix}
= -\begin{bmatrix} -320 \\ -80 \\ -10 \end{bmatrix}
$$

Solving, $u_1 = 11380/11EI$, $u_2 = -1640/11EI$, and $u_3 = 460/11El$

6.8.20 END MOMENTS USING SLOPE DEFLECTION EQUATIONS

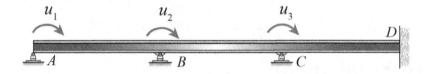

$$u_1 = 11380/11EI;\ u_2 = -1640/11EI;\ u_3 = 460/11EI$$

Span AB:

$$M_{AB} = \frac{4EI}{12}\left(\frac{11380}{11EI}\right) + \frac{2EI}{12}\left(\frac{-1640}{11EI}\right) + (-320) = 0$$

$$M_{BA} = \frac{2EI}{12}\left(\frac{11380}{11EI}\right) + \frac{2EI}{12}\left(\frac{-1640}{11EI}\right) + 160 = 282.7$$

Span BC:

$$M_{BC} = \frac{4EI}{12}\left(\frac{-1640}{11EI}\right) + \frac{2EI}{12}\left(\frac{460}{11EI}\right) + (-240) = -282.7$$

$$M_{CB} = \frac{2EI}{12}\left(\frac{-1640}{11EI}\right) + \frac{4EI}{12}\left(\frac{460}{11EI}\right) + 240 = 229.1$$

Span CD:

$$M_{CD} = \frac{4EI}{8}\left(\frac{460}{11EI}\right) + 0 + (-250) = -229.1$$

$$M_{DC} = \frac{2EI}{8}\left(\frac{460}{11EI}\right) + 0 + 250 = 260.4$$

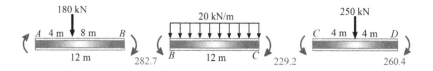

Member	AB	BA	BC	CB	CD	DC
Moment	0.00	282.7	−282.7	229.1	−229.1	260.4

6.8.21 REACTIONS

Load	+120	+60	+120	+120	+125	+125
Moment	−23.56	+23.56	+4.47	−4.47	−3.92	+3.92
Total	+96.44	+208.33		+236.61		+128.92

Example 28

Determine the support reactions of the beam using the stiffness method, given that $I_{AB} = I$ and $I_{BC} = 2I$.

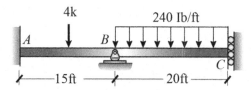

6.8.22 FIXED-END MOMENTS

$$\text{FEM}_{AB} = -PL/8 = -(4 \times 15)/8 \qquad = -7.5$$
$$\text{FEM}_{BA} = +PL/8 = +(4 \times 15)/8 \qquad = +7.5$$
$$\text{FEM}_{BC} = -wL^2/12 = -(0.24 \times 20^2)/12 \quad = -8.0$$
$$\text{FEM}_{CB} = +wL^2/12 = +(0.24 \times 20^2)/12 \quad = +8.0$$
$$\text{FEV}_{CB} = +wL/2 = +(0.24 \times 20)/2 \qquad = +2.4$$

6.8.23 DEGREES OF FREEDOM

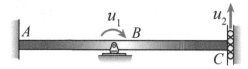

6.8.24 JOINT LOADING

$$P_1 = \text{FEM}_{BA} + \text{FEM}_{BC} = 7.5 - 8.0 \quad = -0.5$$
$$P_2 = \text{FEV}_{CB} \qquad\qquad\qquad = +2.4$$

6.8.25 Unit Rotation at B

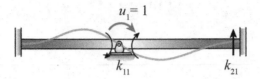

$$k_{11} = \frac{4EI}{15} + \frac{4E(2I)}{20} = \frac{2EI}{3}; \ k_{21} = \frac{6E(2I)}{20^2}$$

Unit displacement at C:

$$k_{12} = \frac{6E(2I)}{20^2}; \ k_{22} = \frac{12E(2I)}{20^3}$$

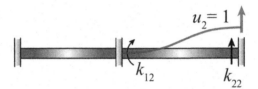

Stiffness equations:

$$\mathbf{KU} = -\mathbf{P}$$

$$\begin{bmatrix} \dfrac{2EI}{3} & \dfrac{6E(2I)}{20^2} \\ \dfrac{6E(2I)}{20^2} & \dfrac{12E(2I)}{20^3} \end{bmatrix} \begin{bmatrix} u_1 \\ u_2 \end{bmatrix} = - \begin{bmatrix} -0.5 \\ 2.4 \end{bmatrix}$$

Solving, $u_1 = 735/11EI$ and $u_2 = -16150/11EI$

6.8.26 End Moments Using Slope Deflection Equations

$$u_1 = 735/11EI; \ u_2 = -16150/11EI$$

Span AB:

$$M_{AB} = 0 + \frac{2EI}{15}\left(\frac{735}{11EI}\right) + (-7.5) = 1.41$$

$$M_{BA} = 0 + \frac{4EI}{15}\left(\frac{735}{11EI}\right) + 7.5 = 25.32$$

Span BC:

$$M_{BC} = \frac{4E \cdot 2I}{20}\left(\frac{735}{11EI}\right) + 0 + \frac{6E \cdot 2I}{20^2}\left(\frac{-16150}{11EI}\right) + (-8) = -25.32$$

$$M_{CB} = \frac{2E \cdot 2I}{20}\left(\frac{735}{11EI}\right) + 0 + \frac{6E \cdot 2I}{20^2}\left(\frac{-16150}{11EI}\right) + 8 = -22.68$$

Member	AB	BA	BC	CB
Moment	+1.41	+25.32	−25.32	−22.68

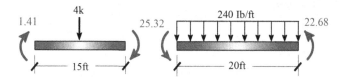

6.8.27 REACTIONS

Load	+2	+2	+2.4	+2.4
Moment	−1.78	+1.78	+2.4	−2.4
Total	+0.22	+8.58	0	

Answer: $R_A = 0.22$ kip; $R_B = 8.58$ kip and $R_C = 0$ kip

6.9   STIFFNESS METHOD FOR FRAME

Example 29

Determine the end moments of the following frame.

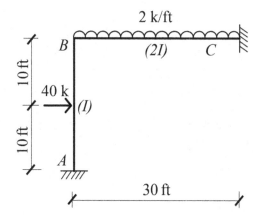

Fixed-end moments:

$$\text{FEM}_{AB} = -\frac{PL}{8} = -\frac{40 \times 20}{8} = -100$$

$$\text{FEM}_{BA} = +\frac{PL}{8} = +\frac{40 \times 20}{8} = +100$$

$$\text{FEM}_{BC} = -\frac{wL^2}{12} = -\frac{2 \times 30^2}{12} = -150$$

$$\text{FEM}_{CB} = +\frac{wL^2}{12} = +\frac{2 \times 30^2}{12} = +150$$

6.9.1 DEGREES OF FREEDOM

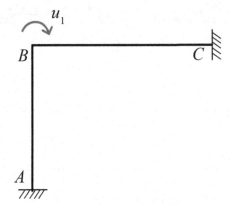

Joint loading:

$$\begin{aligned} P_1 &= \text{FEM}_{BA} + \text{FEM}_{BC} \\ &= 100 - 150 = -50 \end{aligned}$$

Unit rotation at B:

$$k_{11} = \frac{4EI}{20} + \frac{4E(2I)}{30} = \frac{7EI}{15}$$

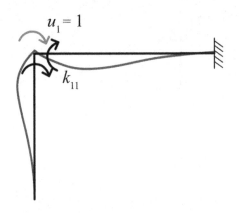

6.9.2 STIFFNESS EQUATIONS

$$[u_1] = -[P_1];$$

$$\left[\frac{7EI}{15}\right][u_1] = -[-50]$$

$$u_1 = 750 / 7EI$$

End moments by slope deflection equations:
Span AB:

$$M_{AB} = 0 + \frac{2EI}{20}\left(\frac{750}{7EI}\right) - 100 = -89.3$$

$$M_{BA} = 0 + \frac{4EI}{20}\left(\frac{750}{7EI}\right) + 100 = 121.4$$

Span BC:

$$M_{BC} = \frac{4E \cdot 2l}{30}\left(\frac{750}{7EI}\right) + 0 - 150 = -121.4$$

$$M_{CB} = \frac{2E \cdot 2l}{30}\left(\frac{750}{7EI}\right) + 0 + 150 = 164.3$$

Free-body diagram:

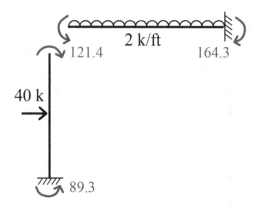

Example 30

Determine the end moments of the following frame.

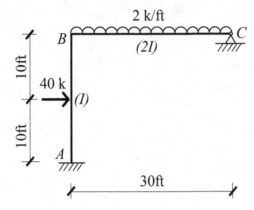

6.9.3   FIXED-END MOMENTS

$$\text{FEM}_{AB} = -\frac{PL}{8} = -\frac{40 \times 20}{8} = -100$$

$$\text{FEM}_{BA} = +\frac{PL}{8} = +\frac{40 \times 20}{8} = +100$$

$$\text{FEM}_{BC} = -\frac{wL^2}{12} = -\frac{2 \times 30^2}{12} = -150$$

$$\text{FEM}_{CB} = +\frac{wL^2}{12} = +\frac{2 \times 30^2}{12} = +150$$

6.9.4   DEGREES OF FREEDOM

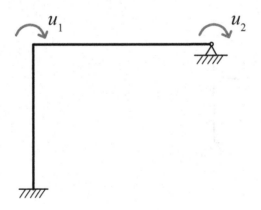

Joint loading:

$$P_1 = \text{FEM}_{BA} + \text{FEM}_{BC}$$
$$= 100 - 150 = -50$$
$$P_2 = \text{FEM}_{CB} = 150$$

Unit rotation at B:

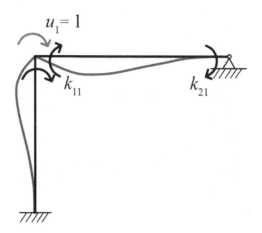

$$k_{11} = \frac{4EI}{20} + \frac{4E(2I)}{30} = \frac{7EI}{15}; \; k_{21} = \frac{2E(2I)}{30}$$

Unit rotation at C:

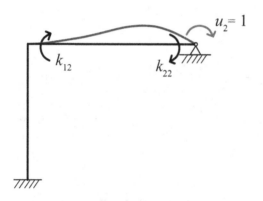

$$k_{12} = \frac{2E(2I)}{30}; \; k_{22} = \frac{4E(2I)}{30}$$

6.9.5 Stiffness Equations

$$\mathbf{KU} = -\mathbf{P}$$

$$\begin{bmatrix} \dfrac{7EI}{15} & \dfrac{2E(2I)}{30} \\ \dfrac{2E(2I)}{30} & \dfrac{4E(2I)}{30} \end{bmatrix} \begin{bmatrix} u_1 \\ u_2 \end{bmatrix} = - \begin{bmatrix} -50 \\ 150 \end{bmatrix}$$

Solving, $u_1 = 625/2EI$ and $u_2 = -2875/4EI$

$$u_1 = 625/2EI; \; u_2 = -2875/4EI$$

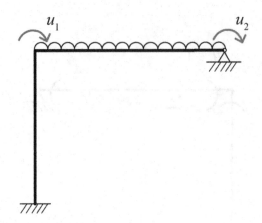

End moments by slope deflection equations:
Span AB:

$$M_{AB} = 0 + \frac{2EI}{20}\left(\frac{625}{2EI}\right) - 100 = -68.75$$

$$M_{BA} = 0 + \frac{4EI}{20}\left(\frac{625}{2EI}\right) + 100 = 162.5$$

Span BC:

$$M_{BC} = \frac{4E \cdot 2I}{30}\left(\frac{625}{2EI}\right) + \frac{2E \cdot 2I}{30}\left(\frac{-2875}{4EI}\right) - 150 = -162.5$$

$$M_{CB} = \frac{2E \cdot 2I}{30}\left(\frac{625}{2EI}\right) + \frac{4E \cdot 2I}{30}\left(\frac{-2875}{4EI}\right) - 150 = 0$$

Free-body diagram:

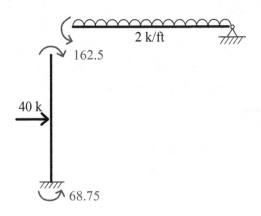

Example 31

Determine the end moments of the following frame.

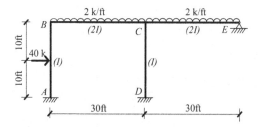

6.9.6 Fixed-End Moments

$$\text{FEM}_{AB} = -\frac{PL}{8} = -\frac{40 \times 20}{8} = -100$$

$$\text{FEM}_{BA} = +\frac{PL}{8} = +\frac{40 \times 20}{8} = +100$$

$$\text{FEM}_{BC} = +\frac{wL^2}{12} = +\frac{2 \times 30^2}{12} = -150$$

$$\text{FEM}_{CB} = +\frac{wL^2}{12} = +\frac{2 \times 30^2}{12} = +150$$

$$\text{FEM}_{CE} = -\frac{wL^2}{12} = -\frac{2 \times 30^2}{12} = -150$$

$$\text{FEM}_{EC} = +\frac{wL^2}{12} = +\frac{2 \times 30^2}{12} = +150$$

$$\text{FEM}_{DC} = \text{FEM}_{CD} = 0$$

6.9.7 Degrees of Freedom

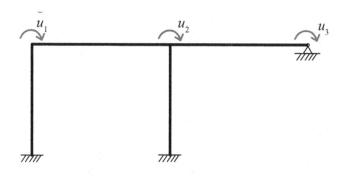

Joint loading:

$$P_1 = \text{FEM}_{BA} + \text{FEM}_{BC} = 100 - 150 = -50$$

$$P_2 = \text{FEM}_{CB} + \text{FEM}_{CE} + \text{FEM}_{CD} = 150 - 150 + 0 = 0$$

$$P_3 = \text{FEM}_{EC} = 150$$

Unit rotation at B:

$$k_{11} = \frac{4EI}{20} + \frac{4E(2I)}{30} = \frac{7EI}{15}; \; k_{21} = \frac{2EI}{30};$$
$$k_{31} = 0$$

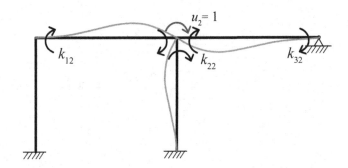

Unit rotation at C:

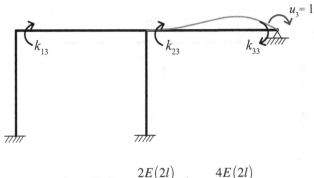

$$k_{12} = \frac{2E(2I)}{30}; \; k_{22} = \frac{4E(2I)}{30} + \frac{4E(2I)}{30} + \frac{4EI}{20} = \frac{11EI}{15}; \; k_{32} = \frac{2E(2I)}{30}$$

Unit rotation at E:

$$k_{13} = 0; \; k_{23} = \frac{2E(2I)}{30}; \; k_{33} = \frac{4E(2I)}{30}$$

6.9.8 Stiffness Equations

$$\begin{bmatrix} \dfrac{7EI}{15} & \dfrac{2E(2l)}{30} & 0 \\[2mm] \dfrac{2E(2l)}{30} & \dfrac{11EI}{15} & \dfrac{2E(2l)}{30} \\[2mm] 0 & \dfrac{2E(2l)}{30} & \dfrac{4E(2l)}{30} \end{bmatrix} \begin{bmatrix} u_1 \\ u_2 \\ u_3 \end{bmatrix} = -\begin{bmatrix} -50 \\ 0 \\ 150 \end{bmatrix}$$

Solving, $u_1 = 875/11EI$, $u_2 = 2125/22EI$, $u_3 = -26875/44EI$

6.9.9 End Moments by Slope Deflection Equations

Span AB:

$$M_{AB} = 0 + (2EI/20)u_1 - 100 = -92.04$$
$$M_{BA} = 0 + (4EI/20)u_1 + 100 = 115.9$$

Span BC:

$$M_{BC} = (4E \cdot 2I/30)u_1 + (2E \cdot 2I/30)u_2 - 150 = -115.9$$
$$M_{CB} = (2E \cdot 2I/30)u_1 + (4E \cdot 2I/30)u_2 + 150 = 186.4$$

Span DC:

$$M_{DC} = 0 + (2EI/20)u_2 + 0 = 9.7$$
$$M_{CD} = 0 + (4EI/20)u_2 + 0 = 19.3$$

Span CE:

$$M_{CE} = (4E \cdot 2I/30)u_2 + (2E \cdot 2I/30)u_3 - 150 = -205.6$$
$$M_{EC} = (2E \cdot 2I/30)u_2 + (4E \cdot 2I/30)u_3 + 150 = 0$$

6.10 STIFFNESS METHOD FOR TRUSS

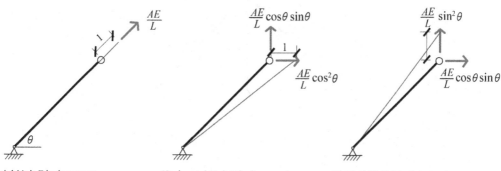

Axial Unit Displacement   Horizontal Unit Displacement   Vertical Unit Displacement

Example 32

Determine the bar forces of the truss using the stiffness method. Cross-sectional areas, in square inch, are in parenthesis.

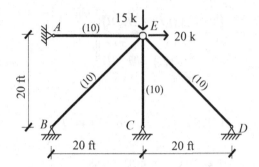

6.10.1 DEGREES OF FREEDOM

Since only the E joint is free to move, the truss has only 2° of freedom.

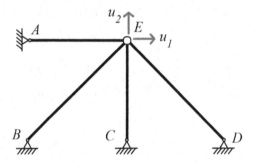

Joint loading:

$$P_1 = 20$$
$$P_2 = -15$$

Horizontal unit displacement at E:

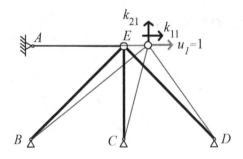

$$k_{11} = \frac{10E}{20} + \frac{10E}{20\sqrt{2}}\cos^2 45 + 0 + \frac{10E}{20\sqrt{2}}\cos^2 135 = \frac{\left(2+\sqrt{2}\right)E}{4}$$

$$k_{21} = 0 + \frac{10E}{20\sqrt{2}}\sin45\cos45 + 0 + \frac{10E}{20\sqrt{2}}\sin135\cos135 = \frac{E}{2}$$

Vertical unit displacement at E:

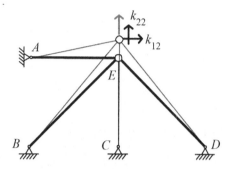

$$k_{12} = 0 + \frac{10E}{20\sqrt{2}} \sin 45 \cos 45 + 0 + \frac{10E}{20\sqrt{2}} \sin 135 \cos 135 = 0$$

$$k_{22} = 0 + \frac{10E}{20\sqrt{2}} \sin^2 45 + \frac{10E}{20} + \frac{10E}{20\sqrt{2}} \sin^2 135 = \frac{\left(2+\sqrt{2}\right)E}{4}$$

6.10.2 STIFFNESS EQUATIONS

$$\mathbf{KU} = \mathbf{P}$$

$$\begin{vmatrix} \dfrac{\left(2+\sqrt{2}\right)E}{4} & 0 \\[2ex] 0 & \dfrac{\left(2+\sqrt{2}\right)E}{4} \end{vmatrix} \begin{bmatrix} u_1 \\ u_2 \end{bmatrix} = \begin{bmatrix} 20 \\ -15 \end{bmatrix}$$

Solving, we find, $u_1 = 40\left(2 - \sqrt{2}\right)/E$, $u_2 = 30\left(-2 + \sqrt{2}\right)/E$.

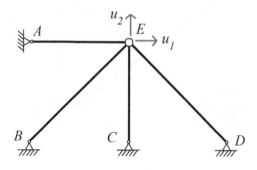

$$F_{AE} = \left[\left(b_x - a_x\right)\cos\theta + \left(b_y - a_y\right)\sin\theta\right]\frac{AE}{L}$$

6.10.3 BAR FORCES

$$F_{AE} = \left[(u_1 - 0)\cos 0 + (u_2 - 0)\sin 0\right]\frac{1}{E} \cdot \frac{10 \times E}{20} = +11.72 \text{ kip}(T)$$

$$F_{BE} = \left[(u_1 - 0)\cos 45 + (u_2 - 0)\sin 45\right]\frac{1}{E} \cdot \frac{10 \times E}{20\sqrt{2}} = +1.46 \text{ kip}(T)$$

$$F_{CE} = \left[(u_1 - 0)\cos 90 + (u_2 - 0)\sin 90\right]\frac{1}{E} \cdot \frac{10 \times E}{20} = -8.78 \text{ kip}(C)$$

$$F_{DE} = \left[(u_1 - 0)\cos 135 + (u_2 - 0)\sin 135\right]\frac{1}{E} \cdot \frac{10 \times E}{20\sqrt{2}} = -10.21 \text{ kip}(C)$$

Example 33

Question: Determine the bar forces of the truss using the stiffness method. Cross-sectional areas, in square inch, are in parenthesis.

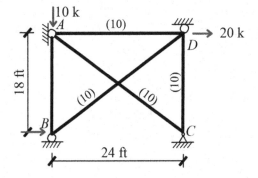

6.10.4 DEGREES OF FREEDOM

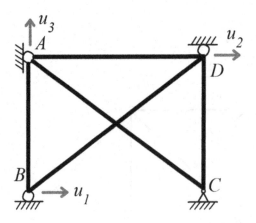

Joint loading:

$$P_1 = 10$$
$$P_2 = 20$$
$$P_3 = -10$$

Horizontal unit displacement at B:

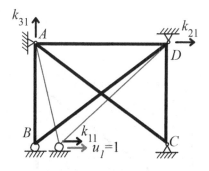

$$k_{11} = \frac{10E}{18}\cos^2 90° + \frac{10E}{30}\left(\frac{24}{30}\right)^2 = \frac{16E}{75}$$

$$k_{21} = -\frac{10E}{30}\left(\frac{24}{30}\right)^2 = -\frac{16E}{75}$$

$$k_{31} = \frac{10E}{18}\cos 90° \sin 90° = 0$$

Horizontal unit displacement at D:

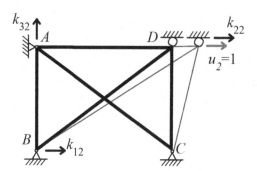

$$k_{22} = \frac{7 \cdot 5E}{24}\cos^2 0° + \frac{10E}{30}\left(\frac{24}{30}\right)^2 + \frac{10E}{18}\cos^2 90 = \frac{631}{1200}$$

$$k_{12} = -\frac{10E}{30}\left(\frac{24}{30}\right)^2 = -\frac{16E}{75}$$

$$k_{32} = \frac{7 \cdot 5E}{24}\cos 0° \sin 0° = 0$$

Vertical unit displacement at A:

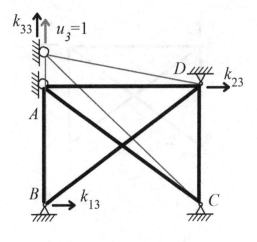

$$k_{33} = \frac{10E}{18}\sin^2 90° + \frac{10E}{30}\sin^2\left[180° + \sin^{-1}\left(\frac{24}{30}\right)\right]$$

$$+ \frac{7.5E}{24}\sin^2 0° = \frac{152E}{225}.$$

$$k_{13} = \frac{10E}{18}\cos 90°\sin 90° = 0.$$

$$k_{23} = \frac{7.5E}{24}\cos 0°\sin 0° = 0.$$

6.10.5 STIFFNESS EQUATIONS

$$\mathbf{KU} = \mathbf{P}$$

$$\begin{bmatrix} \dfrac{16E}{75} & -\dfrac{16E}{75} & 0 \\[2ex] -\dfrac{16E}{75} & \dfrac{631E}{1200} & 0 \\[2ex] 0 & 0 & \dfrac{152E}{225} \end{bmatrix} \begin{bmatrix} u_1 \\ u_2 \\ u_3 \end{bmatrix} = \begin{bmatrix} 10 \\ 20 \\ -10 \end{bmatrix}$$

$$\therefore u_1 = \frac{1143}{8E}$$

$$u_2 = \frac{96}{E}$$

$$u_3 = -\frac{1125}{76E}.$$

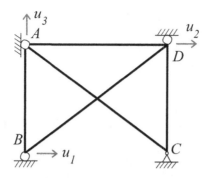

6.10.6   BAR FORCES

Bar forces:

$$F_{xy} = \left[\left(b_x - a_x \right) \cos\theta + \left(b_y - a_y \right) \sin\theta \right] \frac{AE}{L}$$

$$F_{AD} = \left[\left(u_2 - 0 \right) \cos 0° + \left(0 - u_3 \right) \sin 0° \right] \frac{7 \cdot 5E}{24} = +30 \text{ kip } (T)$$

$$F_{BA} = \left[\left(0 - u_1 \right) \cos 9\, 0° + \left(u_3 - 0 \right) \sin 90° \right] \frac{10E}{18} = -8.22 \text{ kip } (C)$$

$$F_{CD} = \left[\left(u_2 - 0 \right) \cos 9\, 0° + \left(0 - 0 \right) \sin 90° \right] \frac{10E}{18} = 0 \text{ kip}$$

$$F_{CA} = \left[\left(0 - 0 \right) \cos\left\{ 90° + \cos^{-1}\left(\frac{18}{30} \right) \right\} + \left(u_3 - 0 \right) \sin\left\{ 90° + \sin^{-1}\left(\frac{24}{3} \right) \right\} \right]$$
$$\times \frac{10E}{30} = -2.96 \text{ kip } (C)$$

$$F_{BD} = \left[\left(u_2 - u_1 \right) \left(\frac{24}{30} \right) + \left(0 - 0 \right) \left(\frac{18}{30} \right) \right] \frac{10E}{30} = -12.5 \, kip(C)$$

$$F_{AD} = +30 \text{ kip} (T) \qquad F_{CA} = -2.96 \text{ kip } (C)$$
$$F_{BA} = -8.22 \text{ kip}(C) \qquad F_{BD} = -12.5 \text{ kip } (C)$$
$$F_{CD} = 0 \text{ kip}$$

6.11 DIRECT STIFFNESS MATRIX FOR TRUSS

6.11.1 COMPUTER APPLICATION OF STIFFNESS METHOD

Example 34

Determine the reactions and bar forces of the truss using the direct stiffness matrix method.

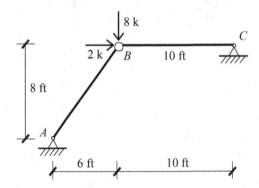

6.11.2 DEGREES OF FREEDOM

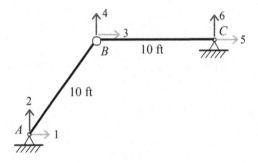

Joint load matrix and displacement matrix:

$$
\mathbf{P} = \begin{bmatrix} P_1 \\ P_2 \\ P_3 \\ P_4 \\ P_5 \\ P_6 \end{bmatrix} = \begin{bmatrix} ? \\ ? \\ 2 \\ -8 \\ ? \\ ? \end{bmatrix}; \quad \mathbf{U} = \begin{bmatrix} u_1 \\ v_2 \\ u_3 \\ v_4 \\ u_5 \\ v_6 \end{bmatrix} = \begin{bmatrix} 0 \\ 0 \\ ? \\ ? \\ 0 \\ 0 \end{bmatrix}
$$

$$
\mathbf{k} = \frac{AE}{L} \begin{bmatrix} c^2 & cs & -c^2 & -cs \\ cs & s^2 & -cs & -s^2 \\ -c^2 & -cs & c^2 & cs \\ -cs & -s^2 & cs & s^2 \end{bmatrix}
$$

	L	$\sin\theta$	$\cos\theta$
AB	10	0.8	0.6
BC	10	0	1

6.11.3 MEMBER STIFFNESS MATRIX

$$\mathbf{k}_{AB} = AE \begin{bmatrix} 0.036 & 0.048 & -0.036 & -0.048 \\ 0.048 & 0.064 & -0.048 & -0.064 \\ -0.036 & -0.048 & 0.036 & 0.048 \\ -0.048 & -0.064 & 0.048 & 0.064 \end{bmatrix};$$

$$\mathbf{k}_{BC} = AE \begin{bmatrix} 0.1 & 0 & -0.1 & -0 \\ 0 & 0 & -0 & -0 \\ -0.1 & -0 & 0.1 & 0 \\ -0 & -0 & 0 & 0 \end{bmatrix}$$

6.11.4 STRUCTURE STIFFNESS MATRIX

$$\mathbf{P = KU}$$

$$\begin{bmatrix} p_1 \\ p_2 \\ 2 \\ -8 \\ p_5 \\ p_6 \end{bmatrix} = AE \begin{bmatrix} 0.036 & 0.048 & -0.036 & -0.048 & 0 & 0 \\ 0.048 & 0.064 & -0.048 & -0.064 & 0 & 0 \\ -0.036 & -0.048 & 0.136 & 0.048 & -0.1 & 0 \\ -0.048 & -0.064 & 0.048 & 0.064 & 0 & 0 \\ 0 & 0 & -0.1 & 0 & 0.1 & 0 \\ 0 & 0 & 0 & 0 & 0 & 0 \end{bmatrix} \begin{bmatrix} 0 \\ 0 \\ u_3 \\ v_4 \\ 0 \\ 0 \end{bmatrix}$$

From stiffness matrix, we write:

$$2 = AE(0.136u_3 + 0.048v_4)$$
$$-8 = AE(0.048u_3 + 0.064v_4)$$

Solving, we find $u_3 = 80 / AE$ and $v_4 = -185 / AE$.

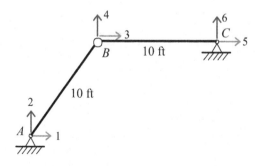

6.11.5 SUPPORT REACTIONS

$$\mathbf{P = KU}$$

$$
\begin{bmatrix} p_1 \\ p_2 \\ 2 \\ -8 \\ p_5 \\ p_6 \end{bmatrix} = AE
\begin{bmatrix}
0.036 & 0.048 & -0.036 & -0.048 & 0 & 0 \\
0.048 & 0.064 & -0.048 & -0.064 & 0 & 0 \\
-0.036 & -0.048 & 0.136 & 0.048 & -0.1 & 0 \\
-0.048 & -0.064 & 0.048 & 0.064 & 0 & 0 \\
0 & 0 & -0.1 & 0 & 0.1 & 0 \\
0 & 0 & 0 & 0 & 0 & 0
\end{bmatrix}
\begin{bmatrix} 0 \\ 0 \\ 80/AE \\ -185/AE \\ 0 \\ 0 \end{bmatrix}
=
\begin{bmatrix} 6\ \text{kip} \\ 8\ \text{kip} \\ 2 \\ -8 \\ -8\ \text{kip} \\ 0\ \text{kip} \end{bmatrix}
$$

Member forces:

$$
F_{AB} = \frac{AE}{L}\begin{bmatrix} -c & -s & c & s \end{bmatrix}\begin{bmatrix} u_1 \\ v_2 \\ u_3 \\ v_4 \end{bmatrix} = \frac{AE}{10}\begin{bmatrix} -0.6 & -0.8 & 0.6 & 0.8 \end{bmatrix}\begin{bmatrix} 0 \\ 0 \\ 80/AE \\ -185/AE \end{bmatrix} = -10\ \text{kip}
$$

$$
F_{BC} = \frac{AE}{L}\begin{bmatrix} -c & -s & c & s \end{bmatrix}\begin{bmatrix} u_3 \\ v_4 \\ u_5 \\ v_6 \end{bmatrix} = \frac{AE}{10}\begin{bmatrix} -1.0 & 0.0 & 1.0 & 0.0 \end{bmatrix}\begin{bmatrix} 80/AE \\ -185/AE \\ 0 \\ 0 \end{bmatrix} = -8\ \text{kip}
$$

6.12 DIRECT STIFFNESS MATRIX FOR BEAM

6.12.1 COMPUTER APPLICATION OF STIFFNESS METHOD

Example 35

Determine all the reactions of the beam using the direct stiffness matrix method.

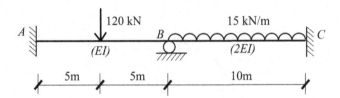

6.12.2 DEGREES OF FREEDOM

6.12.3 Member Load Matrix

$$\mathbf{p}_{AB} = \begin{bmatrix} p_1 \\ p_2 \\ p_3 \\ p_4 \end{bmatrix} = \begin{bmatrix} 120/2 \\ 120 \times 10/8 \\ 120/2 \\ -120 \times 10/8 \end{bmatrix} = \begin{bmatrix} 60 \\ -150 \\ 60 \\ 150 \end{bmatrix}$$

$$\mathbf{p}_{BC} = \begin{bmatrix} p_3 \\ p_4 \\ p_5 \\ p_6 \end{bmatrix} = \begin{bmatrix} 15 \times 10/2 \\ 15 \times 10^2/12 \\ 15 \times 10/2 \\ -15 \times 10^2/12 \end{bmatrix} = \begin{bmatrix} 75 \\ -125 \\ 75 \\ 125 \end{bmatrix}$$

6.12.4 Displacement Matrix

$$\mathbf{u}_{AB} = \begin{bmatrix} \delta_1 \\ \theta_2 \\ \delta_3 \\ \theta_4 \end{bmatrix} = \begin{bmatrix} 0 \\ 0 \\ 0 \\ ? \end{bmatrix}; \ \mathbf{u}_{BC} = \begin{bmatrix} \delta_3 \\ \theta_4 \\ \delta_5 \\ \theta_6 \end{bmatrix} = \begin{bmatrix} 0 \\ ? \\ 0 \\ 0 \end{bmatrix}$$

6.12.5 Member Stiffness Matrix

$$\mathbf{k} = \frac{EI}{L^3} \begin{bmatrix} 12 & 6L & -12 & 6L \\ 6L & 4L^2 & -6L & 2L^2 \\ -12 & -6L & 12 & -6L \\ 6L & 2L^2 & -6L & 4L^2 \end{bmatrix}$$

For $AB, L = 10$ m, $I = I$ and for $BC, L = 10$ m, $I = 2I$, we find:

$$\mathbf{k}_{AB} = EI \begin{bmatrix} 0.012 & 0.06 & -0.012 & 0.06 \\ 0.06 & 0.4 & -0.06 & 0.2 \\ -0.012 & -0.06 & 0.012 & -0.06 \\ 0.06 & 0.2 & -0.06 & 0.4 \end{bmatrix}$$

$$\mathbf{k}_{BC} = EI \begin{bmatrix} 0.024 & 0.12 & -0.024 & 0.12 \\ 0.12 & 0.8 & -0.12 & 0.4 \\ -0.024 & -0.12 & 0.024 & -0.12 \\ 0.12 & 0.4 & -0.12 & 0.8 \end{bmatrix}$$

Structure stiffness matrix:

$$\mathbf{P} = \mathbf{KU}$$

$$\begin{bmatrix} 60 \\ -150 \\ 135 \\ 25 \\ 75 \\ 125 \end{bmatrix} = EI \begin{bmatrix} 0.012 & 0.06 & -0.012 & 0.06 & 0 & 0 \\ 0.06 & 0.4 & -0.06 & 0.2 & 0 & 0 \\ -0.012 & -0.06 & 0.036 & 0.06 & -0.024 & 0.12 \\ 0.06 & 0.2 & 0.06 & 1.2 & -0.12 & 0.4 \\ 0 & 0 & -0.024 & -0.12 & 0.024 & -0.12 \\ 0 & 0 & 0.12 & 0.4 & -0.12 & 0.8 \end{bmatrix} \begin{bmatrix} 0 \\ 0 \\ 0 \\ \theta_4 \\ 0 \\ 0 \end{bmatrix}$$

From the stiffness matrix, we write:

$$25 = EI(1.2\theta_4); \quad \theta_4 = 125/6EI$$

6.12.6 INTERNAL FORCE EQUATIONS

Member AB:

$$\mathbf{q}_{AB} = \mathbf{k}_{AB}\mathbf{u}_{AB} + \mathbf{p}_{AB}$$

$$\begin{bmatrix} V_1 \\ M_2 \\ V_3 \\ M_4 \end{bmatrix} = EI \begin{bmatrix} 0.012 & 0.06 & -0.012 & 0.06 \\ 0.06 & 0.4 & -0.06 & 0.2 \\ -0.012 & -0.06 & 0.012 & -0.06 \\ 0.06 & 0.2 & -0.06 & 0.4 \end{bmatrix} \begin{bmatrix} 0 \\ 0 \\ 0 \\ 125/6EI \end{bmatrix} + \begin{bmatrix} 60 \\ 150 \\ 60 \\ -150 \end{bmatrix} = \begin{bmatrix} 61.25 \\ 154.17 \\ 58.75 \\ -141.67 \end{bmatrix}$$

Member BC:

$$\mathbf{q}_{BC} = \mathbf{k}_{BC}\mathbf{u}_{BC} + \mathbf{p}_{BC}$$

$$\begin{bmatrix} V_3 \\ M_4 \\ V_5 \\ M_6 \end{bmatrix} = EI \begin{bmatrix} 0.024 & 0.12 & -0.024 & 0.12 \\ 0.12 & 0.8 & -0.12 & 0.4 \\ -0.024 & -0.12 & 0.024 & -0.12 \\ 0.12 & 0.4 & -0.12 & 0.8 \end{bmatrix} \begin{bmatrix} 0 \\ 125/6EI \\ 0 \\ 0 \end{bmatrix} + \begin{bmatrix} 75 \\ 125 \\ 75 \\ -125 \end{bmatrix} = \begin{bmatrix} 77.5 \\ 141.67 \\ 72.5 \\ -116.67 \end{bmatrix}$$

6.12.7 FREE-BODY DIAGRAM

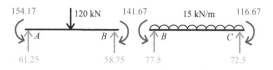

Example 36

Determine all the reactions of the beam using the direct stiffness matrix method.

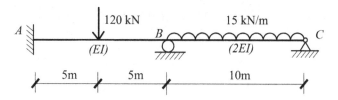

6.12.8 DEGREES OF FREEDOM

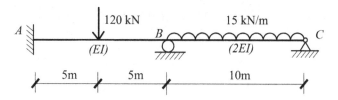

6.12.9 MEMBER LOAD MATRIX

$$\mathbf{p}_{AB} = \begin{bmatrix} p_1 \\ p_2 \\ p_3 \\ p_4 \end{bmatrix} = \begin{bmatrix} 120/2 \\ 120 \times 10/8 \\ 120/2 \\ -120 \times 10/8 \end{bmatrix} = \begin{bmatrix} 60 \\ -150 \\ 60 \\ 150 \end{bmatrix}$$

$$\mathbf{p}_{BC} = \begin{bmatrix} p_3 \\ p_4 \\ p_5 \\ p_6 \end{bmatrix} = \begin{bmatrix} 15 \times 10/2 \\ 15 \times 10^2/12 \\ 15 \times 10/2 \\ -15 \times 10^2/12 \end{bmatrix} = \begin{bmatrix} 75 \\ -125 \\ 75 \\ 125 \end{bmatrix}$$

6.12.10 DISPLACEMENT MATRIX

$$\mathbf{u}_{AB} = \begin{bmatrix} \delta_1 \\ \theta_2 \\ \delta_3 \\ \theta_4 \end{bmatrix} = \begin{bmatrix} 0 \\ 0 \\ 0 \\ ? \end{bmatrix}; \quad \mathbf{u}_{BC} = \begin{bmatrix} \delta_3 \\ \theta_4 \\ \delta_5 \\ \theta_6 \end{bmatrix} = \begin{bmatrix} 0 \\ ? \\ 0 \\ ? \end{bmatrix}$$

6.12.11 MEMBER STIFFNESS MATRIX

For $AB, L = 10$ m, $I = I$ and for $BC, L = 10$ m, $I = 2I$, we find:

$$\mathbf{k}_{AB} = EI \begin{bmatrix} 0.012 & 0.06 & -0.012 & 0.06 \\ 0.06 & 0.4 & -0.06 & 0.2 \\ -0.012 & -0.06 & 0.012 & -0.06 \\ 0.06 & 0.2 & -0.06 & 0.4 \end{bmatrix}$$

$$\mathbf{k}_{BC} = EI \begin{bmatrix} 0.024 & 0.12 & -0.024 & 0.12 \\ 0.12 & 0.8 & -0.12 & 0.4 \\ -0.024 & -0.12 & 0.024 & -0.12 \\ 0.12 & 0.4 & -0.12 & 0.8 \end{bmatrix}$$

6.12.12 STRUCTURE STIFFNESS MATRIX

$$\begin{bmatrix} 60 \\ -150 \\ 135 \\ 25 \\ 75 \\ 125 \end{bmatrix} = EI \begin{bmatrix} 0.012 & 0.06 & -0.012 & 0.06 & 0 & 0 \\ 0.06 & 0.4 & -0.06 & 0.2 & 0 & 0 \\ -0.012 & -0.06 & 0.036 & 0.06 & -0.024 & 0.12 \\ 0.06 & 0.2 & 0.06 & 1.2 & -0.12 & 0.4 \\ 0 & 0 & -0.024 & -0.12 & 0.024 & -0.12 \\ 0 & 0 & 0.12 & 0.4 & -0.12 & 0.8 \end{bmatrix} \begin{bmatrix} 0 \\ 0 \\ 0 \\ \theta_4 \\ 0 \\ \theta_6 \end{bmatrix}$$

From the stiffness matrix, we write:

$$25 = EI(1.2\theta_4) + EI(0.4\theta_6)$$
$$125 = EI(0.4\theta_4) + EI(0.8\theta_6)$$

Solving, we find $\theta_4 = -75/2EI$ and $\theta_6 = 175/EI$.

6.12.13 INTERNAL FORCE EQUATIONS

Member AB:

$$\mathbf{q}_{AB} = \mathbf{k}_{AB}\mathbf{u}_{AB} + \mathbf{p}_{AB}$$

$$\begin{bmatrix} V_1 \\ M_2 \\ V_3 \\ M_4 \end{bmatrix} = EI \begin{bmatrix} 0.012 & 0.06 & -0.012 & 0.06 \\ 0.06 & 0.4 & -0.06 & 0.2 \\ -0.012 & -0.06 & 0.012 & -0.06 \\ 0.06 & 0.2 & -0.06 & 0.4 \end{bmatrix} \begin{bmatrix} 0 \\ 0 \\ 0 \\ -75/2EI \end{bmatrix} + \begin{bmatrix} 60 \\ 150 \\ 60 \\ -150 \end{bmatrix} = \begin{bmatrix} 57.75 \\ 142.5 \\ 62.25 \\ -165 \end{bmatrix}$$

Member BC:

$$\mathbf{q}_{BC} = \mathbf{k}_{BC}\mathbf{u}_{BC} + \mathbf{p}_{BC}$$

$$\begin{bmatrix} V_3 \\ M_4 \\ V_5 \\ M_6 \end{bmatrix} = EI \begin{bmatrix} 0.024 & 0.12 & -0.024 & 0.12 \\ 0.12 & 0.8 & -0.12 & 0.4 \\ -0.024 & -0.12 & 0.024 & -0.12 \\ 0.12 & 0.4 & -0.12 & 0.8 \end{bmatrix} \begin{bmatrix} 0 \\ -75/2EI \\ 0 \\ 175/EI \end{bmatrix} + \begin{bmatrix} 75 \\ 125 \\ 75 \\ -125 \end{bmatrix} = \begin{bmatrix} 91.5 \\ 165 \\ 58.5 \\ 0 \end{bmatrix}$$

Free-body diagram:

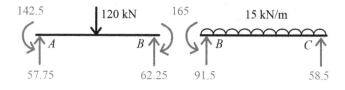

7 Reinforced Concrete Design

7.1 INTRODUCTION

Reinforced concrete is a composite material made up of concrete and reinforcing steel. The reinforcing steel, usually in the form of bars or mesh, is placed within the concrete to increase its tensile strength and resist cracking. This combination results in a material that is strong in compression and tension, making it well-suited for use in construction. Reinforced concrete is widely used in building construction for its strength, durability, and versatility. It is used for a variety of structures, including buildings, bridges, and sidewalks. The reinforcing steel provides the necessary strength to resist bending and shear forces, while the concrete acts as a compression member. In addition to its strength and durability, reinforced concrete is also fire-resistant and able to withstand earthquakes, making it an ideal choice for building construction in areas prone to natural disasters.

The use of reinforced concrete also offers a number of environmental benefits. Concrete is made from natural materials, such as sand, gravel, and cement, and it can be recycled when no longer needed. Additionally, the carbon footprint of reinforced concrete is relatively low compared to other building materials. Reinforced concrete is used in construction for several reasons:

1. *Strength and durability.* Reinforced concrete is a strong and durable material that can withstand heavy loads and resist the effects of weather and natural disasters, making it suitable for long-term use in buildings, bridges, and other structures.
2. *Versatility.* Reinforced concrete can be molded into various shapes and sizes, making it a versatile material that can be used in a wide range of construction projects.
3. *Cost-effectiveness.* Reinforced concrete is relatively inexpensive compared to other building materials, especially in the long run, as it requires low maintenance and has a long service life.
4. *Fire resistance.* Reinforced concrete has good fire resistance and can provide a degree of protection in the event of a fire.
5. *Earthquake resistance.* Reinforced concrete structures are able to withstand earthquakes due to their ability to absorb shock and resist failure.
6. *Sustainability.* Reinforced concrete is made from natural materials and can be recycled when no longer needed, making it a sustainable choice for construction projects.

Reinforced concrete is a versatile and cost-effective building material that offers strength, durability, fire resistance, and environmental benefits, making it a popular choice for construction projects. Reinforced concrete plays a major role in modern architecture, offering a range of benefits for architects and builders alike. Some of the ways that reinforced concrete is used in architecture include:

1. *Structural support.* Reinforced concrete is used as the primary structural material in buildings, bridges, and other structures, providing the necessary strength to resist bending and shear forces.
2. *Flexibility in design.* Reinforced concrete can be molded into various shapes and sizes, making it a versatile material for architects to work with. This flexibility allows for the creation of unique and innovative building designs.
3. *Speed of construction.* Reinforced concrete can be cast on-site, making it a fast and efficient building material. This allows architects and builders to complete projects quickly and move on to the next one.

DOI: 10.1201/9781032638072-7

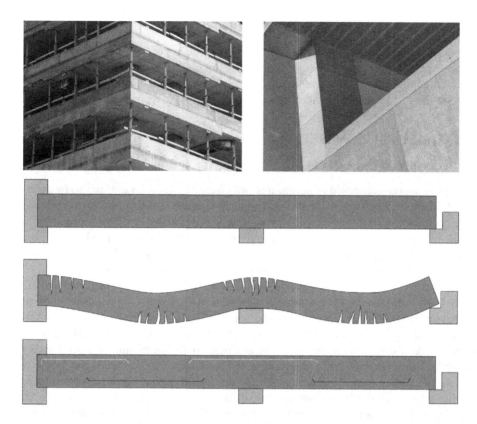

FIGURE 7.1   Typical reinforced concrete building and use of reinforcement in concrete.

4. *Cost-effectiveness*. Reinforced concrete is a cost-effective building material, especially in the long run, as it requires low maintenance and has a long service life.
5. *Fire resistance*. Reinforced concrete structures are fire-resistant, providing a degree of protection for the building and its occupants in the event of a fire.
6. *Earthquake resistance*. Reinforced concrete structures are able to withstand earthquakes, making it a suitable choice for buildings in seismic areas.

Reinforced concrete is a versatile and cost-effective material that offers architects and builders a range of benefits, including structural support, flexibility in design, speed of construction, fire resistance, and earthquake resistance. It is a popular choice for contemporary architecture and continues to play a major role in modern building design.

7.2   ADVANTAGES OF CONCRETE STRUCTURES

- It offers greater compressive strength than other structures.
- Resistance to fire and water is significantly higher.
- Structures are very rigid.
- It has longer service life, usually 50 to 100 years.
- It is economical for foundations, floor slabs, basement walls, etc.
- A special feature of concrete is its ability to be cast into an extraordinary variety of shapes.

7.3 DISADVANTAGES OF CONCRETE STRUCTURES

- Concrete has a very low tensile strength, requiring the use of tensile reinforcing.
- Forms are required to hold the concrete in place until it hardens sufficiently.
- Low strength per unit of weight of concrete leads to heavy members.
- Properties of concrete vary widely because of variations in its proportioning and mixing.

7.4 MATERIALS OF CONCRETE

7.4.1 PROPERTIES OF FRESH AND HARDENED CONCRETE

Concrete is a material composed of aggregates (which may be gravel, sand, and so forth) cemented together. Cement is mixed with water to form a paste. This mixture coats and surrounds the aggregates. A chemical reaction between the cement and water, called hydration, produces heat and causes the mixture to solidify and harden, binding the aggregates into a rigid mass. The cement used for most structural concrete is Portland cement. A portion of the Portland cement is sometimes replaced by fly ash, silica fume, or other supplemental cementitious material.

The properties of the hardened concrete can be affected by a number of factors, but the most important is the ratio of water to cementitious materials. More water is always added to the mix than is necessary for the chemical reaction with the concrete, so that the fresh concrete has a workable consistency. The excess water eventually evaporates, causing shrinkage and making the concrete more porous. As the water content of the cement paste is increased, then the workability of the fresh concrete is also increased, but the strength and durability of the hardened concrete is reduced. Several chemical admixtures, called *plasticizers*, are available that can improve fresh concrete's workability without increasing its water–cement ratio. An alternative use is to reduce the water needed in a mix while maintaining workability, and plasticizers are thus often called water reducers or water-reducing admixtures.

Several other chemical admixtures are used to alter the properties of either the fresh or the hardened concrete. Calcium chloride, for example, may be added to accelerate hydration. But calcium chloride is a source of free chloride ions, and these can cause the steel reinforcement in the concrete to deteriorate. For this reason, design specifications either forbid or limit the amount of calcium chloride that can be used. As the cement paste is mixed, small bubbles of air are trapped in it. Sometimes a chemical admixture called an air entrainer, or air-entraining agent, is added to the paste to increase the creation of these air bubbles. This entrained air makes the resulting concrete more resistant to freeze–thaw deterioration. A secondary benefit is that it also improves the workability of the fresh concrete. Typical air content, by volume, ranges from about 1% to 2% in non-air-entrained concrete to as much as 6% in air-entrained concrete.

Cement hardens more rapidly at higher temperatures. Retarders are sometimes added in hot weather and other cases where it is desirable to slow the rate of hydration. This can help prevent partial hardening of the concrete before pouring is complete. A variety of materials is used for aggregates. (An *aggregate* can refer to all the material, such as gravel, or to one piece of the material, such as a single stone.) The general requirement is that aggregates must be sound, durable, and nonreactive with other constituents in the concrete. Since aggregates are relatively inexpensive, it is desirable that they occupy as much of the concrete's volume as possible. This is accomplished by controlling the gradation of the aggregates in such a way that the voids between the larger aggregates are filled by progressively smaller particles.

Aggregates are classified as fine or coarse. *Fine* aggregates will pass through a sieve of 1/4 in mesh openings, and *coarse* aggregates will not. It is desirable to use the largest aggregates that can be placed without causing segregation or the uneven distribution of coarse aggregates in the mixture as the fresh concrete flows around reinforcing steel, inserts, or other items embedded in the concrete element. For most buildings, bridges, and comparable structures, the maximum size of the aggregates is within a range of about 3 / 4 in to 11 / 2 in. Every aspect of batching, mixing, placing, consolidating, and curing concrete can significantly influence its behavior. It is particularly important to ensure that the temperature of the concrete during curing is within a tolerable range and that the mixture remains moist.

7.4.2   SPECIFYING CONCRETE

Structural concrete is specified in terms of two basic parameters: unit weight, *wc*, and compressive strength, *fc'*.

7.4.2.1   Unit Weight

The *unit weight* of concrete is defined as the weight of a cubic foot of hardened concrete. It is denoted by the symbol *wc*. The type of aggregates used in the concrete controls its unit weight. Unit weights range from 90 lbf/ft³ for structural lightweight concrete up to about 160 lbf/ft³ for normal-weight concrete.

Special applications, such as insulation, require extremely lightweight concrete, but concretes lighter than 90 lbf/ft³ are not permitted for structural applications. Heavyweight concrete uses iron ore or steel slugs for aggregate and yields concrete with a unit weight in excess of 200 lbf/ft³. But heavyweight concrete is rarely encountered in routine design. In the ACI code, distinction is made between all-lightweight and sand-lightweight concrete. All-lightweight concrete contains only lightweight aggregates, whereas sand-lightweight concrete contains lightweight coarse aggregate along with natural sand for the fine aggregate. All-lightweight concrete results in unit weights near the lower bound of 90 lbf/ft³, and sand-lightweight concrete has a unit weight approaching 115 lbf/ft³. The unit weight of plain normal-weight concrete is approximately 145 lbf/ft³. The unit weight of a reinforced concrete member is estimated as 150 lbf/ft³, a slightly heavier unit weight that allows for the fact that the steel in the member is heavier than the concrete it displaces. These values are used throughout this book whenever normal-weight concrete is specified.

7.4.2.2   Specified Compressive Strength

The *specified compressive strength* of concrete is the expected compressive stress at the failure of a cylinder of a standard size that is cast, cured, and tested in accordance with ASTM specifications. The symbol for specified compressive strength is *ft*. The concrete cylinder typically has a diameter of either 4 in or 6 in and a height equal to twice its diameter. Test cylinders are cast using properly selected samples of fresh concrete and cured under controlled temperature and humidity until tested at a specified age, which is usually 28 days. When information on strength gain is needed at earlier or later ages, additional cylinders are made from the same batch sample, and these are tested at intervals (such as at 7 days, 14 days, and so on) to determine the rate of strength gain.

Most structural concrete produced today has a compressive strength from 3000 psi to 6000 psi. Job-cast elements, such as footings, slabs, beams, walls, and so forth, typically use 3000 psi to 4000 psi concrete, while plant-produced precast concrete elements typically use higher-strength 5000 psi to 6000 psi concrete. Strengths significantly above 6000 psi can be achieved, and such concrete is sometimes used in columns and walls of high-rise buildings and other applications when higher performance is needed.

Many equations in ACI 318 refer to the quantity $\sqrt{f_c'}$. By convention, this means the square root of only the numerical value of f_c' as expressed in pounds per square inch (psi). The units themselves are not changed by the operation so that the result is also in psi. If f_c' is given in kips per square inch (ksi), convert to psi before taking the square root. For example, if f_c' equals 4 ksi, f_c' equals $\sqrt{4000}$ psi, or 63.2 psi.

7.5 CEMENT

Cement is a binding material used in construction that hardens and adheres to other materials to form a strong and durable structure. It is typically made from a mixture of limestone, clay, and other materials that are heated in a kiln and then ground into a fine powder. The powder is then mixed with water to create a paste that can be used to bind bricks, stones, and other building materials together. Cement is made by burning a mixture of **limestone ($CaCO_3$)** and **clay** and pulverizing the resulting **clinker** into a very fine powder.

In 1824, Joseph Aspdin patented cement which he called Portland cement, because the render made from it was in a color similar to the prestigious Portland stone on the Isle of Portland, England. However, Aspdin's cement was not like modern Portland cement but was a first step in its development, which was later developed by his son William Aspdin.

Cement is typically classified into two main types: hydraulic and non-hydraulic cement. Hydraulic cement, such as Portland cement, can harden even when submerged in water and is the most commonly used type of cement in construction. Non-hydraulic cement, on the other hand, hardens only through a chemical reaction with carbon dioxide in the air and is typically used in specialty applications. Cement is used in a wide variety of construction applications, including the production of concrete, mortar, and stucco. It is also used in the construction of roads, bridges, dams, and other structures. The production and use of cement can have significant environmental impacts, including the emission of greenhouse gases and the depletion of natural resources. As a result, there is ongoing research into developing more sustainable and eco-friendly cement alternatives.

In the production of cement, the raw materials are first mined and then crushed and blended in specific proportions to create a homogenous mixture. This mixture is then heated in a kiln to a high temperature (about 1,450°C) to cause chemical reactions that result in the formation of clinker, a material that is then ground into a fine powder to produce cement. The use of cement has many advantages in construction, including its ability to form strong and durable structures, its versatility in different applications, and its ability to be produced in large quantities. However, cement production is also associated with several environmental concerns, including greenhouse gas emissions, land use changes, and water use. Efforts are being made to reduce the environmental impacts of cement production through the development of more sustainable production methods, the use of alternative materials, and the implementation of energy-efficient practices. Overall, cement plays a critical role in modern construction, and its properties and uses continue to be studied and refined to meet the changing needs of the construction industry.

The chemical composition of cement can vary depending on the specific type of cement being produced. However, the most common type of cement used in construction is Portland cement, which typically consists of the following compounds:

- Tricalcium silicate (Ca_3SiO_5)
- Dicalcium silicate (Ca_2SiO_4)

| Limestone | Clay | Clinker | Cement |

FIGURE 7.2 Primary ingredients of cement.

- Tricalcium aluminate ($Ca_3Al_2O_6$)
- Tetracalcium aluminoferrite ($Ca_4AlFe_2O_{10}$)

In addition to these compounds, Portland cement may also contain small amounts of other minerals, such as gypsum, limestone, and iron ore. The specific chemical composition of cement can affect its properties, such as its strength, setting time, and durability, which can vary depending on the specific application and environmental conditions.

7.5.1 Types of Cement

- *Type I.* **Normal Portland cement** is used for general construction, having none of the distinguishing qualities of the other types.
- *Type II.* **Moderate Portland cement** is used in general construction, where resistance to moderate sulfate action is required or where heat buildup can be damaging, as in the construction of large piers and heavy retaining walls.
- *Type III.* **High-early-strength Portland cement** cures faster and gains strength earlier than normal Portland cement; it is used when the early removal of formwork is desired, or in cold weather construction, to reduce the time required for protection from low temperatures.
- *Type IV.* **Low-heat Portland cement** generates less heat of hydration than normal Portland cement; it is used in the construction of massive concrete structures, as in gravity dams, where a large buildup in heat can be damaging.
- *Type V.* **Sulfate-resisting Portland cement** is used where resistance to severe sulfate action is required.

7.5.2 Aggregates

Aggregates are a type of material used in construction that are composed of a combination of sand, gravel, crushed stone, slag, or other mineral materials. They are typically mixed with cement, water, and other materials to create concrete, which is used to build a wide range of structures, from buildings and bridges to roads and sidewalks. *Aggregates* are the broad category of coarse particulate materials used in construction, including sand, gravel, crushed stone, etc. Aggregates can be classified into two main categories: fine and coarse aggregates. *Fine* aggregates are typically materials with a particle size smaller than 4.75 mm (0.19 in), while coarse aggregates are typically materials with a particle size between 4.75 mm and 19.0 mm (0.75 in).

Coarse aggregate consists of crushed stone or gravel having a particle size larger than ¼ in.

- *Crushed stones.* Angular in shape since they are obtained by crushing of natural stones.
- *Gravels.* Rounded in shape because of wearing action of wind or water by geological process.

Fine aggregate consists of sand having a particle size less than 1/4 in.

- *Coarse sand.* Grain sizes are larger, usually red yellowish in color.
- *Fine sand.* Grain sizes are much smaller, usually silvery in color.

The properties of aggregates, such as their size, shape, and texture, can affect the properties of the resulting concrete, including its strength, workability, and durability. Therefore, the selection and quality of aggregates are important factors to consider in the design and construction of concrete structures. Aggregates can also be used in other construction applications, such as the production of asphalt, which is used to pave roads, parking lots, and other surfaces. In this application,

Crushed Stones   Gravels

Coarse Sand   Fine Sand

FIGURE 7.3   Typical types of aggregates.

aggregates are combined with bitumen, a sticky, black, and highly viscous liquid, to create a material that is resistant to wear and tear, weathering, and other environmental stresses.

In addition to their importance in construction, the production and transportation of aggregates can have environmental impacts, such as the depletion of natural resources, land use changes, and emissions of particulate matter and greenhouse gases. Therefore, there is a growing interest in using alternative materials or sustainable practices to reduce the environmental impacts of aggregates in construction.

7.5.3   CONCRETE

A solid rock-like material that is made by mixing cement, coarse aggregates, and fine aggregates with sufficient water to cause the cement to set and bind the entire mass.

- Cement acts as a binder material for the concrete.
- Coarse aggregates primarily carry the load that is applied on concrete.
- Fine aggregates act as a filler material of voids of coarse aggregates.
- Water invokes the chemical reaction known as hydration that promotes the bond between all materials of concrete.

Unit weight. Normal concrete has unit weight of 150 lb/ft^3, where lightweight concrete can be up to 120 lb/ft^3.

7.5.4   ADMIXTURES

Admixtures are a type of construction chemical that are added to concrete to improve or alter its properties. They are generally classified into five types:

- *Air-entraining.* Disperses microscopic air bubbles into concrete to improve resistance against cracking in freezing environment.

Casting of Concrete

Hardened concrete

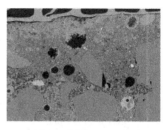

Optical microscopic image of concrete

FIGURE 7.4 Concrete surface.

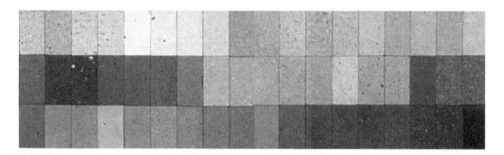

FIGURE 7.5 Concrete surface with different pigmentation.

- *Accelerating.* Promotes early strength development that reduces times required for curing and protection of the concrete and the earlier removal of forms.
- *Retarding.* Prolongs setting time of the concrete for ease of transportation.
- *Plasticizers.* Increases workability of concrete, allowing it to be placed more easily, with less effort.
- *Water reducers.* Reduce the amount of water required to achieve a desired slump and improve the workability of concrete.

Some additional types of admixtures used in concrete for special cases are as follows:

- *Pozzolanic admixtures.* Include materials such as fly ash, silica fume, and slag that react with calcium hydroxide to form additional cementitious compounds and improve the durability and strength of concrete.
- *Corrosion inhibitors.* Help protect reinforcing steel in concrete from corrosion caused by exposure to chlorides and other aggressive chemicals.
- *Shrinkage reducers.* Reduce the amount of drying shrinkage that occurs in concrete as it cures, which can help prevent cracking.
- *Coloring admixtures.* Add pigments to concrete to change color of concrete, for aesthetics.
- *Fiber reinforcement.* Include materials such as steel, glass, and synthetic fibers that improve the toughness, ductility, and crack resistance of concrete.

7.6 MECHANICAL PROPERTIES OF CONCRETE

The design of concrete structures requires an understanding of the behavior of concrete under various states of stress and strain. Of particular importance are the uniaxial compressive stress–strain relationship, the tensile strength, and the volume changes that occur in hardened concrete.

7.6.1 COMPRESSIVE STRESS–STRAIN RELATIONSHIP

The compressive stress–strain relationship is determined from a uniaxial compression test performed on a cylinder of hardened concrete. This requires a stiff test machine, that is, one that will not itself be deflected by the test that is capable of measuring strain beyond the peak compressive stress, f'_c. Figure 7.6 shows representative stress–strain curves for normal-weight concretes having compressive strengths of 3,000 psi, 4,000 psi, and 5,000 psi. The following characteristics are evident.

- Behavior is essentially linearly elastic up to a stress of about $0.65 f'_c$ and becomes distinctly nonlinear beyond that stress.
- The slope of the linear portion (that is, the modulus of elasticity) increases as f'_c increases.
- The compressive strength is reached at a strain of approximately 0.002.
- There is a descending branch of the curve beyond f'_c, reaching an ultimate strain of at least 0.003.

Based on similar tests involving a wide range of compressive strengths and unit weights, the ACI code adopts the following criteria for the design of structural concrete.

The modulus of elasticity (in psi) is defined in ACI Sec. 8.5 by the equation

$$E_c = 33w_c^{1.5} \sqrt{f'_C}$$

In this equation, w_c is the unit weight, in lbf/ft^3, and f'_C is the compressive strength, in psi. The equation was derived empirically, however, and the canceling of units should be disregarded.

7.6.2 ULTIMATE STRAIN OF CONCRETE, ε_c

The ultimate strain in concrete in compression is $\varepsilon_c = 0.003$.

- **Figure 7.7(a):** Fresh concrete is placed into a standard cylindrical mold.
- **Figure 7.7(b):** After initial setting, the concrete cylinder that comes out of the mold has size of 6 in in diameter and 12 in in height. The cylinder is then placed in a water tank for curing.
- **Figure 7.7(c):** After 28 days of initial casting, the test cylinder is then placed in a universal testing machine (UTM). Compressive force is applied from both sides (top and bottom) of the cylinder.
- **Figure 7.7(d):** Due to the high pressure exerted by the UTM, the concrete cylinder crushes.

The compressive stress at which concrete crushes is known as the *compressive strength* (ultimate strength) of concrete, which is denoted by f'_C, and for normal concrete, it is around 4000 psi.

Compressive strength below 3000 psi is not recommended for regular construction, and it could be as high as 20,000 psi.

7.6.3 TENSILE STRENGTH

Concrete is a brittle material, and its tensile strength is small compared to its compressive strength only about a tenth. There are several ways to measure tensile strength, but the most important in design is the *modulus of rupture*. This is the flexural tensile stress at failure in a prism of plain concrete when subjected to pure bending. For convenience, ACI uses an empirical equation, ACI Eq. 9–10, that relates the modulus of rupture, f_r, to the specified compressive strength, f'_C (both in psi).

$$f_r = 7.5\lambda\sqrt{f'_C}$$

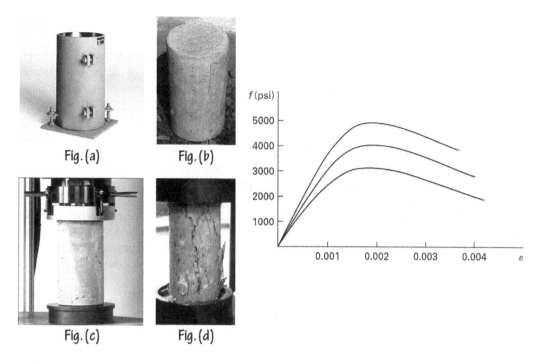

FIGURE 7.6   Representative stress–strain curves for concrete in uniaxial compression.

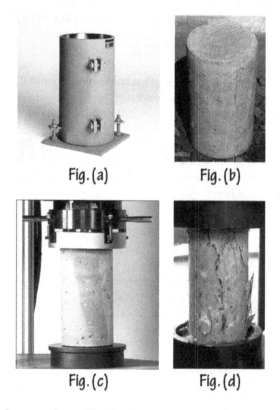

FIGURE 7.7   Strength of concrete is specified by the compressive strength it develops within 28 days after placement. Uniaxial compression test is used to determine the compressive strength of concrete.

The parameter λ is 1.0 for normal-weight concrete, 0.85 for sand-lightweight concrete, and 0.75 for all-lightweight concrete.

The *splitting tensile strength*, f_{ct}, is an alternative measure of tensile strength that may be determined by laboratory test. The splitting tensile strength is defined in terms of the principal tensile stress at failure of a test cylinder of a standard size that is loaded in compression along a main diameter, as shown in Figure 7.7.

7.6.4   Volume Changes

Changes in volume will occur for various reasons in a completed structure. The design of concrete structures, then, must account for the effects of these changes. Depending on circumstances, these effects can be either beneficial or detrimental.

For example, if the deflection of a reinforced concrete beam increases significantly over time, attached finish items may bend and crack. On the other hand, similar changes in volume may "soften" the effects of support settlements, reducing the internal stresses that generally occur. There are three primary sources of volume changes to consider: temperature change, creep, and shrinkage.

7.6.4.1   Temperature Change

The axial deformation of concrete caused by a temperature change ΔT is:

$$\Delta L = C_T(\Delta T)L$$

ΔT is the change in temperature, L is the original unstrained length in the direction under consideration, and C_T is the coefficient of thermal expansion. The coefficient of thermal expansion varies with the aggregate type and the other ingredients in the concrete, but for practical design purposes, it is assumed to be the same as for steel, about 0.000006 in/in-°F.

7.6.4.2   Creep

In the discussion of the compressive stress–strain relationship earlier in this chapter, the duration of loading was assumed to be short. In reality, concrete that must sustain stress for long periods will undergo additional strain, which is termed *creep*. This strain occurs more quickly at first, and the rate of creep gradually diminishes over time.

Several equations have been proposed to predict the amount of creep at a specified time after loading. In the case of plain concrete, the ACI code uses a multiplier, ξ, which is applied to the immediate deformation to predict the cumulative additional deformation at a later time. In the ACI approach, the multiplier is as follows.

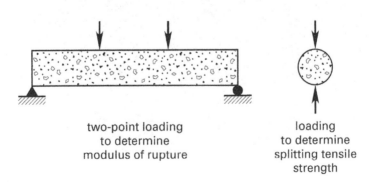

two-point loading
to determine
modulus of rupture

loading
to determine
splitting tensile
strength

FIGURE 7.7a   Test arrangements to determine concrete tensile strength.

$$\xi = 1.0 \text{ for stress sustained 3 months}$$

$$\xi = 1.2 \text{ for stress sustained 6 months}$$

$$\xi = 1.4 \text{ for stress sustained 12 months}$$

$$\xi = 2.0 \text{ for stress sustained 60 months or longer}$$

7.6.4.3 Shrinkage

Shrinkage is strain associated with the evaporation of the excess water in the concrete mixture. The ultimate shrinkage strain depends on the mixture's properties. Typical values range from 0.0004 to 0.0008 in/in. The rate at which the shrinkage occurs depends on many factors, the two most important being the average ambient relative humidity and the volume-to-surface ratio of the structural member.

For structural members of usual proportions, the ultimate shrinkage strain requires several years to develop. Because the expected life of a structure is considerably longer, the expected ultimate strain is usually the controlling value used in design.

7.6.5 PROPERTIES OF REINFORCING STEEL

Concrete is brittle, prone to creep, and relatively weak in tension. Most structural applications, then, require ways of overcoming these deficiencies.

There are two common approaches. The conventional method is to embed steel reinforcement bars, or rebars. The rebars bond with the hardened concrete and reinforce it. An alternative method is to prestress the concrete. This is accomplished by inducing compressive stresses in those regions that will experience tensile stresses when loads are applied.

The current ACI code attempts to unify many of the concepts for these two approaches. Nevertheless, there are many differences in means, methods, and materials of construction, and the two approaches are considered separately in this book. This section describes the essential material properties of steel used for reinforcement. The properties of materials used in prestressing are discussed in a later chapter devoted to prestressed concrete.

7.6.5.1 Reinforcing Bars

Reinforcing bars, or *rebars*, are round steel bars produced by hot rolling. Raised ribs on the surface of the bars, called *deformations*, create a mechanical interlock between the steel and the hardened concrete, helping to maintain the bond between the two. An ASTM specification controls the percentage of the cross section that must comprise the deformations. Figure 7.7b shows schematically a typical reinforcing bar.

Reinforcing bars are designated by a number that gives the number of eighths of an inch in the nominal diameter. For example, a no. 5 bar has a nominal diameter of 5/8 in (0.625 in). Because of

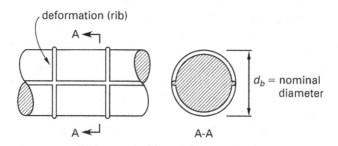

FIGURE 7.7b   Elevation and cross section of typical reinforcing bar.

TABLE 7.1

Properties of Standard Reinforcing Bars (No. 14 and No. 18 Omitted)

bar no.	nominaldiameter, d_b (inch)	nominal area, A_b (in²)	weight, $\left(\dfrac{lbf}{ft}\right)$
3	0.375	0.11	0.38
4	0.500	0.20	0.67
5	0.625	0.31	1.04
6	0.750	0.44	1.50
7	0.875	0.60	2.04
8	1.000	0.79	2.67
9	1.128	1.00	3.40
10	1.270	1.27	4.30
11	1.410	1.56	5.31

the irregularities caused by the deformations, the actual diameter differs from the nominal diameter. For the same reason, the cross-sectional area of the bar, which is an important design property, does not match precisely the area of a circle of the same nominal diameter.

In current practice, deformed bars ranging from no. 3 to no. 11 reinforce structures of usual proportions. Bars of special size, no. 14 and no. 18, are used in exceptionally large or heavily loaded members. Table 7.1 gives the properties of the standard sizes of deformed bars.

7.6.5.2 Smooth Bars and Wire Fabric

Prior to 1971, the ACI Building Code permitted the use of *smooth bars* without deformations as reinforcing bars. These were used primarily in situations where low bond stresses between steel and concrete could be tolerated. The current codes permit smooth bars only in the form of continuous spirals so that bond is not a consideration.

Another type of reinforcement is *welded wire fabric*. This consists of longitudinal and transverse wires that are machine-welded to produce a rectangular grid. The wires may be either smooth or deformed. They are designated by a letter (either *W* for smooth wire or *D* for deformed) followed by a number indicating the cross-sectional area of the wires in hundreds of a square inch. For example, the designation *W5* indicates a smooth wire with a cross-sectional area of 0.05 in²; D30 indicates a deformed wire with a cross-sectional area of 0.30 in².

The designation of wire fabric gives the wire spacing first in the longitudinal direction and then in the transverse direction, followed by the wire sizes for the longitudinal and transverse directions. For example, the designation 12 × 6 W1.4 × W2.5 indicates a fabric with wires spaced 12 in on centers longitudinally and 6 in on centers transversely and providing 0.014 in² per foot of reinforcement in the longitudinal direction and 0.025 in² per 6 in (or 0.050 in² per foot) in the transverse direction. Appendix E of ACI 318 includes a table of commonly used patterns of wire fabric.

7.6.5.3 Mechanical Properties

Reinforcement is specified by a designation that refers first to an appropriate ASTM specification and then to a grade which corresponds to the yield stress of the steel in kips per square inch. A commonly specified reinforcement, for example, is ASTM A615 grade 60, which has a minimum yield stress of 60 ksi. The distinction between specifications has to do primarily with whether the steel will be welded or not. From a design standpoint, the most important item specified is the yield stress.

Unlike concrete, steel does not creep under sustained stress at normal temperatures. Fortunately, the coefficients of thermal expansion for steel and concrete are nearly the same (about 0.000006

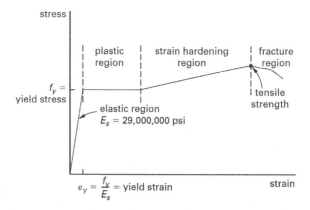

FIGURE 7.8 Idealized uniaxial stress–strain curve for grade 60 reinforcing steel.

in/in-°F), which means that embedded reinforcement can expand and contract with temperature changes without breaking its bond with the surrounding concrete.

The most important properties for reinforcing steel are associated with the uniaxial stress–strain relationship. Figure 7.8 shows the idealized stress–strain relationship for grade 60 reinforcement, the most commonly used grade in modern construction. Curves like this show some important characteristics.

- Steel is linearly elastic practically up to its yield stress.
- The slope of the stress–strain curve, which is the modulus of elasticity, is constant for all steel grades and is equal to 29,000,000 psi or 29,000 ksi. There is a well-defined yield plateau at the yield stress, indicating large plastic deformation.
- Beyond the yield plateau is a region over which stress increases with strain. This increase is called *strain hardening*, and it produces a tensile strength that is significantly larger than the yield stress.
- Reinforcing steel is very ductile and can stretch to about 25 times its original length before fracturing.

7.6.5.3.1 Salient Features

- The stress–strain curve of steel has a definitive yield point, unlike concrete.
- All steels start to yield approximately at 0.002 strain, which is known as *yield strain* $\left(\varepsilon_y\right)$
- The behavior of steel is linear before yield strain, but beyond that, it becomes nonlinear.
- Steel can undergo very large deformation prior to failure, unlike concrete.
- ACI318 code recommends that the tensile strain $\left(\varepsilon_t\right)$ of steel in a concrete beam should have strain of 0.005 or more before failure.

The tensile strength (yield strength) of steel is denoted by f_y, and grade 60 steel ($f_y = 60$ ksi) is very common in construction. Tensile strength is determined by the tension test of steel.

7.7 WHY REINFORCEMENT IS REQUIRED IN CONCRETE STRUCTURE

Concrete is strong in resisting compression but very weak in resisting *tension*. In reinforced concrete analysis, it is assumed that concrete cannot resist any tension at all.

- **Figure 7.9(a), (b):** A continuous beam, shown in Figure 7.9(a), deflects like the shape shown in Figure 7.9(b) due to its self-weight. Crack appears at the bottom at midspan as well as at the top near supports.

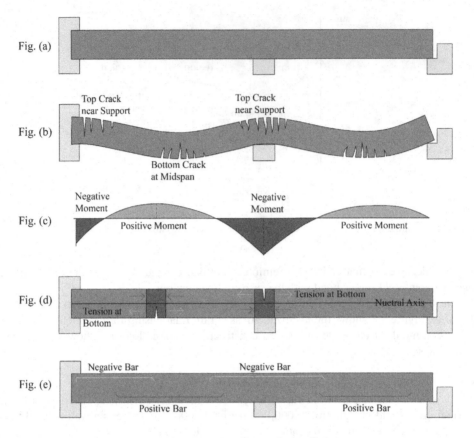

FIGURE 7.9 Reinforcement placement in beam according to the implied moment.

- **Figure 7.9(c):** This happens because positive moment occurs at midspan of the beam and negative moment occurs near support. The moment diagram of the beam is shown in Figure 7.9(c).
- **Figure 7.9(d):** The positive moment causes tension *below* the neutral axis, and the negative moment causes tension *above* the neutral axis. Since concrete is very weak in resisting tension, cracks appear at these tensile zones.
- **Figure 7.9(e):** To help concrete resist tension, steel reinforcements are provided at the crack-prone zones (tensile zones). The reinforcement placed at the bottom of a section at midspan is known as a *positive bar.* The reinforcement placed at the top near the support is known as the *negative bar.*

7.8 ELASTIC ANALYSIS OF BEAM (IF SECTION IS UNCRACKED)

The *elastic analysis* of a beam is a method used to determine the stress and deformation in a beam under different loading conditions. The analysis involves applying the principles of mechanics and the theory of elasticity to calculate the stresses and deflections of a beam when subjected to loads.

To perform an elastic analysis of a beam, the following steps can be followed:

1. *Determine the geometry of the beam.* The geometry of the beam, including its length, cross-sectional shape and size, and material properties, are important factors in the analysis. These parameters are used to calculate the second moment of area of the cross section, which is a measure of the beam's resistance to bending.

2. *Apply the loads.* Determine the magnitude and location of the loads acting on the beam, including point loads, distributed loads, and moments.
3. *Determine the reactions.* Calculate the reactions at the supports of the beam using the equations of statics.
4. *Draw the shear force and bending moment diagrams.* These diagrams show the variation in shear force and bending moment along the length of the beam and are used to determine the internal stresses in the beam.
5. *Calculate the deflections.* Use the differential equation of the deflection curve and the boundary conditions to determine the deflection of the beam at any point.
6. *Calculate the stresses.* Calculate the normal stress, shear stress, and bending stress at any point in the beam using the equations of mechanics and the theory of elasticity.

The elastic analysis of a beam is a complex process that requires a good understanding of mechanics and elasticity theory. Software tools are available that can perform the analysis automatically, but a basic understanding of the underlying principles is still important.

7.8.1 DEVELOPMENT OF TENSILE CRACKS IN RC BEAM

Cracking moment (M_{cr}). The moment at which tensile cracks begin to form, that is, when the tensile stress in the bottom of the beam equals the modulus of rupture (f_r), is referred to as the *cracking moment*.

The *cracking moment* is a term commonly used in structural engineering and refers to the moment at which a reinforced concrete section experiences cracking due to the stresses imposed on it. When a concrete beam is subjected to bending, the tension forces on the lower portion of the beam can exceed the tensile strength of the concrete, causing it to crack. At this point, the beam will continue to deform under load, but the cracks will allow for some degree of rotation at the cracked section, leading to a redistribution of stresses and a change in the beam's behavior.

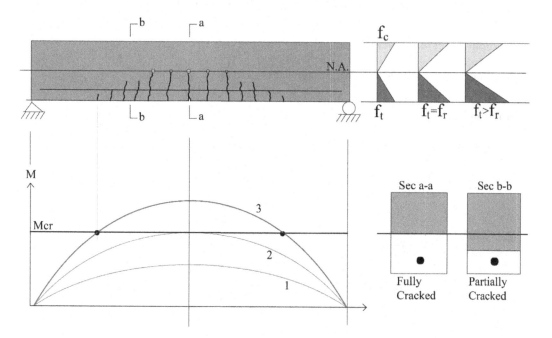

FIGURE 7.10   Visualization of tensile crack in RC beam.

The cracking moment is an important parameter in the design of reinforced concrete beams and is typically calculated based on the beam's dimensions, the properties of the reinforcing steel, and the design loadings. It is important to ensure that the cracking moment is sufficiently high to prevent excessive cracking and ensure adequate structural performance.

Modulus of rupture (f_r). The bending tensile stress at which the concrete begins to crack.

Modulus of rupture is a material property used to measure the flexural strength of a material, such as concrete or ceramics. It is also referred to as flexural strength or bending strength. The *modulus of rupture* is defined as the maximum bending stress that a material can withstand before it fractures. In practical terms, it is the measure of the material's ability to resist bending or flexure under a load. The test to determine the modulus of rupture typically involves placing a test specimen in a three-point bend test machine and applying a load to the center of the specimen until it fractures. The amount of load required to cause the fracture is recorded, and the modulus of rupture is calculated based on the dimensions of the specimen and the maximum load applied.

The modulus of rupture is an important property in material selection and design, particularly in applications where the material will be subjected to bending or flexural stresses, such as in the construction of bridges, buildings, and other structures. It can be used to compare the strength of different materials or to assess the quality of a particular batch of material.

- At small load, the bending moment diagram (curve 1) is far below the cracking moment $a - a$, and the maximum tensile stress (f_t) in the beam is also smaller than the modulus of rupture (f_r).
- As load increases, the moment diagram (curve 2) touches M_{cr}, and at that instant, f_t becomes equal to (f_r); thus, crack begins to form at section $a - a$.
- If load is increased further, the moment diagram (curve 3) rises beyond M_{cr}, and more cracks are formed on both sides of section *a-a*. Theses cracks propagate vertically until they touch the neutral axis (*NA*).
- If the crack propagates fully up to the neutral axis, the section is called a *fully cracked* section. If the crack starts but cannot reach the neutral axis, it is called a *partially cracked* section. If the crack does not start at all, it is called an *uncracked* section.

Example 1 (Section Uncracked)

Question: Determine if the beam (Figure 7.11) has cracked or not, given that $f_c' = 4\,ksi$, $f_y = 60\,ksi$.

Solution

1. Modulus of rupture (tensile stress at which concrete cracks):

$$f_r = 7.5\sqrt{f_c'} = 7.5\sqrt{4000} = 474\,psi$$

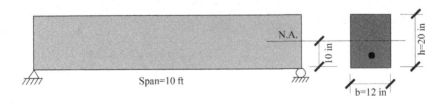

FIGURE 7.11   Beam section for example 1.

Note: While using the following formula, $E_c\sqrt{f_c'}$, the f_c' must be in psi. That is why $f_c' = 4$ ksi is written as 4,000 psi.

2. Moment of inertia of gross section:

$$I_g = \frac{bh^3}{12} = \frac{12 \times 20^3}{12} = 8000\,in^4$$

3. Loading intensity due to self-weight:

$$w = \rho A = \frac{150lb}{ft^3} \times \frac{12 \times 20}{144}ft^2 = \frac{250lb}{ft}$$

Note: The ρ is the density of concrete, which usually equals to 150 lb / f^3. The 1,144 is the conversion factor from in² to ft².

4. Moment at midspan due to self-weight:

$$M = \frac{wL^2}{8} = \frac{\dfrac{250lb}{ft} \times 10^2\,ft^2}{8} = 3125lb - ft$$

5. Stress at the bottommost point of the midspan:

$$\sigma = \frac{My}{I_g} = \frac{(3125 \times 12lb - in) \times 10\,in}{8000\,in^4} = 46.87\,psi$$

6. Since $\sigma(46.87) < f_r(474)$, no crack has occurred yet.

Example 2 (Section Uncracked)

Question: If 1.25 k/ft live load is applied on the beam of example 1, determine whether there is any flexural crack now. If flexural crack still does not occur, then determine the additional load that the beam can carry just before the crack occurs, given that $\sqrt{f_c'} = 4$ ksi, $fy = 60$ ksi.

Solution

Check if concrete has cracked.

1. Modulus of rupture (tensile stress at which concrete cracks):

$$f_r = 7.5\sqrt{f_c'} = 7.5\sqrt{4000} = 474\,psi$$

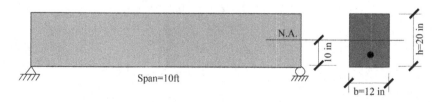

FIGURE 7.12 Beam section for example 2.

2. Moment of inertia of gross section:

$$I_g = \frac{bh^3}{12} = \frac{12 \times 20^3}{12} = 8000\,in^4$$

3. Self-weight:

$$w_{self} = 150 \times \frac{12 \times 20}{144} = \frac{250\,lb}{ft} = \frac{0.25k}{ft}$$

4. Total load per unit length of beam:

$$w_{total} = w_{self} + w_{live} = 0.25 + 1.25 = \frac{1.50k}{ft}$$

5. Moment at midspan:

$$M = \frac{wL^2}{8} = \frac{1.50 \times 10^2}{8} = 18.75k - ft$$

6. Stress at the bottommost point of the midspan:

$$\sigma = \frac{My}{I_g} = \frac{(18.75 \times 12) \times 10}{8000} = 0.281\,ksi = 281\,psi$$

7. Since $\sigma\,(281) < f_r\,(474)$, no crack has occurred yet.

7.8.1.1 Determine Additional Load

8. Cracking moment:

$$M_{cr} = \frac{f_r I}{y} = \frac{474\,psi \times 8000\,in^4}{10in} = 379{,}200\,lb - in = 31.6\,k - ft$$

Note: This is the same formula, $\sigma = \frac{M_y}{I}$. If $M = M_{cr}$, then $\sigma = f_r$, and the formula is rewritten for M_{cr}. The initial result is divided by 1,000 for pound-to-kip conversion, and an additional division by 12 for inch-to-foot conversion.

9. Total load required for cracking:

$$w = \frac{8M_{cr}}{L^2} = \frac{8 \times 31.6k - ft}{10^2 ft^2} = \frac{2.528k}{ft}$$

Note: This is the same formula, $M = wL^2/8$, just rewritten for w.

10. Total live load required:

$$w_{live} = w_{total} - w_{self} = 2.528 - 0.25 = \frac{2.278k}{ft}$$

11. Additional live load required:

$$w_{add.live} = w_{total\,live} - w_{existing\,live} = 2.278 - 1.25 = \frac{1.028\,k}{ft}$$

(Ans)

7.9   ELASTIC ANALYSIS OF BEAM (IF SECTION IS CRACKED)

When a beam with an I-section is cracked, the stiffness of the section is reduced and its behavior under load changes. In this case, elastic analysis can be used to determine the maximum load that the cracked beam can support.

To perform elastic analysis of a cracked I-section beam, the following steps can be followed:

1. Determine the properties of the uncracked section, such as the moment of inertia, cross-sectional area, and modulus of elasticity.
2. Calculate the moment of inertia of the cracked section. This can be done by subtracting the moment of inertia of the crack from the moment of inertia of the uncracked section.
3. Calculate the section modulus of the cracked section. This is the ratio of the moment of inertia to the distance from the neutral axis to the farthest edge of the section.
4. Calculate the maximum bending moment that the cracked beam can withstand. This can be done using the formula $M = (f\_y * Z) / y$, where f_y is the yield strength of the material, Z is the section modulus of the cracked section, and y is the distance from the neutral axis to the extreme fiber.
5. Compare the maximum bending moment to the actual bending moment on the beam. If the actual bending moment is less than the maximum bending moment, the beam is safe. If the actual bending moment is greater than the maximum bending moment, the beam will fail.

It is important to note that elastic analysis assumes that the material of the beam behaves elastically, meaning that it returns to its original shape after being deformed. The material may exhibit plastic behavior, where it does not return to its original shape after being deformed. In such cases, a more advanced analysis, such as plastic analysis, may be required.

7.9.1   BENDING STRESSES IN RC BEAM

Bending stresses in a reinforced concrete beam are caused by the load or loads applied to the beam. When a load is applied to a reinforced concrete beam, the beam will bend, causing tension stresses to develop on the bottom of the beam and compression stresses to develop on the top of the beam. The amount of bending stress depends on the magnitude and distribution of the applied load, as well as the size and shape of the beam.

To resist these bending stresses, steel reinforcing bars (rebars) are placed in the bottom of the beam, where tension stresses develop. The rebars are embedded in the concrete and act to resist the tension stresses that would otherwise cause the concrete to crack and fail. The concrete itself resists the compression stresses that develop on the top of the beam. The amount of reinforcing steel required in a reinforced concrete beam depends on the design loads, the beam dimensions, and the desired strength and stiffness of the beam. The design of reinforced concrete beams typically follows established codes and standards, such as the ACI (American Concrete Institute) Building Code Requirements for Structural Concrete.

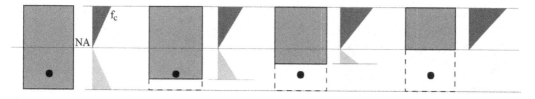

FIGURE 7.13   Visualization of stress in concrete.

Calculating the bending stresses in a reinforced concrete beam involves using the basic principles of mechanics of materials, including the concepts of stress and strain, moment of inertia, and the bending moment equation. The bending moment equation relates the bending moment, beam length, beam depth, and the properties of the cross section of the beam to the maximum bending stress that develops in the beam. The bending stresses in a reinforced concrete beam are resisted by the concrete in compression and the reinforcing steel in tension. The design of reinforced concrete beams involves calculating the required amount of reinforcing steel and determining the maximum bending stress that the beam can safely resist.

7.9.1.1 Stress in Concrete

- *Case (a):* If concrete is uncracked, the flexural stress diagram exists in both the compressive zone (shaded top) and the tensile zone (shaded bottom).
- *Case (b):* As load is increased, the crack propagates upward, and the depth of the effective concrete section (bottom half area) reduces. Since there is no concrete in the cracked zone (dotted line), the stress diagram shrinks in size in tensile zone. The stress diagram in compressive zone increases.
- *Case (c):* The crack continues to propagate upward upon addition of more loads. The stress diagram in the tensile zone shrinks further, and the stress diagram in the compressive zone increases even more.
- *Case (d):* When concrete is fully cracked, that is, the crack has reached the neutral axis, the stress diagram in the tensile zone completely vanishes and only the stress in the compressive zone remains.

7.9.1.2 Stress in Steel

For all cases, from (a) to (d), the stress in steel exists. But its magnitude increases as load is increased.

7.9.2 Transformed Section

- Since the concrete beam is made of composite material, that is, it consists of both concrete and steel, it is therefore necessary to convert it into one homogeneous material before any analysis.
- Usually, the steel is converted into its equivalent concrete, but the other way is also possible, and the resulting section is known as a *transformed section*.
- To convert the steel area $\left(A_s\right)$ into its equivalent area of fictitious concrete, it should be multiplied by the modular ratio (n). The resulting area $\left(nA_s\right)$ is known as the *transformed area*, and it must be at the same depth of the actual steel.
- Though real concrete cannot resist any tension, it is assumed that the fictitious concrete of the transformed area can resist tension.
- The modular ratio is defined as $n = E_s / E_c$, where Es is the elastic modulus of steel and E_c is the elastic modulus of concrete.

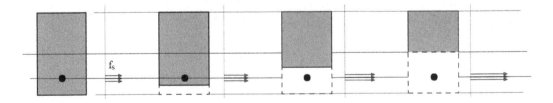

FIGURE 7.14 Visualization of stress in steel.

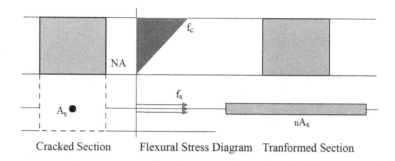

Cracked Section   Flexural Stress Diagram   Tranformed Section

FIGURE 7.15 Visualization of transformed section.

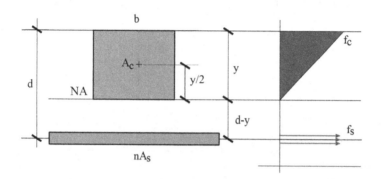

Transformed Section   Flexural Stress Diagram

FIGURE 7.16 Visualization of the neutral axis.

7.9.3 DETERMINATION OF STRESS IN CONCRETE AND STEEL

7.9.3.1 Location of the Neutral Axis

The depth of the neutral axis (y) is unknown before analysis. But the *moment of compression area* (A_c) about the neutral axis always equals the *moment of tensile area* (nAs) about the neutral axis. This property is used to determine y.

Moment of compressive area = moment of tensile area:

$$A_c\left(\frac{y}{2}\right) = (nA_s)(d-y)$$

7.9.3.2 Moment of Inertia

Following is the general formula for the moment of inertia of the transformed section:

$$I = \left[\frac{bh^3}{3}\right]_{conc.} + \left[\frac{bh^3}{12} + Ad^2\right]_{steel}$$

According to the figure, $h = y$ for concrete. Since the height of steel is very small, $bh^3/12$ is very close to zero, thus could be ignored. For steel, $A = nA_s$ and $d = d - y$. Therefore, the general formula becomes:

$$I = \left[\frac{by^3}{3}\right]_{conc.} + \left[(nA_s)(d-y)^2\right]_{steel}$$

7.9.3.3 Stress in Concrete and Steel

$$f_C = \frac{My}{I}, \qquad f_s = n\frac{M(d-y)}{I}$$

Notice the presence of n in the equation of fs. Since the steel was transformed to its equivalent concrete, the concrete stress $M (d\text{-}y)/I$ must be multiplied by n to determine the steel stress.

Example 3 (Section Cracked)

Question: Calculate the maximum bending stresses in concrete and stress in steel of the following beam section by using the transformed area method, given that $n = 9$ and $M = 70\ k - ft$.

Solution

TABLE 7.2
Description of the Symbols

Symbol	Description
y	Depth of neutral axis
$y/2$	Distance between the center of effective concrete area and the neutral axis
A_c	Area of concrete, $Ac = by$
A_s	Area of steel
nA_s	Transformed area of steel
d	Effective depth of beam
$d\text{-}y$	Distance between steel and $n.a.$
f_c	Maximum stress in concrete
f_s	Stress in steel

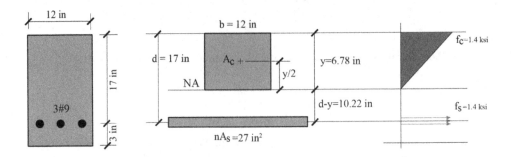

FIGURE 7.17   Beam section for example 3.

7.9.3.4   Determine the Steel Area

Each #9 steel bar has an area of 1.00 in^2. Therefore, three #9s have three times of that amount.

$$A_s = 3 \times 1.00 = 3.00 \text{ in}^2$$

7.9.3.5   Location of the Neutral Axis

$$A_C \left(\frac{y}{2} \right) = (nA_s)(d - y)$$

$$(12 \cdot y) \left(\frac{y}{2} \right) = (9 \times 3.00)(17 - y)$$

$$6y^2 + 27y - 459 = 0$$

Solving, we find $y = 6.78$ in; -11.28 in. Since only the positive value is accepted for length, $y = 6.78$ in.

7.9.3.6   Moment of Inertia of the Transformed Section

$$I = \left[\frac{by^3}{3} \right]_{\text{conc.}} + \left[(nA_s)(d - y)^2 \right]_{\text{steel}}$$

$$= \frac{12 \times 6.78^3}{3} + (9 \times 3.00)(17 - 6.78)^2$$

$$= 4067 \text{ in}^4$$

7.9.3.7   Bending Stresses

$$f_C = \frac{My}{I} = \frac{(70 \times 12)6.78}{4067} = 1.40 \, ksi$$

$$f_s = n \frac{M(d - y)}{I} = 9 \times \frac{(70 \times 12)10.22}{4067} = 19.0 \, ksi$$

7.9.4.   ELASTIC ANALYSIS OF T-BEAM

An *elastic analysis* of a T-beam involves calculating the stresses and deflections of the beam using the principles of elasticity. T-beams are commonly used in construction and are composed of a horizontal flange and a vertical web. The flange resists the bending moment, while the web resists shear forces.

7.9.4.1   Location of the Neutral Axis

To ease the analysis procedure, the T-beam is considered into two parts, the *web part* (shaded in dark gray) and the *flange part* (shaded in light gray).

   Moment of compressive area = moment of tensile area:

$$\left[A_w \left(\frac{y}{2} \right) \right]_{\text{web}} + 2 \left[\left(\frac{A_f}{2} \right) \left(y - \frac{h_f}{2} \right) \right]_{\text{flange}} = (nA_s)(d - y)$$

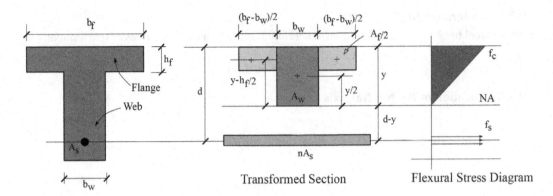

FIGURE 7.18   Visualization of the neutral axis for the T-beam.

TABLE 7.3
Descrption of the Symbols

Symbol	Description
b_w	Width of web
b_f	Width of flange
A_w	Area of web, $A_w = b_w y$
A_f	Area of flange, $A_f = (b_w\text{-}b_f)\, h_f$

Notice the presence of 2 in the flange part. Since only one side of the flange is considered in the expression, it must be multiplied by 2 to account for the other side of the flange.

7.9.4.2   Moment of Inertia

$$I = \left[\frac{b_w y^3}{3}\right]_{web} + 2\left[\frac{\dfrac{(b_f - b_w)}{2}\cdot h_f^3}{12} + \left(\frac{A_f}{2}\right)\left(y - \frac{h_f}{2}\right)^2\right]_{flange} + \left[(nA_s)(d - y)^2\right]_{steel}$$

Though the expressions of the T-beam look scary, in numeric problems, they turn out to be quite simple.

Example 4 (Section Cracked)

Question: Calculate the bending stresses in the T-beam by using the transformed area method, $n = 9$ and $M = 250\ k - ft$.

Solution

Transformed steel area:

$$nA_s = 9 \times (6 \times 1.00) = 72.00\ in^2$$

Width of each flange block:

$$\frac{b_f - b_w}{2} = \frac{60 - 10}{2} = 25 \text{ in}$$

Therefore, the area of each flange block (shaded light gray) is:

$$\frac{A_f}{2} = 25 \text{x} 5 = 125 \; in^2$$

7.9.4.3   Location of the Neutral Axis

$$\frac{b_f - b_w}{2} = \frac{60 - 10}{2} = 25 \text{ in}$$

$$(10 \cdot y)\left(\frac{y}{2}\right) + 2(125)(y - 2.5) = 9(6.00)(28 - y)$$

$$5y^2 + 304y - 2137 = 0$$

So y = 6.36 in.

Distance from the flange center to the neutral axis:

$$y - \frac{h_f}{2} = 6.36 - \frac{5}{2} = 3.86 \text{ in}$$

7.9.4.4   Moment of Inertia of the Transformed Section

$$I = \frac{10 \times 6.36^3}{3} + 2\left[\frac{25 \times 5^3}{12} + 125 \times 3.86^2\right] + 72.00 \times 21.64^2$$

$$= 30,391 \text{ in}^4$$

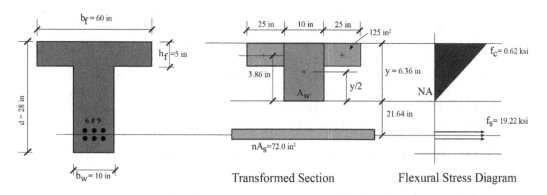

FIGURE 7.19   T-beam section for example 4.

7.9.4.5   Bending Stresses

$$f_c = \frac{My}{l} = \frac{(250 \times 12)6.36}{30{,}391} = 0.62\,\text{ksi}$$

$$f_s = n\frac{My}{l} = 9 \times \frac{(250 \times 12)21.64}{30{,}391} = 19.22\,\text{ksi}$$

7.10   ULTIMATE STRENGTH OF THE RECTANGULAR BEAM

7.10.1   ULTIMATE STRENGTH OF THE BEAM

The concept of the ultimate strength of a rectangular beam is based on the idea that every material has a maximum amount of stress it can withstand before it fails. In the case of a rectangular beam, the *ultimate strength* is the maximum bending moment that the beam can withstand before it breaks or permanently deforms.

The ultimate strength of a rectangular beam is affected by several factors, including the material properties, the dimensions of the beam, and the type of loading applied. The yield stress of the material is an important factor in determining the ultimate strength of the beam. The *yield stress* is the maximum stress that a material can withstand before it starts to deform plastically, meaning that it does not return to its original shape when the load is removed. The dimensions of the beam also play a significant role in determining the ultimate strength. The height and width of the beam affect the cross-sectional area and the moment of inertia, which determines how much load the beam can carry before it fails.

The type of loading applied to the beam also affects its ultimate strength. A beam subjected to pure bending will have a different ultimate strength than a beam subjected to a combination of bending and shear. In engineering practice, the ultimate strength of a rectangular beam is typically calculated using mathematical models that take into account the material properties, dimensions, and type of loading. These models can help engineers design beams that are strong enough to withstand the expected loads and avoid failure.

Case (1): At small load, both f_y and f_c are smaller. Flexural diagrams of concrete exist in both tensile (lower) and compressive (upper) zone. No crack in the beam at this stage.

Case (2): As load increases, f_c becomes equal to f_r (point 2 in Figure 7.20b). Therefore, the first crack appears at midspan.

Case (3): Further increase of load causes the crack to propagate upward. The stress diagram in the tensile zone of the concrete shrinks.

Case (4): The crack reaches the neutral axis, resulting in the tensile stress zone of the concrete to completely disappear.

Case (5): More loads cause f_s to become equal to f_y (point 5 in Figure 7.20a). Therefore, the steel has yielded at this stage near 0.002 strain.

Case (6): Additional load causes concrete strain to enter in the nonlinear zone (point 6 in Figure 7.20b). This results in the stress block becoming parabolic in shape instead of triangular.

Case (7): Further load forces the concrete to reach its maximum stress f_c' (point 7 in Figure 7.20b) at 0.003 strain. Only the topmost level of concrete experiences f_c'.

Case (8): At this ultimate stage, the level of f_c' sinks a bit. The concrete reaches its failure stage (point 8 in Figure 7.20b), and the beam fails by the crushing of concrete.

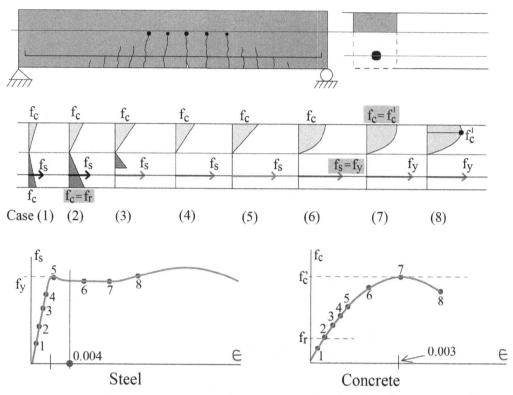

Fig. (a): Stress strain diag. of Steel   Fig. (b): Stress strain diag. of Concrete

FIGURE 7.20   Ultimate strength of a beam visualized.

TABLE 7.4

Description of the Symbols

Symbol	Description
f_c	Stress in concrete
f_r	Modulus of rupture of concrete
f_c'	Compressive strength of concrete
f_s	Stress in steel
f_y	Yield strength of steel

7.10.2   Whitney's Stress Block

Figure 7.21(b): The actual stress diagram of concrete (shaded light gray) is somewhat difficult to analyze due to its parabolic shape.

Figure 7.21(c): To overcome this, C. S. Whitney proposed to replace the parabolic area by a simple *fictitious rectangular area* which is known as *Whitney's stress block*, given that the center of gravity of both parabolic and rectangular area must coincide. The width of this block is found to be equal to $0.85\ f_c'$ from empirical results, and the height (a) is unknown before analysis.

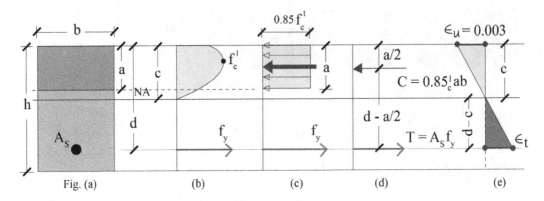

FIGURE 7.21 Whitney's stress block for beam visualized.

TABLE 7.5
Description of the Symbols

Symbol	Description
b	Width of beam
h	Depth of beam
d	Effective depth of beam
A_s	Cross-sectional area of steel
f'_c	Compressive strength of concrete
f_y	Yield strength of steel
c	Depth of neutral axis
a	Depth of Whitney's block
C	Compressive force resisted by concrete
T	Tensile force resisted by steel
ϵ_u	Ultimate strain in concrete
ϵ_t	Strain in tensile reinforcement

The height of Whitney's stress block is determined by force equilibrium condition of a section. The compressive force resisted by concrete (C) must equal the tensile force resisted by steel (T). Therefore:

$$C = T; \qquad 0.85 f'_c ab = A_s f_y; \qquad a = \frac{0.85 f'_c b}{A_s f_y}$$

Note that the depth of Whitney's block (a) is slightly smaller than the depth of the neutral axis (c).

Figure 7.21(d): The compressive force C is halfway the depth of Whitney's block. Therefore, the distance between C and T becomes ($d - a/2$). To determine the nominal moment capacity of the beam, both compressive force (C) or tensile force T could be used.

By using compressive force:

Nominal moment capacity = compressive force × moment arm

$$M_n = C\left(d - \frac{a}{2}\right)$$

$$M_n = 0.85 f_c' ab\left(d - \frac{a}{2}\right)$$

By using tensile force:

Nominal moment capacity = tensile force × moment arm

$$M_n = T\left(d - \frac{a}{2}\right)$$

$$M_n = A_s fy\left(d - \frac{a}{2}\right)$$

Both formulas give the same results, but the latter one is widely used for determining the nominal moment capacity.

Figure 7.21(e): The ultimate strain of concrete ε_u is equal to 0.003, which is found by empirical results.

Since the depth of the neutral axis is c, the distance between the steel and the neutral axis is equal to $(d-c)$. The triangles of Figure 7.21(e) (light gray and dark gray) are similar triangles.

Therefore, we write:

$$\frac{\epsilon_t}{\epsilon_u} = \frac{d-c}{c}$$

$$\epsilon_t = \left(\frac{d-c}{c}\right)\epsilon_u$$

7.10.3 THE β_1 AND $\varnothing$ FACTORS

β_1 factor relates the **depth of neutral axis** (c) to the compressive strength of concrete (f_c').

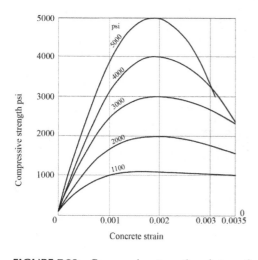
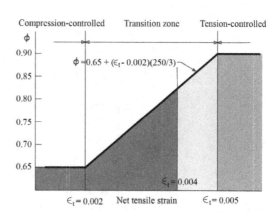

FIGURE 7.22 Compressive strength and strength reduction factor of concrete.

If $f_c' \leq 4000$ psi:

$$\beta_1 = 0.85$$

If $f_c' > 4000$ psi, the following formula should be used, but the resulting β_1 must be greater than or equal to 0.65:

$$\beta_1 = 0.85 - 0.05\left(\frac{f_c' - 4000}{1000}\right) \geq 0.65$$

$\varnothing$ factor accounts for **how far the steel has yielded**.

$$\text{If } \epsilon_t \geq 0.005: \quad \varnothing = 0.90$$

$$\text{If } 0.004 \leq \epsilon_t < 0.005: \quad \phi = 0.65 + \frac{250}{3}\left(\epsilon_t - 0.002\right)$$

Note: The preceding formula is valid only for grade 60 steel (fy = 60 ksi).
    ACI code does not permit $\epsilon_t \leq 0.004$ for beam; that zone is only reserved for column.

Example 5 (Nominal Capacity)

Question: Determine the nominal moment capacity of the beam, given that $f_c' = 4$ ksi, $f_y = 60$ ksi.

Solution

1. Set the compression to equal the tension to find the depth of Whitney's stress block.

$$C = T$$
$$0.85f_c'ab = A_sf_y$$
$$a = \frac{A_sf_y}{0.85f_c'b}$$
$$= \frac{3.00 \times 60}{0.85 \times 4 \times 12}$$
$$= 4.41\,\text{in}$$

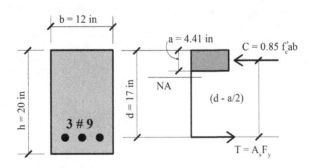

FIGURE 7.23   Beam section for example 5.

2. Determine the nominal moment that the beam can carry.

$$M_n = A_s f_y \left(d - \frac{a}{2} \right)$$
$$= 3.00 \times 60 \left(17 - \frac{4.41}{2} \right)$$
$$= 2663 \, k\text{-in}$$
$$= 221.9 \, k - ft$$

Answer: 221.9 k-ft

Example 6

Design capacity when $\epsilon_t > 0.005$.

Question: Determine the ultimate moment capacity of the following beam, given that $f_C' = 4 \, ksi$; $f_y = 60 \, ksi$.

Solution

1. Find the depth of Whitney's stress block.

$$a = \frac{A_s f_y}{0.85 f_c' b} = \frac{3.00 \times 60}{0.85 \times 4 \times 12} = 4.41 \, in$$

2. Since $f_c' \leq 4000$ psi, $\beta_1 = 0.85$.
3. Find the location of the neutral axis.

$$c = \frac{a}{\beta_1} = \frac{4.41}{0.85} = 5.19 \, in$$

4. Check if the tensile steel has yielded beyond the strain of 0.005.

$$\epsilon_t = \left(\frac{d-c}{c} \right) \epsilon_u = \left(\frac{17-5.19}{5.19} \right) 0.003 = 0.00683$$

5. Since $\epsilon_t \geq 0.005$, $\varnothing = 0.90$.

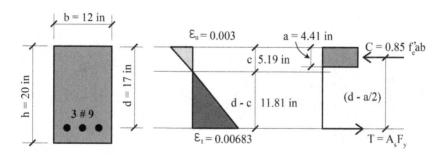

FIGURE 7.24 Beam section for example 6.

6. Determine the ultimate moment capacity.

$$\phi M_n = \phi A_s F_y \left(d - \frac{a}{2} \right)$$

$$= 0.90 \times 3.00 \times 60 \left(17 - \frac{4.41}{2} \right)$$

$$= 2396\, \text{k-in}$$

$$= 199.7\, \text{k} - \text{ft}$$

Answer: 199.7 k-ft

Example 7

Design capacity when $0.004 \le \epsilon_t < 0.005$.
   Question: Determine the ultimate strength of the section, given that $f_C' = 4$ ksi, $f_y = 60$ ksi.

Solution

1. Find the depth of the stress block:

$$a = \frac{A_s f_y}{0.85 f_c' b} = \frac{3.00 \times 60}{0.85 \times 4 \times 10} = 5.29\, \text{in}$$

2. Since $f_C' \le 4000$ psi, $\beta_1 = 0.85$.
3. Find the location of the neutral axis.

$$c = \frac{a}{\beta_1} = \frac{5.29}{0.85} = 6.23\, \text{in}$$

4. Check if the tensile steel has yielded beyond the strain of 0.005.

$$\epsilon_t = \left(\frac{d-c}{c} \right) \epsilon_u = \left(\frac{15 - 6.23}{6.23} \right) 0.003 = 0.00423$$

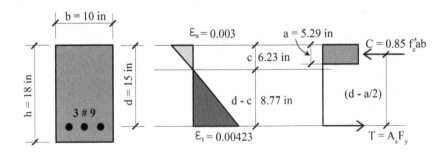

FIGURE 7.25   Beam section for example 7.

5. Since $\epsilon_t < 0.005$, $\varnothing$ does not equal to 0.90 anymore.

$$\phi = 0.65 + (\epsilon_t - 0.002)\frac{250}{3}$$

$$= 0.65 + (0.00423 - 0.002)\frac{250}{3} = 0.836$$

6. Determine the ultimate moment capacity.

$$\phi M_n = \phi A_s F_y \left(d - \frac{a}{2}\right)$$

$$= 0.834 \times 3.00 \times 60\left(15 - \frac{5.29}{2}\right)$$

$$= 1859 k - in$$

$$= 154.9 k - ft$$

Answer: 154.9 k-ft

Example 8

Design capacity when ϵ_t 0.004.
   Question: Determine the ultimate moment capacity of the following beam, given that $f_c' = 4$ ksi, $f_y = 60$ ksi.

Solution

1. Find the depth of the stress block.

$$a = \frac{A_s f_y}{0.85 f_c' b} = \frac{3.00 \times 60}{0.85 \times 4 \times 10} = 5.29 \text{ in}$$

2. Since $f_c' \leq 4000$ psi, $\beta_1 = 0.85$.
3. Find the location of the neutral axis.

$$c = \frac{a}{\beta_1} = \frac{5.29}{0.85} = 6.23 \text{ in}$$

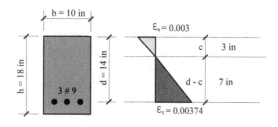

FIGURE 7.26   Beam section for example 8.

4. Check if the tensile steel has yielded beyond the strain of 0.005.

$$\epsilon_t = \left(\frac{d-c}{c}\right)\epsilon_u = \left(\frac{14-6.23}{6.23}\right)0.003 = 0.00374$$

5. Since $\epsilon_t < 0.004$, the section is not ductile. Therefore, the beam **cannot be used**. The size of the beam (specifically depth) should be increased, or the amount of reinforcement should be decreased, so that $\epsilon_t \geq 0.005$ becomes true.

Answer: Increase beam size or reduce reinforcement.

7.11   T-BEAM ANALYSIS

7.11.1   T-Beam

Reinforced concrete floor systems normally consist of slabs and beams that are placed monolithically. As a result, the two parts (slab and beam) act together to resist loads, which is known as T-beam. In effect, the beams have extra widths at their tops, called flanges, and the part of a T-beam below the slab is referred to as the web or stem. Since slabs and beams are cast monolithically, the flange width is unknown before an analysis.

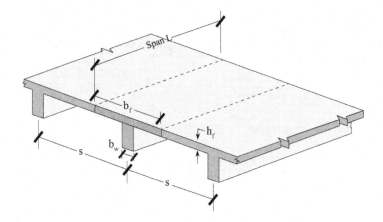

7.11.2   Effective Width, B_f

ACI code suggests using the following guideline to determine the effective width of the T-beam.

$$b_f = \text{Minimum of} \begin{cases} L/4 \\ 16h_f + b_w \\ s \end{cases}$$

Symbol	Description
b_f	Width of flange
h_f	Depth of flange (or slab)
b_w	Width of web
s	Center-to-center spacing of web
L	Span of beam

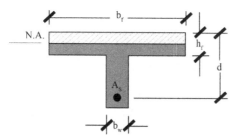

Fig. (a) Rectangular Beam

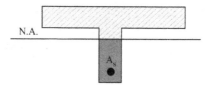

Fig. (b) T- Beam

- Do not be deceived by the looks. A T-beam may behave like a rectangular beam, depending on the location of the neutral axis.
- The neutral axis (*N.A.*) for T-beams can fall either in the flange or in the web, depending on the proportions of the slabs and webs.
- If the neutral axis falls in the flange, Figure 7.28(a), it is effectively a rectangular beam, even though it looks like a T-beam. Because the concrete below the neutral axis is assumed to be cracked and its shape has no effect on the flexure calculations (other than weight), the section above the neutral axis is rectangular that makes it a rectangular beam.
- If the neutral axis is below the flange, Figure 7.28(b), the compression concrete above the neutral axis no longer consists of a single rectangle. Therefore, it is now a true T-beam, and the normal rectangular beam expressions do not apply anymore.

Example 9

Determine the design strength of the T-beam of 30 ft span, given that $f_c' = 4$ ksi, $f_y = 60$ ksi.

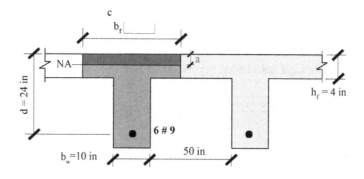

Solution

Find the effective width:

$$
b_f = \text{Min. of} \begin{cases} L/4 = (30 \times 12)/4 = 90 \text{ in} \\ 16h_f + b_w = 16 \times 4 + 10 = 74 \text{ in} \\ s = 50 + 10 = 60 \text{ in (governs)} \end{cases}
$$

7.11.3 Check the Location of the Neutral Axis

Compressive area:

$$
C = T;
$$
$$
0.85 f_c' A_C = A_s f_y \left[0.85 f_c' ab \text{ is written as } 0.85 f_c' A_C \right]
$$
$$
A_C = \frac{A_s f_y}{0.85 f_c'} = \frac{6.00 \times 60}{0.85 \times 4} = 105.8 \text{ in}^2
$$

Flange area:

$$
A_f = b_f h_f = 60 \times 4 = 240 \text{ in}^2
$$

Since $A_C \left(105.8 \text{ in}^2 \right) < A_f \left(240 \text{ in}^2 \right)$, the neutral axis is in the flange. Therefore, the beam is not a true T-beam.

7.11.4 Find the Capacity Like a Rectangular Beam

1. Depth of stress block:

$$
a = \frac{A_s f_y}{0.85 f_c' b} = \frac{6.00 \times 60}{0.85 \times 4 \times 60} = 1.76 \text{ in}
$$

2. Since $f_c' \leq 4000$ psi, $\beta_1 = 0.85$.
3. Location of the neutral axis:

$$
c = \frac{a}{\beta_1} = \frac{1.76}{0.85} = 2.07 \text{ in}
$$

4. Check if the tensile steel has yielded beyond strain of 0.005.

$$
\epsilon_t = \left(\frac{d-c}{c} \right) \epsilon_u = \left(\frac{24 - 2.07}{2.07} \right) 0.003 = 0.0318
$$

5. Since $\epsilon_t \geq 0.005$, $\phi = 0.90$.

6. Design the moment capacity:

$$\phi M_n = \phi A_s F_y \left(d - \frac{a}{2} \right)$$

$$= 0.90 \times 6.00 \times 60 \left(24 - \frac{1.76}{2} \right)$$

$$= 7491 \text{k} - \text{in} = 624.3 \text{k} - \text{ft (Ans.)}$$

7.11.5 ANALYSIS OF T-BEAM

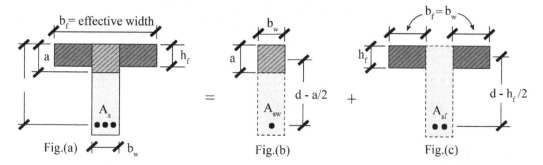

Fig.(a)   Fig.(b)   Fig.(c)

- Figure 7.30(a): To ease the analysis procedure, a T-beam is divided into a set of rectangular parts consisting of the overhanging parts of the flange (chcheckered light gray) and the compression part of the web (chcheckered light gray).
- Figure 7.30(b): The compressive area (small gray area on the top of the web) is just like the compressive area of a rectangular beam. Therefore, the compressive force in the web is:

$$C_w = 0.85 f'_c a b_w$$

- Figure 7.30(c): The compressive areas (chcheckered light gray) are nothing but two rectangular compressive areas that are split apart. Imagine that they are joined together, thus forming a single rectangular block. Therefore, the compressive force in the flange is:

$$C_f = 0.85 f'_c \left(b_f - b_w \right) h_f$$

- The nominal moment, M_n, is determined by multiplying C_w and C_f by their respective lever arms from their centroids to the centroid of the steel.

$$M_n = C_w \left(d - \frac{a}{2} \right) + C_f \left(d - \frac{h_f}{2} \right)$$

- Therefore, the design capacity of the T-beam is:

$$\phi M_n = \phi \left[C_w \left(d - \frac{a}{2} \right) + C_f \left(d - \frac{h_f}{2} \right) \right]$$

Note: The expression of the design moment capacity, ϕM_n, could also be developed using the steel area that balances the web compression (A_{sw}) and the steel area that balances the flange compression (A_{sf}), but the preceding form is simpler to use.

Example 10

Determine the design strength of the T-beam of 10ft span, given that $f'_c = 3000$ psi and $f_y = 60,000$ psi.

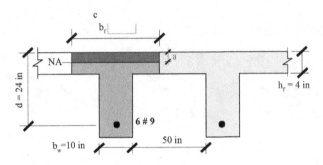

Solution

Find effective width:

$$b_f = \text{Min. of} \begin{cases} L/4 = (10 \times 12)/4 = 30 \text{ in (governs)} \\ 16h_f + b_w = 16 \times 4 + 10 = 74 \text{ in} \\ s = 50 + 10 = 60 \text{ in} \end{cases}$$

Check location of the neutral axis:

$$A_c = \frac{A_s f_y}{0.85 f'_c} = \frac{6 \times 60}{0.85 \times 3} = 141.17 \text{ in}^2$$

$$A_f = b_f h_f = 30 \times 4 = 120 \text{ in}^2$$

Since $A_c (141.17 \text{ in}^2) > A_f (120 \text{ in}^2)$, the neutral axis is in the web. Therefore, the beam is indeed a true T-beam.

Find the depth of the neutral axis:

$$y = (A_c - A_f)/b_w = (141.17 - 120)/10 = 2.12 \text{ in}$$

$$a = h_f + y = 4 + 2.12 = 6.12 \text{ in}$$

7.11.6   CHECK STRAIN

$$c = \frac{a}{\beta_1} = \frac{6.12}{0.85} = 7.20 \text{ in}$$

$$\epsilon_t = \left(\frac{d - c}{c}\right)\epsilon_u = \left(\frac{18 - 7.2}{7.2}\right)0.003 = 0.0045$$

Since $\epsilon_t < 0.005, \phi$ should be determined.

$$\phi = 0.65 + \frac{250}{3}\left(\epsilon_t - 0.002\right)$$

$$= 0.65 + \frac{250}{3}\left(0.0045 - 0.002\right) = 0.858$$

Find the compressive forces:

$$C_f = 0.85 f_c'\left(b_f - b_w\right)h_f$$

$$= 0.85 \times 3 \times \left(30 - 10\right) \times 4 = 204 \text{ kip}$$

$$C_w = 0.85 f_c' a b_w$$

$$= 0.85 \times 3 \times 6.12 \times 10 = 156.1 \text{ kip}$$

7.11.7 DETERMINE CAPACITY

$$\phi M_n = \phi\left[C_w\left(d - \frac{a}{2}\right) + C_f\left(d - \frac{h_f}{2}\right)\right]$$

$$= 0.858\left[156.1\left(18 - \frac{6.12}{2}\right) + 204\left(18 - \frac{4}{2}\right)\right]$$

$$= 4801\,\text{k-in} = 400.1\,\text{k} - \text{ft (Ans.)}$$

7.12 DOUBLY REINFORCED BEAM

7.12.1 DESIGN CAPACITY OF DOUBLY REINFORCED BEAM

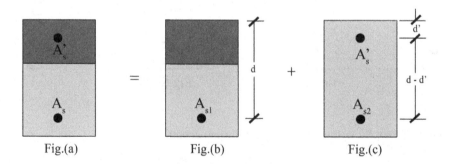

Fig.(a)   Fig.(b)   Fig.(c)

- Figure 7.32(a): Beams with both tensile and compressive steel are referred to as doubly reinforced beams. The steel that is used on the compression sides of the beams is called the compression steel $\left(A_s'\right)\left(A_s\right)$
- The nominal resisting moment of the beam is assumed to consist of two parts, shown in Figure 7.32(b) and Figure 7.32(c).
- Figure 7.32 (b): The first part is the moment $\left(M_{n1}\right)$ resisted by compression concrete (shaded dark gray) and the balancing tensile reinforcing $\left(A_{s1}\right)$.

$$M_{n1} = A_{s1} f_y\left(d - \frac{a}{2}\right)$$

- $\left(M_{n2}\right)\left(A'_s\right)$ and the balancing amount of the additional tensile steel $\left(A_{s2}\right)$.

$$M_{n2} = A_{s2}f'_s\left(d-d'\right)$$

- The total design moment capacity is found by adding the two nominal capacities and multiplying by ϕ.

$$\phi M_n = \phi\left(M_{n1} + M_{n2}\right)$$

$$\phi M_n = \phi\left[A_{s1}f_y\left(d-\frac{a}{2}\right) + A_{s2}f'_s\left(d-d'\right)\right]$$

Symbol	Description	Remarks
A_s	Total tensile steel area, $A_s = A_{s1} + A_{s2}$	Known
A'_s	Compression steel area	Known
A_{s1}	Tensile steel area that balances concrete compression	Unknown
A_{s2}	Tensile steel area that balances A'_s	Unknown
f_y	Stress in tensile steel	Known
f'_s	Stress in compressive steel	Unknown

7.12.2   DETERMINATION OF A_{s1} AND A_{s2} (IF COMPRESSION STEEL YIELDS)

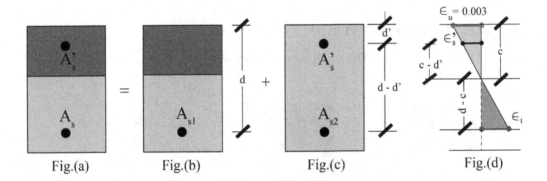

Fig.(a)   Fig.(b)   Fig.(c)   Fig.(d)

- Figure 7.33(c): By writing the force equilibrium equation:

$$C = T;\ A_{s2}f_y = A'_s f'_s$$

- The tensile steel always yields $\left(f_y\right)$, but the stress in the compression steel $\left(f'_s\right)$ may reach yield strength or not.
- Figure 7.33(d): To determine f'_s, we must know the strain in the compression steel $\left(\epsilon'_s\right)$. From the two similar triangles in the compressive zone (shaded light gray on top), we write:

$$\frac{\epsilon'_s}{c-d'} = \frac{\epsilon_u}{c};\ \epsilon'_s = \left(\frac{c-d'}{c}\right)\epsilon_u$$

- If $\epsilon_s' \geq \epsilon_y$, then the compression steel has yielded; if $\epsilon_s' < \epsilon_y$, then the compression steel has not yielded. Here, ϵ_y is the yield strain of steel.
- If $\epsilon_s' \geq \epsilon_y$, then f_s' becomes equal to f_y. Therefore:

$$A_{s2}f_y = A_s'f_s'; \; A_{s2}f_y = A_s'f_y; \; A_{s2} = A_s'$$

- The A_{s1} is now easily found using the following equation:

$$A_s = A_{s1} + A_{s2}; \; A_{s1} = A_s - A_{s2}$$

Note: The case of compression steel not yielded is beyond the scope of this text.

Example 11

Determine the design moment capacity of the doubly reinforced beam, given that compressive steel has yielded and $\epsilon_t > 0.005$. Take $f_c' = 3$ ksi, $f_y = 60$ ksi.

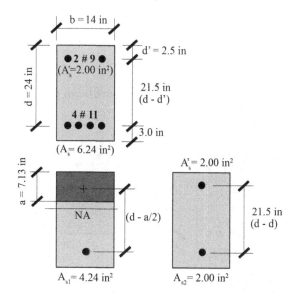

Solution

From the given beam:

$$A_s = 4 \times 1.56 = 6.24 \, \text{in}^2$$
$$A_s' = 2 \times 1.00 = 2.00 \, \text{in}^2$$

Find the depth of Whitney's stress block.
It is given that compression steel has yielded; therefore, $f_s' = f_y = 60$ ksi .

$$T = C; \; T_{\text{steel}} = C_{\text{conc}} + C_{\text{comp. steel}}$$
$$A_s f_y = 0.85 f_c' ab + A_s' f_s'$$
$$6.24 \times 60 = 0.85 \times 3 \times a \times 14 + 2.00 \times 60$$
$$a = 7.13 \, \text{in}$$

7.12.3 Determine Steel Area A_{s1} and A_{s2}

The formula $A_s' = A_{s2}$ applies if compression steel yields.

$$A_{s2} = A_s' = 2.00\,\text{in}^2$$
$$A_{s1} = A_s - A_{s2} = 6.24 - 2.00 = 4.24\,\text{in}^2$$

7.12.4 Determine Capacity

It is given that $\epsilon_t > 0.005$; therefore, $\phi = 0.90$.

$$
\begin{aligned}
\phi M_n &= \phi\left[A_{s1}f_y\left(d - \frac{a}{2}\right) + A_s'f_s'\left(d - d'\right)\right] \\
&= 0.90\left[4.24 \times 60\left(24 - \frac{7.13}{2}\right) + 2 \times 60 \times (24 - 2.5)\right] \\
&= 7000\,\text{k} - \text{in} = 583.4\,\text{k} - \text{ft (Ans.)}
\end{aligned}
$$

Example 12

Determine the design moment capacity of the doubly reinforced beam, given that $f_c' = 3\,\text{ksi}, f_y = 60\,\text{ksi}$.

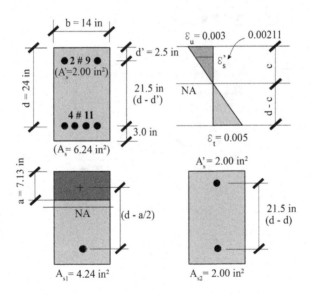

Solution

It is the same beam of the previous problem, example 11. But it is not given that the compression steel yields, and the validity of $\epsilon_t > 0.005$ is also unknown. We must verify these ourselves.

It will be assumed that the compression steel has yielded, $f_s' = f_y = 60$ ksi, and the assumption will be checked later.

7.12.5 FIND THE DEPTH OF WHITNEY'S STRESS BLOCK

$$A_s f_y = 0.85 f_c' ab + A_s' f_s'$$
$$(4 \times 1.56) \times 60 = 0.85 \times 3 \times a \times 14 + 2.00 \times 60; \ a = 7.13 \text{ in}$$

7.12.6 FIND THE DEPTH OF THE NEUTRAL AXIS

$$c = \frac{a}{\beta} = \frac{7.13}{0.85} = 8.39 \text{ in}$$

7.12.7 CHECK THE COMPRESSION STEEL STRAIN

$$\epsilon_s' = \left(\frac{c - d'}{c}\right)\epsilon_u = \left(\frac{8.39 - 2.5}{8.39}\right)0.003 = 0.00211$$

$$\epsilon_y = \frac{f_y}{E} = \frac{60 \text{ ksi}}{29000 \text{ ksi}} = 0.00206$$

Since $\epsilon_s' > \epsilon_y$, the compression steel has yielded. Therefore, the assumption $f_s' = f_y = 60$ ksi was correct.

$$A_{s2} = A_s' = 2.00 \text{ in}^2$$
$$A_{s1} = A_s - A_{s2} = 6.24 - 2.00 = 4.24 \text{ in}^2$$

7.12.8 CHECK THE TENSILE STEEL STRAIN

$$\epsilon_t = \left(\frac{d - c}{c}\right)\epsilon_u = \left(\frac{24 - 8.39}{8.39}\right)0.003 = 0.0055$$

Since $\epsilon_t > 0.005$, therefore, $\phi = 0.90$.

7.12.9 DETERMINE CAPACITY

$$\phi M_n = \phi\left[A_{s1} f_y\left(d - \frac{a}{2}\right) + A_s' f_y\left(d - d'\right)\right]$$

$$= 0.90\left[4.24 \times 60\left(24 - \frac{7.13}{2}\right) + 2 \times 60 \times (24 - 2.5)\right]$$

$$= 7000 \text{ k-in} = 583.4 \text{ k-ft (Ans.)}$$

7.13   RECTANGULAR BEAM DESIGN

7.13.1   GUIDELINE FOR BEAM DESIGN

Designing a beam according to ACl – 318 requires meeting a lot of specifications. Many of these are beyond the scope of this text; therefore, only the very basic requirements are listed here, in simplified form.

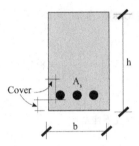

7.13.2   BEAM DEPTH

Beam depth (h) is usually around span/12 as a rule of thumb, but ACl – 318 suggests to use the following table as a guideline for minimum depth. Overall dimensions are selected to whole inches and are incremented by 1 in to 2 in to simplify the construction process by using uniform formwork.

Simply Supported	One End Continuous	Both End Continuous	Cantilever
$\ell/16$	$\ell/18.5$	$\ell/21$	$\ell/8$

Span length is denoted by ℓ.

7.13.3   BEAM PROPORTIONS

Unless architectural or other requirements dictate, the depth width ratio is $1\frac{1}{2}$ to 2 for shorter beams (up to 25ft length). For longer spans, better economy is usually obtained if a deep section, such as 3 to 4 times of the widths, is used.

7.13.4   BAR SIZES

For the usual situations, bars of sizes #11 and smaller are practical. It is usually convenient to use bars of one size only in a beam, although occasionally two sizes of bars are used.

7.13.5   MINIMUM REINFORCEMENT

ACl (10.5.1) specifies a certain minimum amount of reinforcing that must be used at every section of flexural members even if required tensile reinforcing is less than minimum.

$$A_{s,\min} = \frac{3\sqrt{f_c'}}{f_y} b_w d$$

7.13.6 COVER

To protect reinforcement from fire and corrosion, it is located at certain minimum distances from the surface of the concrete, which is known as the *cover*. The code requires a minimum cover of $1\frac{1}{2}$ in for a beam.

7.13.7 REINFORCEMENT RATIO, NOMINAL FLEXURAL RESISTANCE FACTOR

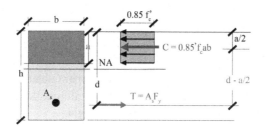

7.13.8 REINFORCEMENT RATIO, (ρ)

Reinforcement ratio is defined as the total amount of tensile area $\left(A_s\right)$ over effective concrete area $\left(bd\right)$.

$$\rho = \frac{A_s}{bd}; \ A_s = \rho bd$$

7.13.9 NOMINAL FLEXURAL RESISTANCE FACTOR, $\left(R_n\right)$

Recalling the expression of depth of Whitney's stress block and substituting $A_s = \rho bd$:

$$a = \frac{A_s f_y}{0.85 f_c' b} = \frac{\left(\rho bd\right) f_y}{0.85 f_c' b} = \frac{\rho d f_y}{0.85 f_c'}$$

Recalling the expression of design moment capacity and substituting A_s and a found earlier:

$$\phi M_n = M_u = \phi A_s f_y \left(d - \frac{a}{2}\right) = \phi \left(\rho bd\right) f_y \left(d - \frac{1}{2} \times \frac{\rho d f_y}{0.85 f_c'}\right)$$

$$M_u = \phi bd^2 \rho f_y \left(1 - \frac{\rho f_y}{1.7 f_c'}\right)$$

Rearranging:

$$\frac{M_u}{\phi bd^2} = \rho f_y \left(1 - \frac{\rho f_y}{1.7 f_c'}\right)$$

The expression on the left-hand side is defined as the nominal flexural resistance factor $\left(R_n\right)$. Therefore:

$$R_n = \frac{M_u}{\phi bd^2}$$

Substituting R_n:

$$R_n = \rho f_y \left(1 - \frac{\rho f_y}{1.7 f_c'}\right)$$

Now, solving ρ, we find:

$$\rho = \frac{0.85 f_c'}{f_y}\left[1 - \sqrt{1 - \frac{2R_n}{0.85 f_c'}}\right]$$

7.13.10 BEAM DESIGN

Beam design means that we need to choose the geometric parameters of the beam, such as width (b), height (h), effective depth (d), and provide adequate reinforcing steel (A_s) so that the beam can carry the imposed load upon it safely, including its self-weight.

7.13.11 STEPS IN BEAM DESIGN

1. Choose the beam depth (h), width (b), and effective depth (d).
2. Estimate the design load.
3. Determine the design moment M_u.
4. Assume $\phi = 0.90$, and determine R_n.
5. Determine ρ by substituting R_n into the expression of ρ.
6. Find $A_s = \rho b d$.
7. Validate the assumption of $\phi = 0.90$.
8. Check if the beam capacity is greater than the design moment, that is, $\phi M_n \geq M_u$.

Example 12 (Simply Supported Beam)

Design a rectangular beam of 22ft length which is required to carry dead load of 1k / ft and live load of 2k / ft, given that $f_c' = 4$ksi, $f_y = 60$ ksi.

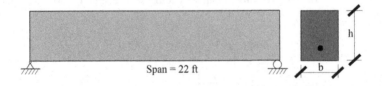

Span = 22 ft

Solution

Estimate the beam size and weight.
   Assume the beam height is span/12.

$$h = 22ft / 12 = 1.833ft = 22\,in$$
$$\therefore h = 22\,in$$

Estimate the effective depth, usually $2.5 \sim 3$ in less than the beam height.

$$d = h - 2.5 = 22 - 2.5 = 19.5 \text{ in}$$

Assume the beam width, usually $50 \sim 65\%$ of the beam height.

$$b = 0.5\,h = 0.50 \times 22 = 11 \text{ in}$$
$$\therefore b = 12 \text{ in.}$$

(Note: We could use $b = 11$ in, but choosing a nice even number is convenient for construction.)
Estimate the beam weight per linear feet.

$$w_{\text{self}} = \left(\frac{12 \times 22}{144}\text{ft}^2\right) \times 150\text{lb}/\text{ft}^3 = 275\text{lb}/\text{ft} = 0.275\text{k}/\text{ft}$$

7.13.12   COMPUTE DESIGN LOAD AND MOMENT

$$w_u = 1.2D + 1.6L$$
$$= 1.2 \times (1 + 0.275) + 1.6 \times 2$$
$$= 4.73\text{k}/\text{ft}$$
$$M_u = \left(w_u L^2\right)/8 = \left(4.73 \times 22^2\right)/8$$
$$= 286.2\text{k} - \text{ft}$$

Determine the reinforcement area (assuming $\phi = 0.90$).

$$R_n = \frac{M_u}{\phi b d^2} = \frac{286.2 \times 12 \text{ k-in}}{0.9 \times 12 \text{ in} \times (19.5 \text{ in})^2}$$
$$= 0.836 \text{ ksi}$$
$$\rho = \frac{0.85 f_c'}{f_y}\left[1 - \sqrt{1 - \frac{2R_n}{0.85 f_c'}}\right]$$
$$= \frac{0.85 \times 4}{60}\left[1 - \sqrt{1 - \frac{2 \times 0.836}{0.85 \times 4}}\right]$$
$$= 0.0163$$
$$A_s = \rho b d = 0.0163 \times 12 \text{ in} \times 19.5 \text{ in}$$
$$= 3.81\text{in}^2$$

We need to provide reinforcement so that it has at least an area of 3.81 in^2. We know that each #9 bar has an area of 1.00 in^2. If we choose four of them, the total area would be 4.00 in^2 which is greater than 3.81 in^2.

Use $4\,\#9$ bars $\left(A_s = 4 \times 1.00 = 4.00\text{in}^2\right)$.

Note: Another bar choice is also possible. For example, each #10 bar has an area of 1.27in^2; therefore, three of them would be $3 \times 1.27 = 3.81\text{in}^2$, which is exactly what we need. Thus, using three #10 bar is also an option. Choosing the appropriate bar size falls into the category of structural detailing, which is far beyond the scope of this text. The author feels that choosing the #9 bar serves the purpose of this example; moreover, it has an area of exactly $1.00\,\text{in}^2$, which makes the calculation easier.

Now, validate the assumption of $\phi = 0.90$.

7.13.13 CHECK STRAIN

$$a = \frac{A_s f_y}{0.85 f_c' b} = \frac{4.00 \times 60}{0.85 \times 4 \times 12} = 5.88 \text{ in}$$

$$c = \frac{a}{\beta_1} = \frac{5.88}{0.85} = 6.92 \text{ in}$$

$$(\beta_1 = 0.85, \text{ since } f_c' \leq 4000 \text{ psi})$$

$$\epsilon_t = \left(\frac{d - c}{c}\right) \epsilon_u$$

$$= \left(\frac{19.5 - 6.92}{6.92}\right) 0.003 = 0.0054$$

Since $\epsilon_t > 0.005$, so $\phi = 0.90$ was okay.

7.13.14 CHECK CAPACITY

$$\phi M_n = \phi A_s F_y \left(d - \frac{a}{2}\right)$$

$$= 0.9 \times 4 \times 60 \times \left(19.5 - \frac{5.88}{2}\right)$$

$$= 3577 \text{ k-in} = 298.1 \text{ k} - \text{ft} > M_u (286.2) \quad \text{O.K.}$$

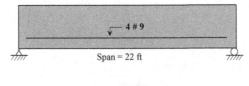

Span = 22 ft

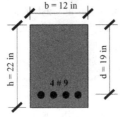

Designed Section

Example 13 (Cantilever Beam)

Design a rectangular cantilever beam of 14 ft length which is required to carry a dead load of $0.5 \text{k}/\text{ft}$ and a live load of $1.0 \text{k}/\text{ft}$, given that $f'_c = 4\text{ksi}$, $f_y = 60$ ksi.

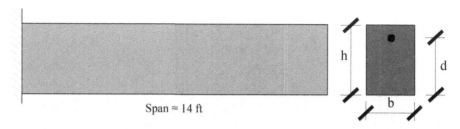

Span = 14 ft

Solution

Since this is a cantilever beam, tensile reinforcement should be at the top of the neutral axis for the entire beam. The effective depth (d) should be measured from the bottom.

7.13.15 ESTIMATE BEAM SIZE AND WEIGHT

Assume beam height span/8 for the cantilever beam.

$$h = 14/8 = 1.75\,\text{ft} = 21\,\text{in}$$
$$\therefore h = 22 \text{ in.}$$

(Note: It is nice to choose an even number for beam sizes.)

Estimate effective depth.

$$d = h - 2.5 = 22 - 2.5 = 19.5 \text{ in}$$

Assume the beam width, usually $50 \sim 60\%$ of the beam height.

$$b = 0.5h = 0.5 \times 22 = 11\,\text{in}$$
$$\therefore b = 12\,\text{in.}$$

Estimate the beam weight per linear feet.

$$w_{\text{self}} = \frac{12 \times 22}{144} \times 150 = 275\,\text{lb}/\text{ft} = 0.275\,\text{k}/\text{ft}$$

7.13.16 COMPUTE DESIGN LOAD AND MOMENT

$$w_u = 1.2\,D + 1.6\,L$$
$$= 1.2 \times (0.5 + 0.275) + 1.6 \times 1.0$$
$$= 2.53\,\text{k}/\text{ft}$$
$$M_u = (w_u L^2)/2 = (2.53 \times 14^2)/2$$
$$= 247.9\,\text{k} - \text{ft}$$

(Note: Unlike a simply supported beam, a cantilever beam has a maximum moment at support which equals $wL^2/2$. The maximum moment in the simply supported beam (like the previous example) occurs at a midspan which equals $wL^2/8$.

Determine reinforcement (assuming, $\phi = 0.90$).

$$R_n = \frac{M_u}{\phi bd^2} = \frac{247.9 \times 12}{0.9 \times 12 \times 19.5^2}$$
$$= 0.724 \, \text{ksi}$$

$$\rho = \frac{0.85 f_c'}{f_y} \left[1 - \sqrt{1 - \frac{2R_n}{0.85 f_c'}} \right]$$
$$= \frac{0.85 \times 4}{60} \left[1 - \sqrt{1 - \frac{2 \times 0.724}{0.85 \times 4}} \right]$$
$$= 0.0137$$

$$A_s = \rho bd = 0.0137 \times 12 \times 19.5$$
$$= 3.21 \, \text{in}^2$$

Use 4 #9 bars $\left(A_s = 4 \times 1.00 = 4.00 \, \text{in}^2 \right)$.
Now, validate the assumption of $\phi = 0.90$.

7.13.17   CHECK STRAIN

$$a = \frac{A_s f_y}{0.85 f_c' b} = \frac{4.00 \times 60}{0.85 \times 4 \times 12} = 5.88 \, \text{in}$$

$$c = \frac{a}{\beta_1} = \frac{5.88}{0.85} = 6.92 \, \text{in}$$

$$\left(\beta_1 = 0.85, \text{ since } f_c' \leq 4000 \, \text{psi} \right)$$

$$\epsilon_t = \left(\frac{d-c}{c} \right) \epsilon_u$$
$$= \left(\frac{19.5 - 6.92}{6.92} \right) 0.003 = 0.00545$$

Since $\epsilon_t > 0.005$, so $\phi = 0.90$ was okay.

7.13.18   CHECK CAPACITY

$$\phi M_n = \phi A_s F_y \left(d - \frac{a}{2} \right)$$
$$= 0.90 \times 4 \times 60 \times \left(19.5 - \frac{5.88}{2} \right)$$
$$= 3576 \, \text{k-in} = 298.1 \, \text{k-ft} > M_u \, (247.9) \quad \text{O.K}$$

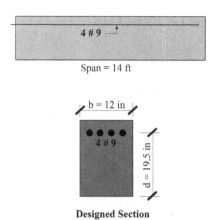

Span = 14 ft

Designed Section

7.14   SHEAR DESIGN

7.14.1   FLEXURAL AND SHEAR CRACKS IN BEAM

Vertical cracks are formed due to flexure, which are known as *flexural crack*.

Inclined cracks are formed due to shear, which are known as *shear crack*.

7.14.2 Typical Reinforcement Layout of Beam

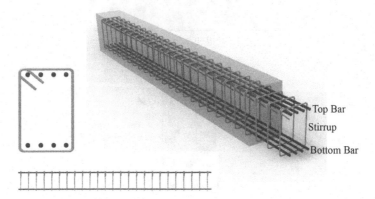

- To resist tensile forces that occur at both the top part and the bottom part of a section, the top bars and bottom bars are provided, respectively. These bars are also known as longitudinal reinforcement.
- Shear forces and torsional forces are resisted by stirrups, which are also called *transverse reinforcement.*
- Unlike top bars and bottom bars, the size of stirrups is selected rather than determined. Usually, #3 and #4 bars are selected for stirrups.
- Based on the selected bar for stirrups, the spacing is then determined.

7.14.3 Shear Strength of Concrete and Shear Force Resisted by Steel

7.14.1.1 Shear of Strength of Concrete (V_C)

- Concrete is very weak in resisting tension, but its shear strength is significantly higher compared to its tensile strength.
- ACl − 318 suggests using the following formula to determine concrete's shear strength,

$$V_c = 2\sqrt{f_c'}bd.$$

- It is a common mistake to include the gross area of concrete (bh) instead of an effective area of concrete (bd) in the preceding formula.
- The unit of f_c' in the preceding formula must be in psi. For example, if $f_c' = 4\,\text{ksi}$ is given, then it should be changed to $f_c' = 4000$ psi while plugging it into the formula.

7.14.4 Shear Force Resisted by Steel (V_S)

- The total shear strength of a concrete beam is denoted by V_n.
- Concrete can provide up to V_c shear force.
- The remaining shear force is provided by steel, which is denoted by V_s. Thus:

$$V_n = V_c + V_s$$

- Multiplying by ϕ on both side:

$$\phi V_n = \phi V_c + \phi V_s$$
$$V_u = \phi V_c + \phi V_s$$
$$V_s = \frac{V_u - \phi V_c}{\phi}$$

- Here, V_u is the design shear force, and $\phi = 0.75$ for shear.
- According to ACI-318, v_u should be calculated at d distance from support, but for the sake of simplicity in this text, V_u will be determined at support.
- Therefore, V_u is nothing but the support reaction. For simply supported beam, $V_u = w_u L / 2$, and for cantilever beam, $V_u = w_u L$.

7.14.5 DETERMINATION OF STIRRUP SPACING

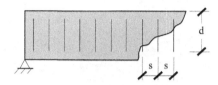

Symbol	Description
V_C	Shear strength of concrete
V_S	Shear force resisted by steel
V_u	Design strength
A_V	Contributing area of a single stirrup to resist shear force
f_y	Yield strength of steel of stirrup
d	Effective depth of beam
s	Spacing of stirrups

1. Find the shear capacity of concrete, $\phi = 0.75$, for shear:

$$\phi V_c = \phi 2\sqrt{f_c'}\, bd$$

2. Find the amount of shear that must be carried by steel:

$$V_s = \frac{V_u - \phi V_c}{\phi}$$

3. Determine the spacing of stirrup:

$$s = \frac{A_V f_y d}{V_s}$$

4. Spacing cannot be greater than $d/2$ or 24 in.
   Note: According to ACI-318 11.4.5.3, there is another $d/4$ or 12 in version of the preceding formula, which is skipped for the sake of simplicity in this text.

The contributing area of stirrup (A_V) is equal to two times the cross-sectional area of the stirrup. Because if an imaginary section (horizontal, vertical, or even inclined) is passed through a stirrup, it will intersect at two cross sections of the stirrup.

For example, a #3 bar has a cross-sectional area of 0.11in^2. If a #3 bar is used for the stirrup, then $A_V = 2 \times 0.11 = 0.22$ in^2.

Example 14

If $V_u = 40$ kip, then determine the spacing of #3 stirrup for the section, given that $f_c' = 3$ ksi and $f_y = 60$ ksi.

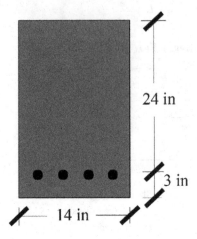

Solution

1. Find the shear capacity of concrete.

$$\phi V_c = \phi 2\sqrt{f_c'}bd$$
$$= 0.75 \times 2\sqrt{3000} \times 14 \times 24 = 27,605\,lb = 27.6\,kip$$

2. Find the amount of shear that must be carried by steel.

$$V_s = \frac{V_u - \phi V_c}{\phi} = \frac{40 - 27.6}{0.75} = 16.53\,kip$$

3. Determine the spacing of stirrup.

$$s = \frac{A_v f_y d}{V_s} = \frac{(2 \times 0.11) \times 60 \times 24}{16.53} = 19.17\,in$$

4. Check the maximum spacing.

$$S_{max} = \begin{cases} d/2 = 24/2 = 12\,in\,(governs) \\ 24\,in \end{cases}$$

Spacing should be the minimum value among 19.17 in, 12 in, and 24 in.

Answer: $s = 12$ in

Example 15

If $V_u = 80\ kip$, then determine the spacing of #3 stirrup for the section, given that $f_c' = 3\ ksi$ and $f_y = 60\ ksi$.

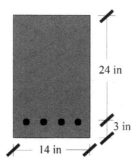

Solution

1. Find the shear capacity of concrete.

$$\phi V_c = \phi 2\sqrt{f_c'}\,bd$$
$$= 0.75 \times 2\sqrt{3000} \times 14 \times 24 = 27,605\,lb = 27.6\,kip$$

2. Find the amount of shear that must be carried by steel.

$$V_s = \frac{V_u - \phi V_c}{\phi} = \frac{80 - 27.6}{0.75} = 69.87\,kip$$

3. Determine the spacing of stirrup.

$$s = \frac{A_v f_y d}{V_s} = \frac{(2 \times 0.11) \times 60 \times 24}{69.87} = 4.53 \text{ in (governs)}$$

4. Check the maximum spacing.

$$s_{max} = \begin{cases} d/2 = 24/2 = 12\text{ in} \\ 24\text{ in} \end{cases}$$

The spacing 4.53 in is the minimum value among 4.53 in, 12 in, and 24 in. Therefore, we could select $s = 4.53$ in. However, measuring such dimension is error-prone and time-consuming in construction. That is why nice, rounded numbers are preferable. For stirrups, it is practiced that spacing should be lower, rounded to the nearest 0.5 in.

Answer: $s = 4.5$ in

Example 16

Design the beam for shear using #3 stirrup. This beam was previously designed for flexure (example 15 of flexural design). Design load $w_u = 4.73k\,/\,ft$ was determined earlier, and f_c' was 4 ksi.

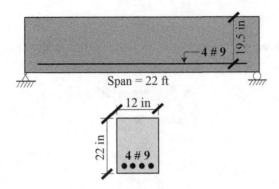

Solution

It is commonly practiced for concrete a beam to divide it into three equal zones and then determine the stirrup spacing for each zone.

- The stirrup spacing in zone 1 will be determined using shear at the support $V_u = w_u L / 2$ (i.e., support reaction).
- For zone 2, it is customary that the stirrup spacing should be double the spacing determined for zone 1.
- For zone 3, the spacing would be the same as zone 1, since the shear diagram is symmetric in this example.

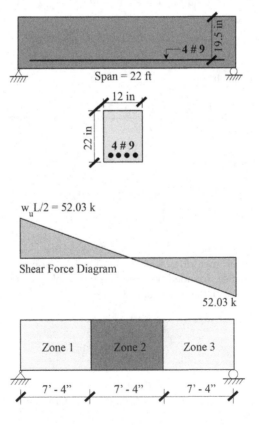

7.14.6   STIRRUP SPACING IN ZONE 1

1. Find the design shear which occurs at support.

$$V_u = \frac{w_u L}{2} = \frac{4.73 \times 22}{2} = 52.03 \text{ kip}$$

2. Find the shear capacity of concrete.

$$\phi V_c = \phi 2 \sqrt{f_c'} bd$$
$$= 0.75 \times 2\sqrt{4000} \times 12 \times 19.5 = 22,199 \text{ lb} = 22.2 \text{ kip}$$

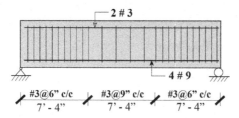

Longitudinal Section

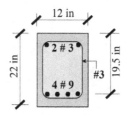

7.14.6.1   Cross Section

3. Find the amount of shear that must be carried by steel.

$$V_s = \frac{V_u - \phi V_c}{\phi} = \frac{52.03 - 22.2}{0.75} = 39.8 \text{ kip}$$

4. Determine the spacing of stirrup.

$$s = \frac{A_v f_y d}{V_s} = \frac{(2 \times 0.11) \times 60 \times 19.5}{39.8} = 6.46 \text{ in}$$

5. Check the maximum spacing.

$$s_{max} = \begin{cases} d/2 = 19.5/2 = 9.75 \text{ in} \\ 24 \text{ in} \end{cases}$$

6. Provide #3 at 6 inch c/c at zone 1.

7.14.7   Stirrup Spacing in Zone 2 and Zone 3

- For zone 2, spacing could be the double of zone 1, that is, $2 \times 6 = 12$ in. But that crosses the limit of maximum spacing, $d/2 = 9.75$ in. Therefore, we cannot provide 12 in. Thus, select 9 in. (Another option would be 9.50 in.)
- In zone 3, the spacing would be the same as zone 1, that is, 6 in.

7.14.8   Hanger Bars

Other than stirrups, we need to provide 2 hanger bars. These bars are not intended to carry any tension or compression but are provided to attach the stirrups with it. Since strength is not a concern, we could just select the thinnest available bar, that is, #3 bar.

7.15   DESIGN OF ONE-WAY SLAB

7.15.1   Slab: One-Way and Two-Way

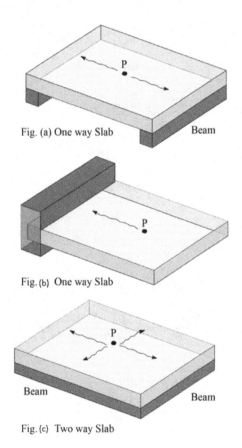

Fig. (a) One way Slab   Beam

Fig. (b) One way Slab

Beam

Beam

Fig. (c) Two way Slab

- *Concrete slabs* are plate structures that are reinforced to span either one or both directions of a structural bay.
- Slab must transfer loads that are placed on it to the support(s). Loads are always transferred to the directions of supports.

- Based on the number of directions of load transfer (flow or propagation), slab could be classified as one-way or two-way.
- Figure 7.51(a): This simply supported slab is carried by two beams from two opposite sides. Any load placed on point P will propagate to the direction of the beams. It is a one-way slab because load propagates only in one direction (one axis). Load cannot propagate to the free edges because there are no beams to carry that load.
- Figure 7.51(b): A cantilever slab, embedded in a wall, is a one-way slab too. Any load placed on point P will propagate to the direction of the wall because it is the only direction where load can go. Load cannot propagate to the three free edges because there are no beams to carry that load.
- Figure 7.51(c): This simply supported slab is supported by four beams on all four edges. Since there are beams on all four sides, any load placed on P will propagate to all four beams. It is a two-way slab because load can flow into two different directions (two different axes).
- Thus, support condition plays an important role to classify slab as one-way or two-way.

7.15.2 SLAB: ONE-WAY AND TWO-WAY

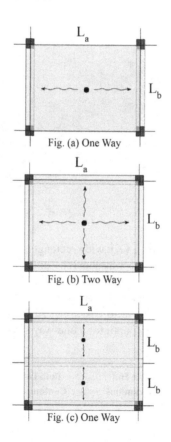

Fig. (a) One Way

Fig. (b) Two Way

Fig. (c) One Way

- But the support conditions alone cannot classify a slab correctly.
- The lengths of slab, L_a and L_b, should be considered, too, if it is supported on all four sides.
- If a rectangular slab is supported on all four sides but the long side is two or more times as long as the short side, for all practical purposes, that slab will act as a one-way slab.

- If the long side is denoted by L_a and the short side is denoted by L_b, the following must be true for a two-way slab, regardless of its support condition:

$$\frac{L_a}{L_b} \le 2$$

- Figure 7.52(a): This simply supported slab is a one-way slab because load is transferred only in one direction. The lengths L_a and L_b are insignificant here.
- Figure 7.52(b): Slab is supported from all four edges; therefore, the length of the sides should be considered. From the figure, it seems that both sides are almost equal in length; therefore, $L_a / L_b < 2$. Thus, it is a two-way slab.
- Figure 7.52(c): This slab is supported from all four edges too; therefore, the length of the sides should be considered. But from the figure, it is clearly visible that $L_a / L_b > 2$; therefore, it is a one-way slab. Actually, there are two one-way slabs in this figure.

7.15.3 DESIGN OF ONE-WAY SLAB

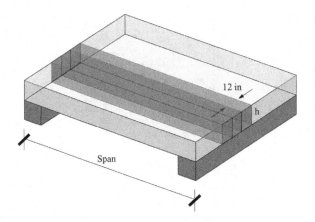

- A one-way slab is assumed to be a very wide rectangular beam.
- To simplify design procedure, a one-way slab is assumed to consist of a series of imaginary beams placed side by side.
- It is customary to consider the width of each of these fictitious beams to be equal to 12 in.
- Slab depth (h) is usually 6 in for most cases, but ACI-318 suggests to use the following table as a guideline for minimum depth for a one-way slab.

Simply Supported	One End Continuous	Both End Continuous	Cantilever
$\ell / 20$	$\ell / 24$	$\ell / 28$	$\ell / 10$

Span length is denoted by ℓ.

- Slab thicknesses are usually rounded off to the nearest $1/4$ in for slabs of 6 in or less in thickness and to the nearest $1/2$ in for slabs thicker than 6 in.
- The span length of one-way slabs is typically between 6ft and 18ft.

7.15.4   MAIN REINFORCEMENT

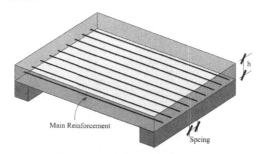

- Main reinforcements are provided to carry the primary loads of one-way slabs.
- They are always aligned to the direction of load propagation.
- For simply supported one-way slab having beam on two opposite directions, the main reinforcement is always from beam to beam.
- Usually, #3 or #4 bars are provided for the main reinforcement.
- Spacing of the main reinforcement cannot be greater than $3h$ or 18 in, where h is the depth of the slab.

7.15.5   TEMPERATURE AND SHRINKAGE REINFORCEMENT

- As concrete hardens, it shrinks. In addition, temperature changes occur that cause expansion and contraction of the concrete.
- The ACI-318 states that shrinkage and temperature reinforcement must be provided in one-way slabs; otherwise, cracks will develop.
- It should be provided in the direction perpendicular to the main reinforcement.
- Following is the total amount of temperature and shrinkage reinforcement (A_s), where b is the slab width and h is the slab depth.

$$A_s = 0.0018\,bh$$

- Usually, #3 bars are provided for temperature and shrinkage reinforcement.
- The spacing of temperature and shrinkage reinforcement cannot be greater than $5h$ or 18 in, where h is the depth of the slab.
- The layer of temperature and shrinkage reinforcement is always placed on top of the layer of the main reinforcement.

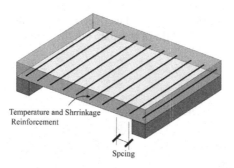

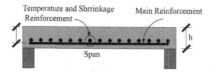

Example 17

Design a simply supported one-way slab of 10 ft span for 200 psf: $f'_c = 4$ ksi, $f_y = 60$ ksi

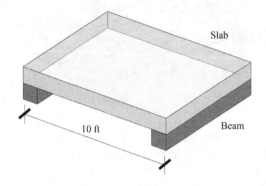

Solution

Estimate slab thickness.
   For simply supported slab, ACI-318 suggests minimum slab thickness of $\ell / 20$.

$$h = \frac{\ell}{20} = \frac{10\text{ ft}}{20} = 0.5\text{ ft} = 6\text{ in}$$

Estimate effective depth.
   It is usually 1 in less than the slab thickness. Therefore:

$$d = h - 1 = 6 - 1 = 5\text{ in}$$

7.15.6 COMPUTE DESIGN LOAD AND MOMENT

$$w_D = h\gamma_{\text{conc.}} = \left(\frac{6}{12}\text{ft}\right)150\text{lb} / \text{ft}^3 = 75\text{ psf}$$

$$w_u = 1.2w_D + 1.6w_L = 1.2 \times 75 + 1.6 \times 200 = 410\text{ psf} = 0.410\text{ ksf}$$

$$M_u = \frac{w_u L^2}{8} = \frac{0.410 \times 10^2}{8} = 5.125\ kft / ft$$

Notice the unit of M_u; it is written as kft/ft. The "per feet" (/ft) part is added to the actual unit because we are designing only 1 ft (12 in) of the slab, not the entire slab.

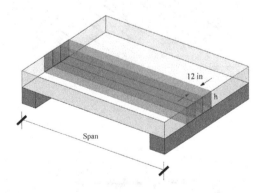

Now, design a simply supported rectangular beam of 12 in width to carry 5.125 kft moment.

7.15.7 Determine Main (Flexural) Reinforcement

$$R_n = \frac{M_u}{\phi b d^2} = \frac{5.125 \times 12}{0.9 \times 12 \times 5^2} = 0.227 \, \text{ksi} / \text{ft}$$

$$\rho = \frac{0.85 \times 4}{60} \left(1 - \sqrt{1 - \frac{2 \times 0.227}{0.85 \times 4}} \right) = 0.00393$$

$$A_s = \rho b d = 0.00393 \times 12 \times 5 = 0.236 \, \text{in}^2 / \text{ft}$$

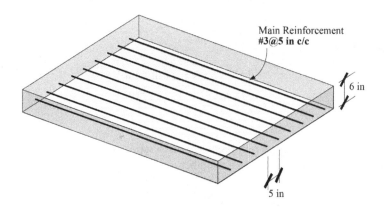

Main Reinforcement
#3@5 in c/c

6 in

5 in

7.15.8 Main Reinforcement

If #3 $\left(0.11 \text{in}^2 \right)$ bar is used, spacing would be:

$$s = \frac{A_{s, \text{provided}}}{A_{s, \text{required}}} = \frac{0.11}{0.236} \times 12 = 5.59 \, \text{in} \approx 5 \, \text{in}$$

7.15.9 Check Spacing

$$s \leq \begin{cases} 3h = 3 \times 6 = 18 \text{ in} & (\text{ok}) \\ 18 \text{ in} & (\text{ok}) \end{cases}$$

Therefore, #3 bars could be used as main reinforcement if they are placed 5 in apart from center to center. More conveniently, it is expressed as:

Use #3 at 5 in c/c.

Note: As an alternate option, we could also choose #4 bar; in that case, spacing would be 10 in.

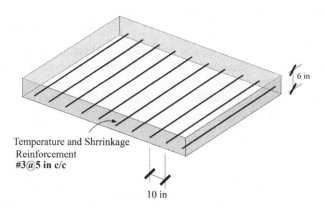

Temperature and Shrrinkage
Reinforcement
#3@5 in c/c

6 in

10 in

7.15.10   Temperature and Shrinkage Reinforcement

$$A_s = 0.0018\,bh = 0.0018 \times 12 \times 6 = 0.129\,\text{in}^2\,/\,\text{ft}$$

If #3 (0.11 in²) bar is used, spacing would be:

$$s = \frac{A_{s,\,\text{provided}}}{A_{s,\,\text{required}}} = \frac{0.11}{0.129} \times 12 = 10.23\,\text{in} \approx 10\,\text{in}$$

7.15.11   Check Spacing

$$s \leq \begin{cases} 5h = 5 \times 6 = 24\,\text{in} & (\text{ok}) \\ 18\,\text{in} & (\text{ok}) \end{cases}$$

Use #3 at 10 in c/c.

The following figure is the actual reinforcement distribution, which looks very congested due to a large amount of reinforcement crossing each other.

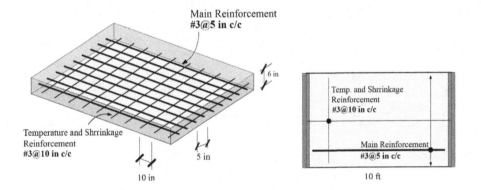

To simplify the drawing of the actual distribution, the following two figures are usually practiced in construction drawing.

Plan: Only one main reinforcement and one temperature and shrinkage reinforcement are shown. Each of the two reinforcements is attached with a double arrow, which shows the direction of distribution of that particular reinforcement.

Longitudinal Section: One main reinforcement and a series of temperature and shrinkage reinforcement on top of it are shown.

Example 18

Design a one-way cantilever slab of 10ft span for 100 psf live load, given that $f_c' = 4\,\text{ksi}, f_y = 60\,\text{ksi}$.

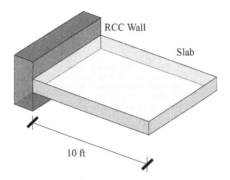

Solution

Estimate the slab thickness.

For a one-way cantilever slab, ACI-318 suggests a minimum slab thickness of $\ell/10$.

$$h = \frac{\ell}{10} = \frac{10\,\text{ft}}{10} = 1\,\text{ft} = 12\,\text{in}$$

Estimate effective depth.

It is usually 1 in less than the slab thickness. Therefore:

$$d = 12 - 1 = 12 - 1 = 11\,\text{in}$$

7.15.12 Compute Design Load and Moment

$$w_D = h\gamma_{\text{conc.}} = (1\,\text{ft})150\,\text{lb}/\text{ft}^3 = 150\,\text{psf}$$

$$w_u = 1.2w_D + 1.6w_L = 1.2 \times 150 + 1.6 \times 100 = 340\,\text{psf}$$

$$M_u = \frac{w_u L^2}{2} = \frac{0.340 \times 10^2}{2} = 17\,\text{kft}/\text{ft}$$

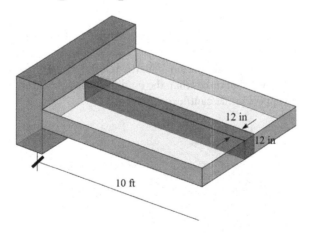

Now, design a rectangular cantilever beam of 12 in width to carry 17kft moment.

7.15.13 Determine Main (Flexural) Reinforcement

$$R_n = \frac{M_u}{\phi b d^2} = \frac{17 \times 12}{0.9 \times 12 \times 11^2} = 0.156\,\text{ksi}/\text{ft}$$

$$\rho = \frac{0.85 \times 4}{60}\left(1 - \sqrt{1 - \frac{2 \times 0.156}{0.85 \times 4}}\right) = 0.00266$$

$$A_s = \rho b d = 0.00266 \times 12 \times 11 = 0.351\,\text{in}^2/\text{ft}$$

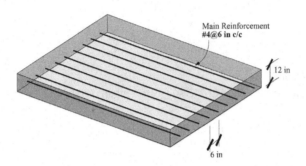

Main Reinforcement
#4@6 in c/c

12 in

6 in

7.15.14 Main Reinforcement

If #4 ($0.20\,\text{in}^2$) bar is used, spacing would be:

$$s = \frac{A_{s,\,provided}}{A_{s,\,required}} = \frac{0.20}{0.351} \times 12 = 6.83\ \text{in} \approx 6\ \text{in}$$

7.15.15 Check Spacing

$$s \leq \begin{cases} 3h = 3 \times 12 = 36\ \text{in} & (\text{ok}) \\ 18\ \text{in} & (\text{ok}) \end{cases}$$

Use #4 at 6 in c/c.

Note: As an alternate option, we could also choose #3 bar; in that case, spacing would be 3 in, which is very congested.

Also notice that, unlike a simply supported slab, the reinforcement layer should be placed near the top surface of the cantilever slab rather than the bottom surface, because flexural cracks form at the top surface near the support in the cantilever slab.

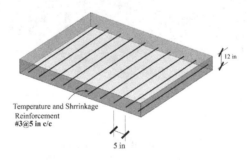

Temperature and Shrrinkage
Reinforcement
#3@5 in c/c

12 in

5 in

7.15.16 TEMPERATURE AND SHRINKAGE REINFORCEMENT

$$A_s = 0.0018\,bh = 0.0018 \times 12 \times 12 = 0.259\,\text{in}^2 \,/\,\text{ft}$$

If #3 (0.11in^2) bar is used, spacing would be:

$$s = \frac{A_{s,\text{ provided}}}{A_{s,\text{ required}}} = \frac{0.11}{0.259} \times 12 = 5.09 \text{ in} \approx 5 \text{ in}$$

7.15.17 CHECK SPACING

$$s \le \begin{cases} 5h = 5 \times 12 = 60 \text{ in} & (\text{ok}) \\ 18 \text{ in} & (\text{ok}) \end{cases}$$

Use #3 at 5 in c/c.

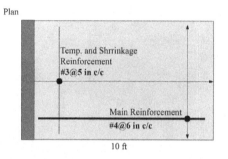

Plan

Temp. and Shrrinkage Reinforcement
#3@5 in c/c

Main Reinforcement
#4@6 in c/c

10 ft

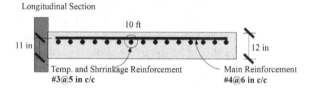

Longitudinal Section

10 ft

11 in

12 in

Temp. and Shrrinkage Reinforcement
#3@5 in c/c

Main Reinforcement
#4@6 in c/c

7.16 DESIGN OF TWO-WAY SLAB

7.16.1 BEAM SLAB

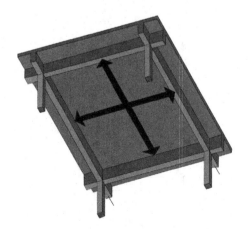

- A two-way slab of uniform thickness may be reinforced in two directions and cast integrally with supporting beams and columns on all four sides of square or nearly square bays.
- Two-way slab and beam construction is effective for medium spans and heavy loads, or when high resistance to lateral forces is required.
- For economy, however, two-way slabs are usually constructed as flat slabs and plates without beams.
- Two-way slabs are most efficient when spanning square or nearly square bays and suitable for carrying intermediate to heavy loads over 15 ft to 40 ft spans.

7.16.2 WAFFLE SLAB

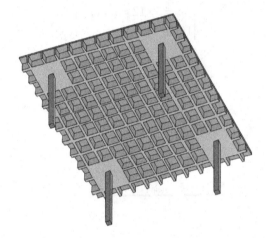

- A *waffle slab* is a two-way concrete slab reinforced by ribs in two directions.
- Waffle slabs are able to carry heavier loads and span longer distances than flat slabs.
- Suitable for spans of 24 ft to 54 ft; longer spans may be possible with posttensioning.
- For maximum efficiency, bays should be square or as nearly square as possible.
- Waffle slabs can be efficiently cantilevered in two directions up to 1/3 of the main span.

7.16.3 FLAT PLATE SLAB

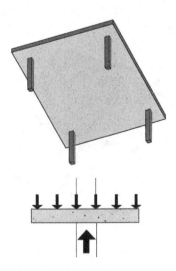

- A *flat plate* is a concrete slab of uniform thickness reinforced in two or more directions and supported directly by columns without beams or girders.
- Simplicity of forming, lower floor-to-floor heights, and some flexibility in column placement make flat plates practical for apartment and hotel construction.
- Suitable for light live to moderate loads over relatively short spans of 12 ft to 24 ft.
- While a regular column grid is most appropriate, some flexibility in column placement is possible.
- Shear at column locations governs the thickness of a flat plate.
- *Punching shear* is the potentially high shearing stress developed by the reactive force of a column on a reinforced concrete slab.

7.16.4 FLAT SLAB

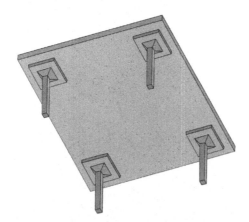

- A *flat slab* is a flat plate thickened at its column supports to increase its shear strength and moment resisting capacity.
- *Drop panel* is the portion of a flat slab thickened around a column head to increase its resistance to punching shear.
- Suitable for relatively heavy loads and spans from 20 ft to 40 ft.

7.16.5 PUNCHING SHEAR FAILURE IN FLAT PLATE

- If flat plate slab is too thin, a column may penetrate through the slab due to high shear stress developed around that column. This shear stress is known as punching shear.
- After failure, a small portion of slab is attached with the column, which is usually larger than the column section in size.
- Punching shear failure occurs when the shear stress developed along the perimeter of this small portion of slab cannot resist the punching shear of the column anymore.
- Punching shear plays a significant role in determining the thickness of a flat plate slab.

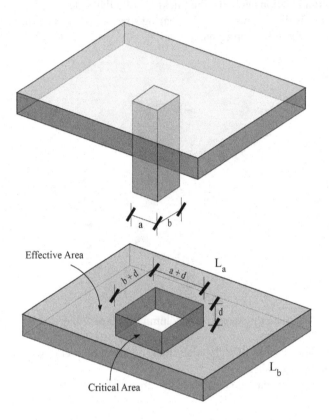

If column dimensions are a and b, then the critical perimeter of a slab for resisting punching shear is $[2(a+d)+2(b+d)]$, where d is the effective thickness of the slab.

- The critical area for resisting punching shear (shown in dark gray in the middle) is:

$$A_{cr} = [2(a+d)+2(b+d)]d$$

- The shear capacity of this critical area is:

$$\phi V_C = \phi 4\sqrt{f_c'}\, A_{cr}$$

Where $\phi = 0.75$. Notice the presence of 4; for beam shear, it was 2.

- If the applied load on the effective area of the slab (shown in light gray) exceeds the shear capacity of the critical area (in the middle), punching shear failure will occur.

- The effective area is gross slab area minus the hole area:

$$A_{\text{eff}} = L_a L_b - (a+d)(b+d)$$

Where L_a and L_b are distances between the column centers in both directions.

- The total load applied on the effective area is:

$$V_u = A_{\text{eff}} q_u$$

Where q_u is the load intensity on the slab.

- To resist punching shear, the following expression must be true; otherwise, the slab will collapse due to punching shear:

$$\phi V_c > V_u$$

Example 19

Determine the minimum thickness of the following interior flat plate slab.

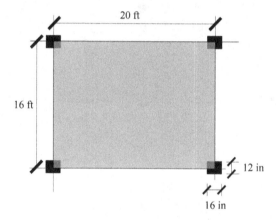

Solution

Assume a service live load equal to 80 psf, a service dead load equal to 110 (including slab weight), $f'_c = 3000$ psi, and $f_y = 60,000$ psi.

7.16.6 Determine Minimum Thickness

For an interior flat plate slab, ACI – 318 Table 9.5(c) suggests to use the following formula for minimum thickness:

$$h = \frac{\ell_n}{33}$$

Where ℓ_n is the larger clear distance between two columns.

Here, we have two clear distances, $(20-16/12)$ ft and $(16-12/12)$ ft, and obviously, the first one is larger.

Therefore:

$$h = \frac{\ell_n}{33} = \frac{20-16/12}{33} = 0.565 \text{ ft} = 6.787 \text{ in};$$

$$\therefore \text{Try 7 inch}$$

But selecting 7 in thickness does not ensure that the slab would resist punching shear failure. Therefore, we need to check punching shear failure by ourselves.

Note: There is another check, known as beam shear check, which does not govern flat plates; therefore, it is skipped in this lecture.

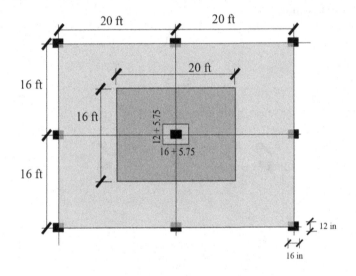

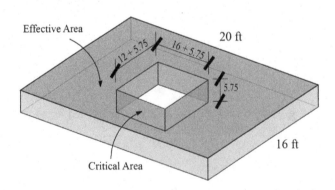

7.16.7 PUNCHING SHEAR CHECK FOR $h = 7$ INCH

1. Effective thickness of the slab is usually 1.25 in less than the total depth for a flat plate slab. Thus:

$$d = h - 1.25 = 7 - 1.25 = 5.75 \text{ in}$$

2. Critical area for resisting punching shear:

$$A_{cr} = [2(a+d)+2(b+d)]d$$
$$= [2(16+5.75)+2(12+5.75)]5.75 = 454.25 \text{ in}^2$$

3. Shear capacity of this critical area:

$$\phi V_c = \phi 4\sqrt{f_C'} A_{cr}$$
$$= 0.75 \times 4 \times \sqrt{3000} \times 454.25 = 74{,}640 = 74.6 \text{ kip}$$

4. Effective area of the slab:

$$A_{eff} = L_a L_a - (a+d)(b+d)$$
$$= 20 \times 16 - (16+5.75)(12+5.75)/144 = 317.32 \text{ ft}^2$$

5. Design loading intensity:

$$q_u = 1.2 \times 110 + 1.6 \times 80 = 260 \text{ psf}$$

6. Total load applied on the effective area:

$$V_u = A_{eff} \cdot q_u = 317.32 \times 260 = 82{,}503 \text{ lb} = 82.5 \text{ kip}$$

Since $\phi V_C (74.6) < V_u (82.5)$, punching shear will occur. Therefore, 7 in thickness is not adequate for the slab. Try $h = 7.5$ in.

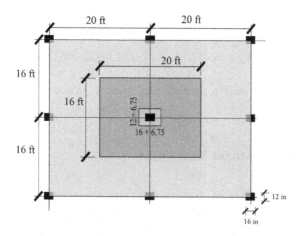

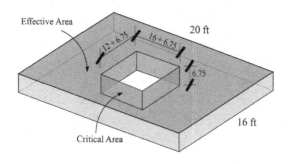

7.16.8 Punching Shear Check for $h = 7.5$ Inch

1. Effective thickness of the slab:

$$d = h - 1.25 = 7.5 - 1.25 = 6.25 \text{ in}$$

2. Critical area for resisting punching shear:

$$A_{cr} = \left[2(a+d)+2(b+d)\right]d$$
$$= \left[2(16+6.25)+2(12+6.25)\right]6.25 = 506.25 \text{ in}^2$$

3. Shear capacity of this critical area:

$$\phi V_C = \phi 4\sqrt{f_C'} A_{cr}$$
$$= 0.75 \times 4 \times \sqrt{3000} \times 506.25 = 83,185 = 83.2 \text{ kip}$$

4. Effective area of slab:

$$A_{eff} = L_a L_b - (a+d)(b+d)$$
$$= 20 \times 16 - (16+6.25)(12+6.25)/144 = 317.18 \text{ ft}^2$$

5. Design loading intensity:

$$q_u = 1.2 \times 110 + 1.6 \times 80 = 260 \text{ psf}$$

6. Total load applied on the effective area:

$$V_u = A_{eff} q_u = 317.18 \times 260 = 82,466 \text{lb} = 82.5 \text{ kip}$$

Now, $\phi V_c (83.2) > V_u (82.5)$; therefore, no punching shear will occur. Thus, $h = 7.5$ in is selected for the interior slab.

7.16.9 Column Strip and Middle Strip

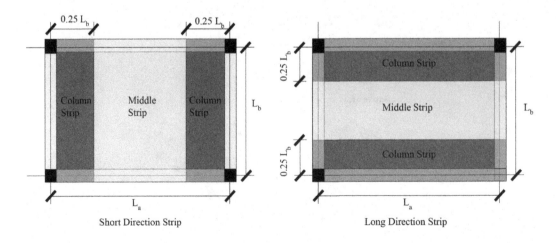

Short Direction Strip   Long Direction Strip

- Unlike a one-way slab, flexural reinforcement is provided in both the long and short directions of a two-way slab.
- The slab is divided into three zones for each direction. The part of the slab that is adjacent to the column is known as the *column strip*, and the other part is known as the *middle strip*.
- The width of the column strip is the same for each direction, that is, $0.25L_b$, where L_b is the smaller dimension of the slab size.

7.16.10   Positive and Negative Moment Zones for Short Direction

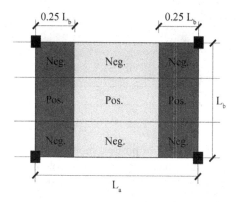

- Each short direction strip is divided into a further three zones. The zones that are in the middle of each strip are positive moment zones, and the rest of the zones are negative moment zones.
- The total static moment in the short direction is $M_s = q_u L_a L_b^2 / 8$, where q_u is the loading intensity on the slab.

Note: $M_s = q_u L_a L_n^2 / 8$ is the actual formula, where L_n is the clear distance in the short direction. For the sake of simplicity in this text, L_b is used instead of L_n.

7.16.11   Distribution of Positive and Negative Moment for Short Direction

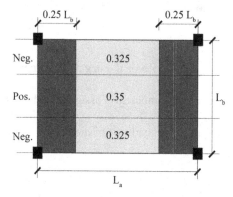

- According to ACI-318 13.6.2, 35% of total static moment is positive for an interior flat plate slab.

- The remaining 65% of total static moment is negative. If we divide that 65% into two negative zones of each strip, each zone gets 32.5%.

7.16.12 Distribution of Positive Moment in Column and Middle Strip

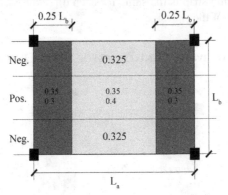

- According to ACI-318 13.6.4.4, 40% of positive moment is assigned to the middle strip.
- The remaining 60% of positive moment is assigned to column strip, i.e., 30% for each side.
- This distribution is written in the second row of the positive moment zone by the values, 0.3, 0.4, and 0.3.

7.16.13 Distribution of Negative Moment in Column and Middle Strip

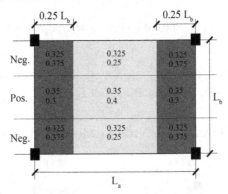

- According to ACl − 318 13.6.4.2, 25% of the negative moment is assigned to the middle strip.
- The remaining 75% of the negative moment is assigned to the column strip, that is, 37.5% for each side.
- This distribution is written in the second row of the negative moment zones by the values 0.375, 0.25, and 0.375.

Example 20

For the flat plate slab in example 1, determine the spacing of reinforcement in the middle strip for the short direction.

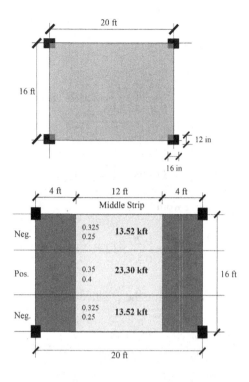

Solution

The width of the column strip is $0.25 L_b = 0.25 \times 16 = 4\,\text{ft}$.

7.16.14   DISTRIBUTION OF MOMENTS

Total moment in short direction:

$$M_s = \frac{q_u L_a L_b^2}{8} = \frac{0.260 \times 20 \times 16^2}{8} = 166.4\,\text{k}-\text{ft}$$

Negative moment in the middle strip for short direction:

$$M_{s,\,\text{mid, neg}} = 0.325 \times 0.25 \times 166.4 = 13.52\,\text{k}-\text{ft}$$

Positive moment in the middle strip for short direction:

$$M_{s,\,\text{mid, pos}} = 0.35 \times 0.4 \times 166.4 = 23.30\,\text{k}-\text{ft}$$

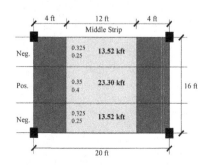

Now, we design the slab by assuming a beam of 12 in width, just like we did for the one-way slab. Determine the flexural reinforcement for the negative moment in the middle strip (assuming $d = 6$ in).

$$R_n = \frac{M_u}{\phi bd^2} = \frac{13.52 \times 12}{0.9 \times 12 \times 6^2} = 0.417 \, \text{ksi} / \text{ft}$$

$$\rho = \frac{0.85 \times 3}{60} \left(1 - \sqrt{1 - \frac{2 \times 0.417}{0.85 \times 3}} \right) = 0.00764$$

$$A_s = \rho bd = 0.00764 \times 12 \times 6 = 0.55 \, \text{in}^2 / \text{ft}$$

If #4 (0.20in^2) bar is used, spacing would be:

$$s = \frac{A_{s,\,provided}}{A_{s,\,required}} = \frac{0.20}{0.55} \times 12 = 4.36 \, \text{in} \approx 4 \, \text{in}$$

Determine the flexural reinforcement for the positive moment in the middle strip (assuming $d = 6$ in).

$$R_n = \frac{M_u}{\phi bd^2} = \frac{23.3 \times 12}{0.9 \times 12 \times 6^2} = 0.719 \, \text{ksi} / \text{ft}$$

$$\rho = \frac{0.85 \times 3}{60} \left(1 - \sqrt{1 - \frac{2 \times 0.719}{0.85 \times 3}} \right) = 0.0144$$

$$A_s = \rho bd = 0.0144 \times 12 \times 6 = 1.04 \, \text{in}^2 / \text{ft}$$

If #4 (0.20 in^2):

$$s = \frac{A_{s,\,provided}}{A_{s,\,required}} = \frac{0.20}{1.04} \times 12 = 2.31 \, \text{in} \approx 2 \, \text{in}$$

8 Design of Steel Structures

Topics to be covered in this chapter are as follows:

- Tension member.
- Compression member.
- Flexural member.
- Connection.

8.1 TENSION MEMBER

8.1.1 EFFECTIVE AREA AND NET AREA

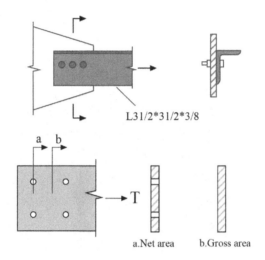

L31/2*31/2*3/8

a.Net area   b.Gross area

Gross Area (A_g): Total available cross-sectional area without any hole.

Net Area (A_n) : Gross area (A_g) minus the whole area A_h.

$$A_n = A_g - A_h$$

Shear Lag: Non-uniform stress distribution, when some part of the section is not connected.

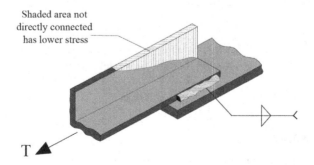

Shaded area not directly connected has lower stress

T

DOI: 10.1201/9781032638072-8

Effective Area (A_e) : Available cross section after applying shear lag factor (U).

$$A_e = A_n U$$

Shear lag factor (U) is determined by the following formula:

$$U = \begin{cases} 1 - \dfrac{\bar{x}}{l} \\ 1.0 \text{ for plate} \end{cases}$$

Here, ℓ is connection length.

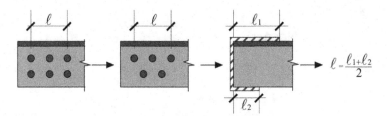

And $\bar{x}$ is the distance between the centroid of the connected part to the connection plane.

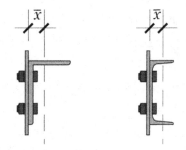

LRFD: load resistance and factor design.
The factored load must be less than or equal to the design strength:

$$\phi P_u \leq \phi_t P_n$$

There are two limit states for tensile member.

8.1.2 YIELDING CRITERIA

If stress on the gross area is large enough, then excessive deformation occurs.

$$\phi_t P_n = 0.90 F_y A_g$$

8.1.3 FRACTURE CRITERIA

If stress on the effective area is large enough, then fracture occurs.

$$\phi_t P_n = 0.75 F_u A_e$$

Example 1

Determine the effective area of the following section.

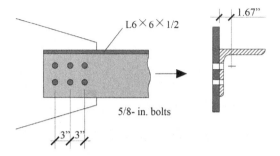

From Table 1–7 of the AISC manual, $A_g = 5.77 \text{ in}^2$, $\bar{y} = \bar{x} = 1.67$ in.

$$A_n = A_g - A_h = 5.77 - \left[\frac{1}{2}\left(\frac{5}{8} + \frac{1}{8}\right)\right]2 = 5.02 \text{ in}^2$$

$$U = 1 - \frac{\bar{x}}{\ell} = 1 - \frac{1.67}{3+3} = 0.7217$$
$$A_e = A_n U = 5.02 \times 0.7217 = 3.63 \text{ in}^2$$

Table 8.1
1-7 Angles Properties (AISC Manual)

L8-L6

Shape	Axis Y-Y						Axis Z-Z			Tan α.	Qs F_y= 36 ksi
	I	S	r		Z	X_p	I	S	r		
	in.⁴	in.³	in.	in.	in.³	in.	in.⁴	in.³	in.		
L6×6×1	35.4	8.55	1.79	1.86	15.4	0.917	14.9	5.70	1.17	1.00	1.00
×⅞	31.9	7.61	1.81	1.81	13.7	0.813	13.3	5.18	1.17	1.00	1.00
×¾	28.1	6.64	1.82	1.77	11.9	0.705	11.6	4.63	1.17	1.00	1.00
×⅝	24.1	5.64	1.84	1.72	10.1	0.594	9.81	4.04	1.17	1.00	1.00
×⁹⁄₁₆	22.0	5.12	1.85	1.70	9.18	0.538	8.90	3.73	1.18	1.00	1.00
×½	19.9	4.59	1.86	1.67	8.22	0.481	8.06	3.40	1.18	1.00	1.00
×⁷⁄₁₆	17.6	4.06	1.86	1.65	7.25	0.423	7.05	3.05	1.18	1.00	0.973
×⅜	15.4	3.51	1.87	1.62	6.27	0.365	6.21	2.69	1.19	1.00	0.912
×⁵⁄₁₆	13.0	2.95	1.88	1.60	5.26	0.306	5.20	2.30	1.19	1.00	0.826

Example 2

A PL $1/2 \times 5$ plate is connected by four $5/8$ in diameter bolts. Determine the tensile strength of the member. Assume A36 steel.

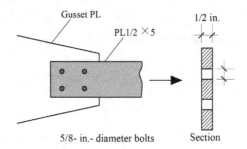

Gusset PL

PL1/2 ✕ 5

1/2 in.

5/8- in.- diameter bolts   Section

$$A_g = 5 \times \frac{1}{2} = 2.5 \, \text{in}^2$$

$$A_n = A_g - A_h = 2.5 - 2\left[\frac{1}{2}\left(\frac{5}{8} + \frac{1}{8}\right)\right] = 1.75 \, \text{in}^2$$

$$A_e = A_n U = 1.75 \times 1.0 = 1.75 \, \text{in}^2 \, [U = 1.0 \text{ for plate}]$$

Design strength based on yielding:

$$\phi_t P_n = 0.90 F_y A_g = 0.90 \times 36 \times 2.5 = 81 \, \text{kip}$$

Design strength based on fracture:

$$\phi_t P_n = 0.75 F_u A_e = 0.75 \times 58 \times 1.75 = 76.1 \, \text{kip}$$

Answer: 76.1 kip

Example 3

Determine the design strength of the following angle member L $31/2 \times 31/2 \times 3/8$, given that bolts are 7/8 in of diameter and are 2 in apart from center to center. Assume A36 steel.

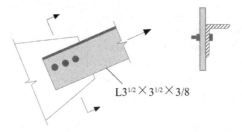

L$3^{1/2} \times 3^{1/2} \times 3/8$

From Table 1–7 of the AISC manual, $A_g = 2.48 \, \text{in}^2$, $\bar{y} = \bar{x} = 1.00$ in.

$$A_n = A_g - A_h = 2.48 - 1\left[\frac{3}{8}\left(\frac{7}{8} + \frac{1}{8}\right)\right] = 2.105 \, \text{in}^2$$

$$U = 1 - \frac{\bar{x}}{\ell} = 1 - \frac{1.0}{2+2} = 0.75$$

$$A_e = A_n U = 2.105 \times 0.75 = 1.578 \, \text{in}^2$$

Design strength based on yielding:

$$\phi_t P_n = 0.90 F_y A_g = 0.90 \times 36 \times 2.48 = 80.35 \, \text{kip}$$

Design strength based on fracture:

$$\phi_t P_n = 0.75 F_u A_e = 0.75 \times 58 \times 1.578 = 68.7 \, \text{kip}$$

Answer: 68.7 kip

8.2   TENSION MEMBER

8.2.1   STAGGERED FASTENER

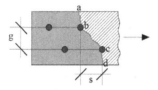

Consider an $s^2 / (4g)$ term for each of the staggering failure line, such as *bc*.
For failure along *abcd* :

$$A_n = A_g - \sum A_h + \sum \frac{s^2}{4g} t$$

Example 4

Determine the net area of the channel C6 × 13. Bolts have 5/8 in diameter.

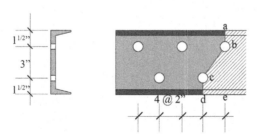

From Table 1–5 of the AISC manual, $A_g = 3.81 \, \text{in}^2, t_w = 7/16$ in.
Line *abe*:

$$A_n = 3.81 - 1 \left[\frac{7}{16} \left(\frac{5}{8} + \frac{1}{8} \right) \right] = 3.48 \, \text{in}^2$$

Line *abcd*:

$$A_n = 3.81 - 2 \left[\frac{7}{16} \left(\frac{5}{8} + \frac{1}{8} \right) \right] + \frac{2^2}{4 \times 3} \times \frac{7}{16} = 3.30 \, \text{in}^2$$

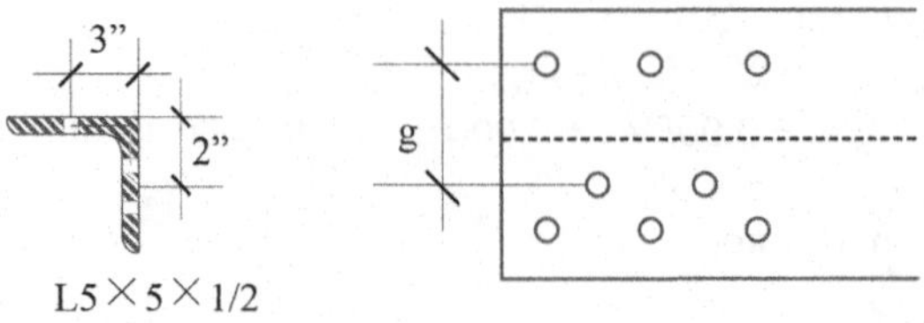

If bolts do not lie in one plane, then the plane should be unfolded to determine g.

$$g = d_1 + d_2 - t$$
$$= 3 + 2 - \frac{1}{2} = 4.5 \text{ in}$$

The term g should be considered along the center line of the holes.

TABLE 8.2
1-5 C-Shapes Dimensions (AISC Manual)

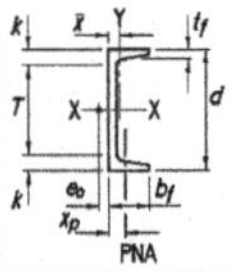

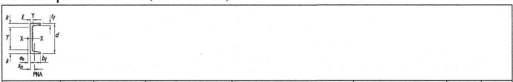

Shape	Area, A	Depth, d		Web: Thickness, t_w		Web: $\frac{t_w}{2}$	Flange: Width, b_f		Flange: Average Thickness, t_f		Distance: K	Distance: T	Distance: Workable Gage	r_{is}	h_0
	in.²	in.		in.			in.	in.	in.		in.	in.	in.	in.	in.
C15×50	14.7	15.0	15	0.716	11/16	3/8	3.72	3 3/4	0.650	5/8	1 7/16	12 1/8	2 1/4	1.17	14.4
×40	11.8	15.0	15	0.520	1/2	1/4	3.52	3 1/2	0.650	5/8	1 7/16	12 1/8	2	1.15	14.4
×33.9	10.0	15.0	15	0.400	3/8	3/16	3.40	3 3/8	0.650	5/8	1 7/16	12 1/8	2	1.13	14.4
C12×30	8.81	12.0	12	0.510	1/2	1/4	3.17	3 1/8	0.501	1/2	1 1/8	9 3/4	1 3/4 g	1.01	11.5
×25	7.34	12.0	12	0.387	3/8	3/16	3.05	3	0.501	1/2	1 1/8	9 3/4	1 3/4 g	1.00	11.5
×20.7	6.08	12.0	12	0.282	5/16	3/16	2.94	3	0.501	1/2	1 1/8	9 3/4	1 3/4 g	0.983	11.5
C10×30	8.81	10.0	10	0.673	11/16	3/8	3.03	3	0.436	7/16	2	8	1 3/4 g	0.924	9.56
×25	7.35	10.0	10	0.526	1/2	1/4	2.89	3 3/8	0.436	7/16	1	8	1 3/4 g	0.911	9.56
×20	5.87	10.0	10	0.379	3/8	3/16	2.74	2 3/4	0.436	7/16	1	8	1 1/2 g	0.894	9.56
×15.3	4.48	10.0	10	0.240	7	1/8	2.60	2 2/5	0.436	7/16	1	8	1 1/2 g	0.868	9.56
C9×20	5.87	9.00	9	0.448	7/16	1/4	2.65	2 5/8	0.413	7/16	1	7	1 1/2 g	0.850	8.59
×15	4.40	9.00	9	0.285	5/16	3/16	2.49	2 1/2	0.413	7/16	1	7	1 3/8 g	0.825	8.59
× 13.4	3.94	9.00	9	0.233	1/4	1/8	2.43	2 3/8	0.413	7/16	1	7	1 3/8 g	0.814	8.59
C8 ×18.75	5.51	8.00	8	0.487	1/2	1/4	2.53	2 1/2	0.390	3/8	15/16	6 1/8	1 1/2 g	0.800	7.61
×13.75	4.03	8.00	8	0.303	5/16	3/16	2.34	2 3/8	0.390	3/8	15/16	6 1/8	1 3/8 g	0.774	7.61
×11.5	3.37	8.00	8	0.220	1/4	1/8	2.26	2 1/4	0.390	3/8	15/16	6 1/8	1 3/8 g	0.756	7.61
C7×14.75	4.33	7.00	7	0.419	7/16	1/4	2.30	2 1/4	0.366	3/8	7/8	5 1/4	1 1/4 g	0.738	6.63
×12.25	3.59	7.00	7	0.314	5/16	3/16	2.19	2 1/4	0.366	3/8	7/8	5 1/4	1 1/4 g	0.722	6.63
×9.8	2.87	7.00	7	0.210	3/16	5/8	2.09	2 1/8	0.366	3/8	7/8	5 1/4	1 1/4 g	0.698	6.63
C6×13	3.82	6.00	6	0.437	7/16	1/4	2.16	2 1/8	0.343	5/16	13/16	4 3/8	1 3/8 g	0.689	5.66
×10.5	3.07	6.00	6	0.314	5/16	3/16	2.03	2	0.343	5/16	13/16	4 3/8	1 1/8 g	0.669	5.66
×8.2	2.39	6.00	6	0.200	3/16	1/8	1.92	1 7/8	0.343	5/16	13/16	4 3/8	1 1/8 g	0.643	5.66

Example 5

The tension member shown in Figure 8.11 is an $L\, 6 \times 31/2 \times 5/16$. The bolts are $3/4$ in in diameter. If A36 steel is used, is the member adequate for a service dead load of 25 kips and a service live load of 25 kips ?

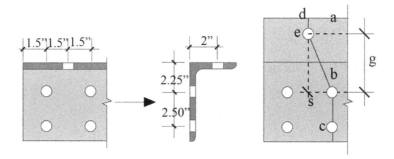

From Table 1–7, pp. 1–42, of the AISC manual, $A_g = 2.87$ in.

8.2.2   NET AREA

Line *abc* :

$$A_n = 2.87 - 2\left[\frac{5}{16}\left(\frac{3}{4}+\frac{1}{8}\right)\right] = 2.343 \text{ in}^2$$

Line *debc*:

$$g = 2.25 + 2 - 5/16 = 3.937 \text{ in}$$

$$A_n = 2.87 - 3\left[\frac{5}{16}\left(\frac{3}{4}+\frac{1}{8}\right)\right] + \frac{1.5^2}{4 \times 3.937} \times \frac{5}{16} = 2.094 \text{ in}^2\,(\leftarrow)$$

8.2.3   EFFECTIVE AREA

From Table 1–7, pp. 1–43, of the AISC manual, $\bar{x} = 0.756$ in.

$$U = 1 - \frac{\bar{x}}{\ell} = 1 - \frac{0.756}{3} = 0.748$$

$$A_e = A_n U = 2.114 \times 0.748 = 1.566 \text{ in}^2$$

8.2.4   CAPACITY

Based on yielding:

$$\phi_t P_n = 0.90 F_y A_g = 0.90 \times 36 \times 2.87 = 92.9 \text{ kip}$$

Based on fracture:

$$\phi_t P_n = 0.75 F_u A_e = 0.75 \times 58 \times 1.566 = 68.1 \text{ kip} (\leftarrow)$$

8.2.5 DEMAND

$$P_u = 1.2 DL + 1.6 LL = 1.2 \times 25 + 1.6 \times 25 = 70 \text{ kip}$$

Since $P_u (70) \phi /_t P_n (68.1)$, the member is not adequate.

Table 8.3

1-7 Angles Properties (AISC Manual)

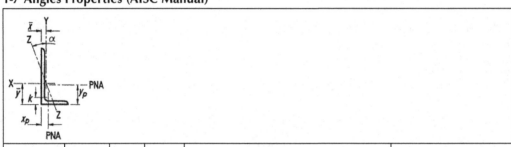

Shape	K	Wt.	Area, A	Axis X-X						Flexural-Torsional Properties		
				I	S	r	y	Z	y_P	J	C_W	r_0
	In.	lb/ft	In.²	In.⁴	In.³	In.	In.	In.³	In.	In.⁴	In.⁶	In.
L6×4×⅞	1 ⅜	27.2	8.00	27.7	7,13	1.86	2.12	12.7	1.43	2.03	4.04	2.82
×¾	1 ¼	23.6	6.94	24.5	6,23	1.88	2.07	11.1	1.37	1.31	2.64	2.85
×⅝	1 ⅛	20.0	5.86	21.0	5,29	1.89	2.03	9.44	1.31	0.775	1.59	2.88
×⁹⁄₁₆	1 ¹⁄₁₆	18.1	5.31	19.2	4,81	1.90	2.00	8.59	1.28	0.572	1.18	2.90
×½	1	16.2	4.75	17.3	4,31	1.91	1.98	7.71	1.25	0.407	0.843	2.91
×⁷⁄₁₆	¹⁵⁄₁₆	14.3	4.18	15.4	3,81	1.92	1.95	6.81	1.22	0.276	0.575	2.93
×⅜	⅞	12.3	3.61	13.4	3,30	1.93	1.93	5.89	1.19	0.177	0.369	2.94
×⁵⁄₁₆	¹³⁄₁₆	10.3	3.03	11.4	2,77	1.94	1.90	4.96	1.15	0.104	0.217	2.96
L6×3½×½	1	15.3	4.50	16.6	4,23	1.92	2.07	7.49	1.50	0.386	0.779	2.88
×⅜	⅞	11.7	3.44	12.9	3,23	1.93	2.02	5.74	1.41	0.168	0.341	2.90
×⁵⁄₁₆	¹³⁄₁₆	9.80	2.89	10.9	2,72	1.94	2.00	4.84	1.38	0.0990	0.201	2.92
L5×5×⅞	1 ⅜	27.2	8.00	17.8	5,16	1.49	1.56	9.31	0.800	2.07	3.53	2.64

	Axis Y-Y						Axis Z-Z				Q_s
Shape	I	S	r	$\bar{x}$	Z	xp	I	S	r	Tan	$F_y = 36$
	In.⁴	In.³	In.	In.	In.³	In.	In.⁴	In.³	In.	α	ksl
L6×4×⅞	9.70	3.37	1.10	1.12	6.26	0.667	5.82	2.91	0.854	0.421	1.00
×¾	8.63	2.95	1.12	1.07	5.42	0.578	5.08	2.51	0.856	0.428	1.00
×⅝	7.48	2.52	1.13	1.03	4.56	0.488	4.32	2.12	0.859	0.435	1.00
×⁹⁄₁₆	6.86	2.29	1.14	1.00	4.13	0.443	3.93	1.92	0.861	0.438	1.00
×½	6.22	2.06	1.14	0.981	3.69	0.396	3.54	1.72	0.864	0.440	1.00
×⁷⁄₁₆	5.56	1.83	1.15	0.957	3.24	0.348	3.14	1.51	0.867	0.443	0.973
×⅜	4.86	1.58	1.16	0.933	2.79	0.301	2.73	1.31	0.870	0.446	0.912
×⁵⁄₁₆	4.13	1.34	1.17	0.908	2.33	0.253	2.31	1.10	0.874	0.449	0.826
L6×3½×½	4.24	1.59	0.968	0.829	2.88	0.375	2.59	1.34	0.756	0.343	1.00
×⅜	3.33	1.22	0.984	0.781	2.18	0.287	2.01	1.02	0.763	0.349	0.912
×⁵⁄₁₆	2.84	1.03	0.991	0.756	1.82	0.241	1.70	0.859	0.767	0.352	0.826
L5×5×⅞	17.8	5.16	1.49	1.56	9.31	0.800	7.60	3.43	0.971	1.00	1.00

L6-L4

8.3 TENSION MEMBER

8.3.1 BLOCK SHEAR

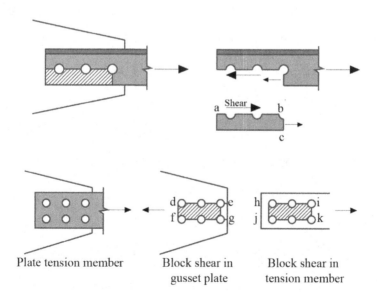

Plate tension member   Block shear in   Block shear in
gusset plate   tension member

Two types of failure occur during block shear.

Tensile failure occurs at cross sections that are transverse to the applied load.

Shear failure occurs at cross sections that are along the direction of the applied load. *Nominal capacity* is the minimum of the following two expressions:

$$R_n = \begin{cases} 0.60 F_y A_{gv} + F_u A_{nt} \\ 0.60 F_u A_{nv} + F_u A_{nt} \end{cases}$$

A_{gv} Gross area for shear

A_{nv} Net area for shear

A_{nt} Net area for tension

F_y Yield Strength

F_u Ultimate Strength

Example 6

Determine the block shear strength of the following plate, which is 1/2 in in thickness. Bolts have a 5/8 in diameter. Assume A36 steel.

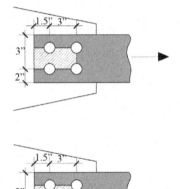

8.3.2 BLOCK SHEAR – MODE 1

Areas:

$$A_{gv} = \frac{1}{2} \times [4.5] \times 2 = 4.5 \text{ in}^2$$

$$A_{nv} = \frac{1}{2} \times \left[4.5 - 1.5 \times \left(\frac{5}{8} + \frac{1}{8} \right) \right] \times 2$$

$$= 3.375 \text{ in}^2$$

$$A_{nt} = \frac{1}{2} \times \left[3 - 1 \times \left(\frac{5}{8} + \frac{1}{8} \right) \right]$$

$$= 1.125 \text{ in}^2$$

8.3.3 Capacity

$$R_n = 0.60 F_y A_{gv} + F_u A_{nt}$$
$$= 0.60 \times 36 \times 4.5 + 58 \times 1.125$$
$$= 162.4 (\leftarrow)$$
$$R_n = 0.60 F_u A_{nv} + F_u A_{nt}$$
$$= 0.60 \times 58 \times 3.375 + 58 \times 1.125$$
$$= 182.7 \text{ kip}$$
$$\phi_t R_n = 0.9 \times 162.4 = 146.1 \text{ kip}$$

8.3.4 Block Shear – Mode 2

8.3.5 Areas

$$A_{gv} = \frac{1}{2} \times [4.5] \times 1 = 2.25 \text{ in}^2$$
$$A_{nv} = \frac{1}{2} \times \left[4.5 - 1.5 \times \left(\frac{5}{8} + \frac{1}{8} \right) \right]$$
$$= 1.687 \text{ in}^2$$
$$A_{nt} = \frac{1}{2} \times \left[5 - 1.5 \times \left(\frac{5}{8} + \frac{1}{8} \right) \right]$$
$$= 1.936 \text{ in}^2$$

8.3.6 Capacity

$$R_n = 0.60 F_y A_{gv} + F_u A_{nt}$$
$$= 0.60 \times 36 \times 2.25 + 58 \times 1.936$$
$$= 160.9 (\leftarrow)$$
$$R_n = 0.60 F_u A_{nv} + F_u A_{nt}$$
$$= 0.60 \times 58 \times 1.687 + 58 \times 1.936$$
$$= 171 \text{kip}$$
$$\phi_t R_n = 0.9 \times 160.9 = 144.8 \text{ kip}$$

Answer: Block shear strength is 144.8 kip.

Example 7

The following angle has a tensile strength of 78.9 kip based on fracture and yielding. Now, determine the tensile strength again, but considering block shear. Bolts have 7 / 8 in diameter. Assume A36 steel.

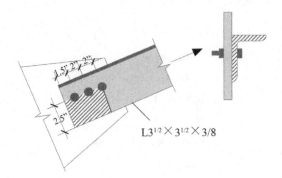

L$3^{1/2} \times 3^{1/2} \times 3/8$

Also, determine the adequacy of the tension member against service dead load of 20 kip and a service live load of 25 kip.

Areas:

$$A_{gv} = \frac{3}{8} \times [5.5] = 2.062 \, \text{in}^2$$

$$A_{nv} = \frac{3}{8} \times \left[5.5 - 2.5 \times \left(\frac{7}{8} + \frac{1}{8} \right) \right] = 1.125 \, \text{in}^2$$

$$A_{nt} = \frac{3}{8} \times \left[2.5 - 0.5 \times \left(\frac{7}{8} + \frac{1}{8} \right) \right] = 0.75 \, \text{in}^2$$

8.3.7   CAPACITY

$$R_n = 0.60 F_y A_{gv} + F_u A_{nt}$$
$$= 0.60 \times 36 \times 2.062 + 58 \times 0.75 = 88 \, \text{kip}$$
$$R_n = 0.60 F_u A_{nv} + F_u A_{nt}$$
$$= 0.60 \times 58 \times 1.125 + 58 \times 0.75 = 82.6 \, \text{kip} (\leftarrow)$$
$$\phi_t R_n = 0.9 \times 82.6 = 74.3 \, \text{kip}$$

Since block shear has lower strength than strength based on fracture and yielding, thus, block shear governs over fracture and yielding.

8.3.8   DEMAND

$$P_u = 1.2 DL + 1.6 LL = 1.2 \times 20 + 1.6 \times 25 = 64 \, \text{kip}$$

Since $P_u (64) \le \phi_t R_n (74.3)$, the member is adequate.

8.4   DESIGN OF TENSION MEMBER

Example 8

Select an unequal-leg angle tension member 15 ft long to resist a service dead load of 35 kip, live load of 70 kip, and wind load of 60 kip. Use A36 steel and 3/4 in bolt. The angle should have minimum thickness of 1/4 in.

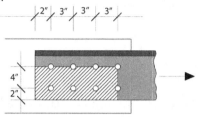

8.4.1   DESIGN LOAD

$$P_u = 1.2DL + 1.6LL$$
$$= 1.2 \times 35 + 1.6 \times 70 = 154 \text{kip}(\leftarrow)$$
$$P_u = 1.2DL + 0.5LL + 1.0WL$$
$$= 1.2 \times 35 + 0.5 \times 70 + 1.0 \times 60$$
$$= 137 \text{kip}$$

Step 1: Determine the gross area based on yielding.

$$A_{g_1} = \frac{P_u}{0.90F_y} = \frac{154}{0.90 \times 36} = 4.75 \text{ in}^2$$

Step 2: Determine the gross area based on fracture.

$$A_{g_2} = \frac{P_u}{0.75F_uU} + \sum A_{holes}$$
$$= \frac{154}{0.75 \times 58 \times 0.80} + t\left(\frac{3}{4} + \frac{1}{8}\right) \times 2 = 4.425 + 1.75\,t$$

If there are four or more bolts in a row, $U = 0.80$ can be used.

Step 3: Determine the minimum radius of gyration.

$$r_{min} = \frac{L}{300} = \frac{15 \times 12}{300} = 0.60 \text{ in}$$

Step 4: Find the appropriate angle from the table.

No.	t	A_{g_1}	A_{g_2}	Angle	r_z
1	1/4	4.75	4.86	None found	
2	5/16	4.75	4.97	None found	
3	3/8	4.75	5.08	None found	
4	7/16	4.75	5.19	L 8×6×7/16	1.31

Step 5: Check for block shear.

$$A_{gv} = \frac{7}{16} \times [11] \times 1 = 4.81 \text{ in}^2$$

$$A_{nv} = \frac{7}{16} \times \left[11 - 3.5 \times \left(\frac{3}{4} + \frac{1}{8} \right) \right] = 3.47 \text{ in}^2$$

$$A_{nt} = \frac{7}{16} \times \left[6 - 1.5 \times \left(\frac{3}{4} + \frac{1}{8} \right) \right] = 2.05 \text{ in}^2$$

$$R_n = 0.60 F_y A_{gv} + F_u A_{nt}$$

$$= 0.60 \times 36 \times 4.81 + 58 \times 2.05 = 222.8 \text{ kip} (\leftarrow)$$

$$R_n = 0.60 F_y A_{gv} + F_u A_{nt}$$

$$= 0.60 \times 58 \times 3.47 + 58 \times 2.05 = 239.7 \text{ kip}$$

$$\phi_t R_n = 0.9 \times 222.8 = 200.5 > P_u (154) \text{ kip O.K.}$$

Answer: Select $L\, 8 \times 6 \times 7/16$.

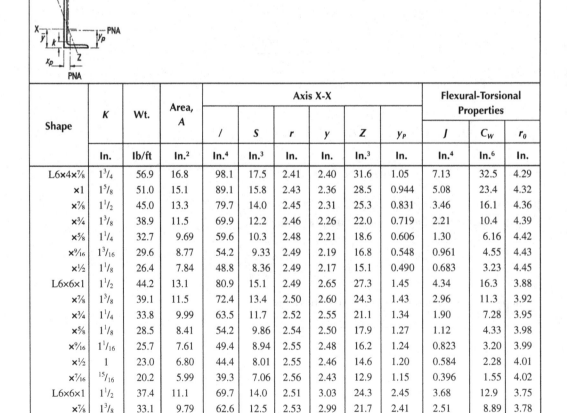

Shape	K	Wt.	Area, A	Axis X-X						Flexural-Torsional Properties		
				I	S	r	y	Z	y_P	J	C_W	r_0
	In.	lb/ft	In.²	In.⁴	In.³	In.	In.	In.³	In.	In.⁴	In.⁶	In.
L6×4×⅞	1³/₄	56.9	16.8	98.1	17.5	2.41	2.40	31.6	1.05	7.13	32.5	4.29
×1	1⁵/₈	51.0	15.1	89.1	15.8	2.43	2.36	28.5	0.944	5.08	23.4	4.32
×⅞	1¹/₂	45.0	13.3	79.7	14.0	2.45	2.31	25.3	0.831	3.46	16.1	4.36
×¾	1³/₈	38.9	11.5	69.9	12.2	2.46	2.26	22.0	0.719	2.21	10.4	4.39
×⅝	1¹/₄	32.7	9.69	59.6	10.3	2.48	2.21	18.6	0.606	1.30	6.16	4.42
×⁹/₁₆	1³/₁₆	29.6	8.77	54.2	9.33	2.49	2.19	16.8	0.548	0.961	4.55	4.43
×½	1¹/₈	26.4	7.84	48.8	8.36	2.49	2.17	15.1	0.490	0.683	3.23	4.45
L6×6×1	1¹/₂	44.2	13.1	80.9	15.1	2.49	2.65	27.3	1.45	4.34	16.3	3.88
×⅞	1³/₈	39.1	11.5	72.4	13.4	2.50	2.60	24.3	1.43	2.96	11.3	3.92
×¾	1¹/₄	33.8	9.99	63.5	11.7	2.52	2.55	21.1	1.34	1.90	7.28	3.95
×⅝	1¹/₈	28.5	8.41	54.2	9.86	2.54	2.50	17.9	1.27	1.12	4.33	3.98
×⁹/₁₆	1¹/₁₆	25.7	7.61	49.4	8.94	2.55	2.48	16.2	1.24	0.823	3.20	3.99
×½	1	23.0	6.80	44.4	8.01	2.55	2.46	14.6	1.20	0.584	2.28	4.01
×⁷/₁₆	¹⁵/₁₆	20.2	5.99	39.3	7.06	2.56	2.43	12.9	1.15	0.396	1.55	4.02
L6×6×1	1¹/₂	37.4	11.1	69.7	14.0	2.51	3.03	24.3	2.45	3.68	12.9	3.75
×⅞	1³/₈	33.1	9.79	62.6	12.5	2.53	2.99	21.7	2.41	2.51	8.89	3.78

L8-L6

Shape	Axis Y-Y						Axis Z-Z			Tan	Q_s
	I	S	r	$\bar{X}$	Z	xp	I	S	r	α	$F_y = 36$
	In.⁴	In.³	In.	In.	In.³	In.	In.⁴	In.³	In.		ksl
L8×8×1⅛	98.1	17.5	2.41	2.40	31.6	1.05	40.7	12.0	1.56	1.00	1.00
×1	89.1	15.8	2.43	2.36	28.5	0.944	36.8	11.0	1.56	1.00	1.00
×⅞	79.7	14.0	2.45	2.31	25.3	0.831	32.7	10.0	1.57	1.00	1.00
×⅛	69.9	12.2	2.46	2.26	22.0	0.719	28.5	8.90	1.57	1.00	1.00
×⅝	59.6	10.3	2.48	2.21	18.6	0.606	24.2	7.72	1.58	1.00	0.997
×⁹⁄₁₆	54.2	9.33	2.49	2.19	16.8	0.548	21.9	7.09	1.58	1.00	0.959
×½	48.8	8.36	2.49	2.17	15.1	0.490	19.8	6.44	1.59	1.00	0.912
L8×6×1	38.8	8.92	1.72	1.65	16.2	0.819	21.3	7.60	1.28	0.542	1.00
×⅞	34.9	7.94	1.74	1.60	14.4	0.719	18.9	6.71	1.28	0.546	1.00
×¾	30.8	6.92	1.75	1.56	12.5	0.624	16.6	5.82	1.29	0.550	1.00
×⅝	26.4	5.88	1.77	1.51	10.5	0.526	14.1	4.91	1.29	0.554	0.997
×⁹⁄₁₆	24.1	5.34	1.78	1.49	9.52	0.476	12.8	4.45	1.30	0.556	0.959
×½	21.7	4.79	1.79	1.46	8.52	0.425	11.5	3.98	1.30	0.557	0.912
×⁷⁄₁₆	19.3	4.23	1.80	1.44	7.50	0.374	10.2	3.51	1.31	0.559	0.850
L8×4×1	11.6	3.94	1.03	1.04	7.73	0.694	7.83	3.48	0.844	0.247	1.00

8.5   COMPRESSION MEMBER

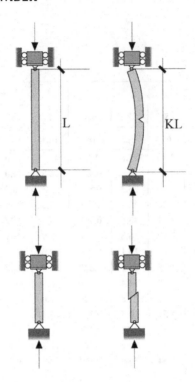

8.5.1   BUCKLING

Buckling is the sudden lateral instability of a slender structural member induced by the action of an axial load before the yield stress of the material is reached.

    L is the physical length of the column.
    K is the effective length factor.
    KL is the effective length.

8.5.2   CRUSHING

Crushing occurs when the direct stress from an axial load exceeds the compressive strength of the material available in the cross section.

    L is the physical length of the column.
    K is the effective length factor.
    KL is the effective length.
    r is the radius of gyration.
    KL/r is the slenderness ratio.
    F_{Cr} is the critical stress.
    Fy is the yield stress.
    E is the modulus of elasticity.
    C_C is the critical coefficient.

$$F_e = \frac{\pi^2 E}{(KL/r)^2} \qquad F_{Cr} = \begin{cases} 0.658^{F_y/F_e} F_y & \text{if } KL/r < C_C \\ 0.877 F_e & \text{if } KL/r > C_C \end{cases}$$

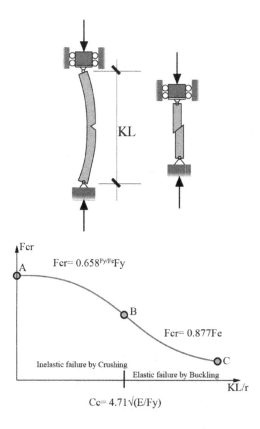

Example 9

A steel column of 25ft length is made of W14×61 shape which is supported by a fixed-hinge joint. Determine the axial capacity of the section. Steel is A992.

Solution

$$K = 0.80 \quad \text{(for fixed-hinge joint)}$$
$$L = 25\,\text{f}$$
$$F_y = 50\,\text{ksi}$$

From Table 1–7 of the AISC manual, $A_g = 17.90$ in and $r_y = 2.45$ in.

8.5.3   CHECK FAILURE MODE

$$KL = 0.8 \times 25 = 20\text{ft}$$
$$\frac{KL}{r} = \frac{20 \times 12}{2.45} = 97.9$$
$$C_C = 4.71\sqrt{\frac{E}{F_y}} = 4.71\sqrt{\frac{29000}{50}} = 113.4$$

Since $KL\,/\,r < C_C$, failure is by crushing.

8.5.4 DETERMINE CAPACITY

$$F_e = \frac{\pi^2 E}{(KL/r)^2} = \frac{\pi^2 \times 29000}{97.9^2} = 29.83 \text{ ksi}$$

$$F_{Cr} = 0.658^{F_y/F_e} F_y = 0.658^{50/29.83} \times 50 = 24.79 \text{ ksi}$$

$$\Phi_C P_n = \Phi_C F_{Cr} A_g = 0.9 \times 24.79 \times 17.9 = 399.3 \text{ kip}$$

Question: If the same column is changed to a length of 35ft, then determine its capacity.

Solution

$$K = 0.80 \quad \text{(for fixed-hinge joint)}$$
$$L = 35 \text{ ft}$$
$$F_y = 50 \text{ ksi}$$

Check failure mode.

$$KL = 0.8 \times 35 = 28 \text{ ft}$$
$$\frac{KL}{r} = \frac{28 \times 12}{2.45} = 137.1$$
$$C_C = 113.4$$

Since $KL/r > C_C$, failure is by buckling.
   Determine capacity.

$$F_e = \frac{\pi^2 E}{(KL/r)^2} = \frac{\pi^2 \times 29000}{137.1^2} = 15.22 \text{ ksi}$$

$$F_{Cr} = 0.877 F_e = 0.877 \times 15.22 = 13.35 \text{ ksi}$$

$$\Phi_C P_n = \Phi_C F_{Cr} A_g = 0.9 \times 13.35 \times 17.9 = 215.1 \text{ kip}$$

Answer: 215.1 *kip*

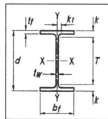

			Web			Flange				Distance				
Shape	Area, A	Depth, d	Thickness, t_w	$\frac{t_w}{2}$	Width, b_f		Thickness, t_f		k		k_1	T	Work-able Gage
									k_{dos}	k_{det}			
	in.²	in.	in.	in.	in.		in.		in.	in.	in.	in.	in.
W14×132	38.8	14.7	14.	0.645	⅝	⁵⁄₁₆	14.7	14¾	1.03	1	1.63	2⁵⁄₁₆	1⁹⁄₁₆	10	5 1/2
×120	35.3	14.5	14½	0.590	⁹⁄₁₆	⁵⁄₁₆	14.7	14⅝	0.940	¹⁵⁄₁₆	1.54	2¼	1½		
×109	32.0	14.3	14⅜	0.525	½	¼	14.6	14⅝	0.860	⅞	1.46	2³⁄₁₆	1½		
×99ᶠ	29.1	14.2	14⅛	0.485	½	¼	14.6	14⅝	0.780	¾	1.38	2¹⁄₁₆	1¹⁄₁₆		
×90ᶠ	26.5	14.0	14	0.440	⁷⁄₁₆	¼	14.5	14½	0.710	1¹⁄₁₆	1.31	2	1¹⁄₁₆		
W14×82	24.0	14.3	14¼	0.510	½	¼	10.1	10⅛	0.855	⅞	1.45	1¹¹⁄₁₆	1¹⁄₁₆	10 7/8	5 1/2
×74	21.8	14.2	14⅛	0.450	⁷⁄₁₆	¼	10.1	10⅛	0.785	1³⁄₁₆	1.38	1⅝	1¹⁄₁₆		
×68	20.0	14.0	14	0.415	⁷⁄₁₆	¼	10.0	10	0.720	¾	1.31	1⁹⁄₁₆	1¹⁄₁₆		
×61	17.9	13.9	13⅞	0.375	⅜	³⁄₁₆	10.0	10	0.645	⅝	1.24	1½	1		
W14×53	15.6	13.9	13⅞	0.370	⅜	³⁄₁₆	8.06	8	0.660	1¹⁄₁₆	1.25	1½	1	10 7/8	5 1/2

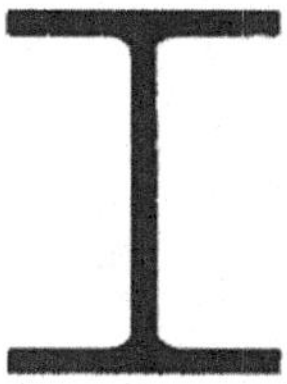

W14-W12

Nom-inal Wt.	Compact Section Criteria		Axis X-X				Axis Y-Y				r_{ts}	h_0	Torsional Properties	
	$\frac{b_f}{2t_f}$	$\frac{h}{t_w}$	I	S	r	Z	I	S	r	Z			J	C_w
lb/ft			in.⁴	in.³	in.	in.³	in.⁴	in.³	in.	in.³	in.	in.	in.⁴	in.⁶
132	7.15	17.7	1530	209	6.28	234	548	74.5	3.76	113	4.23	13.7	12.3	25500
120	7.80	19.3	1380	190	6.24	212	495	67.5	3.74	102	4.20	13.6	9.37	22700
109	8.49	21.7	1240	173	6.22	192	447	61.2	3.73	92.7	4.17	13.4	7.12	20200
99	9.34	23.5	1110	157	6.17	173	402	55.2	3.71	83.6	4.14	13.4	5.37	18000
90	10.2	25.9	999	143	6.14	157	362	49.9	3.70	75.6	4.10	13.3	4.06	16000
82	5.92	22.4	881	123	6.05	139	148	29.3	2.48	44.8	2.85	13.4	5.07	6710
74	6.41	25.4	795	112	6.04	126	134	26.6	2.48	40.5	2.83	13.4	3.87	5990
68	6.97	27.5	722	103	6.01	115	121	24.2	2.46	36.9	2.80	13.3	3.01	5380
61	7.75	30.4	640	92.1	5.98	102	107	21.5	2.45	32.8	2.78	13.3	2.19	4710
53	6.11	30.9	541	77.8	5.89	87.1	57.7	14.3	1.92	22.0	2.22	13.2	1.94	2540

8.6   BIAXIAL BENDING

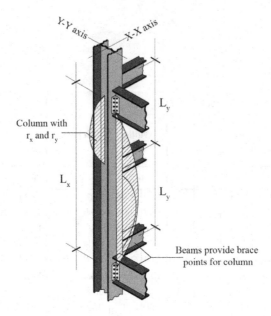

Weak axis buckling is usually about the y-y axis. Slenderness ratio is:

$$\frac{KL_y}{r_y}$$

Strong axis buckling is usually about the x-x axis. Slenderness ratio is:

$$\frac{KL_x}{r_x}$$

The column should be designed with maximum slenderness ratio.

Example 10

Determine the design capacity of the steel column *AB*, which is laterally braced at *C* in *x* direction, given that steel is A36.

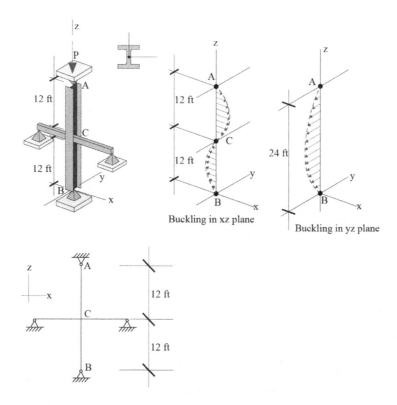

From Table 1–7 of the AISC manual, $A_g = 11.50$ in^2, $r_x = 4.27$ in, and $r_y = 1.98$ in.

8.6.1 DETERMINE GOVERNING SLENDERNESS RATIO

$$\left(\frac{KL}{r_y}\right)_{AC} = \frac{1.0 \times 12 \times 12}{1.98} = 72.7 \, (\leftarrow)$$

$$\left(\frac{KL}{r_y}\right)_{CB} = \frac{1.0 \times 12 \times 12}{1.98} = 72.7$$

$$\left(\frac{KL}{r_X}\right)_{AB} = \frac{1.0 \times 24 \times 12}{4.27} = 67.4$$

The largest slenderness ratio $KL/r = 72.7$ governs the behavior of the ABC column.

8.6.2 CHECK IF COLUMN IS SLENDER OR STOCKY

$$C_C = 4.71 \sqrt{\frac{E}{F_y}} = 4.71 \sqrt{\frac{29000}{36}} = 133.68$$

Since $(KL/r) < C_C$, column is stocky.

8.6.3 DETERMINE CAPACITY

$$F_e = \frac{\pi^2 E}{(KL/r)^2} = \frac{\pi^2 \times 29000}{72.7^2} = 54.1 \text{ ksi}$$

$$F_{Cr} = 0.658^{(F_y/F_e)} F_y = 0.658^{(36/54.1)} \times 36 = 27.3 \text{ ksi}$$

$$\phi_C P_n = \phi_C F_{Cr} A = 0.9 \times 27.3 \times 11.5 = 282.5 \text{ kip}$$

Answer: $P_n = 282.5$ kip

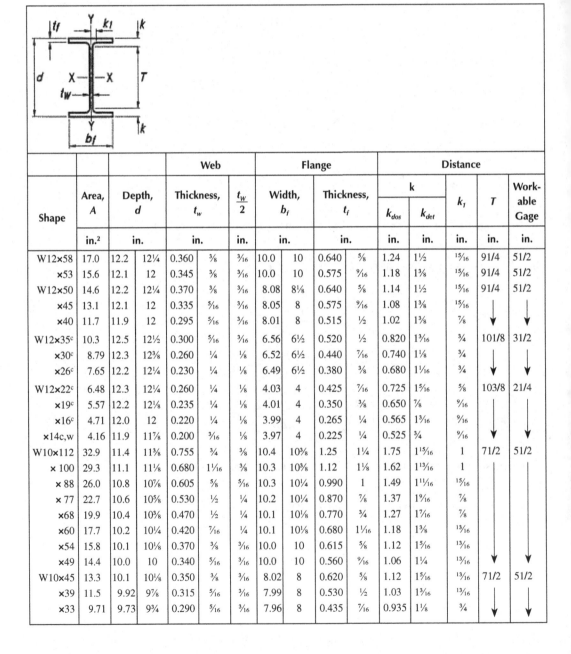

Shape	Area, A	Depth, d		Web Thickness, t_w		$\frac{t_w}{2}$	Flange Width, b_f		Thickness, t_f		Distance k k_{dos}	k_{det}	k_1	T	Work-able Gage
	in.²	in.		in.		in.	in.		in.		in.	in.	in.	in.	in.
W12×58	17.0	12.2	12¼	0.360	⅜	³⁄₁₆	10.0	10	0.640	⅝	1.24	1½	¹⁵⁄₁₆	9¼	5½
×53	15.6	12.1	12	0.345	⅜	³⁄₁₆	10.0	10	0.575	⁹⁄₁₆	1.18	1⅜	¹⁵⁄₁₆	9¼	5½
W12×50	14.6	12.2	12¼	0.370	⅜	³⁄₁₆	8.08	8⅛	0.640	⅝	1.14	1½	¹⁵⁄₁₆	9¼	5½
×45	13.1	12.1	12	0.335	⁵⁄₁₆	³⁄₁₆	8.05	8	0.575	⁹⁄₁₆	1.08	1⅜	¹⁵⁄₁₆	↓	↓
×40	11.7	11.9	12	0.295	⁵⁄₁₆	³⁄₁₆	8.01	8	0.515	½	1.02	1⅜	⅞	↓	↓
W12×35ᶜ	10.3	12.5	12½	0.300	⁵⁄₁₆	³⁄₁₆	6.56	6½	0.520	½	0.820	1³⁄₁₆	¾	10⅛	3½
×30ᶜ	8.79	12.3	12⅜	0.260	¼	⅛	6.52	6½	0.440	⁷⁄₁₆	0.740	1⅛	¾	↓	↓
×26ᶜ	7.65	12.2	12¼	0.230	¼	⅛	6.49	6½	0.380	⅜	0.680	1¹⁄₁₆	¾	↓	↓
W12×22ᶜ	6.48	12.3	12¼	0.260	¼	⅛	4.03	4	0.425	⁷⁄₁₆	0.725	1⁵⁄₁₆	⅝	10⅜	2¼
×19ᶜ	5.57	12.2	12⅛	0.235	¼	⅛	4.01	4	0.350	⅜	0.650	⅞	⁹⁄₁₆	↓	
×16ᶜ	4.71	12.0	12	0.220	¼	⅛	3.99	4	0.265	¼	0.565	1³⁄₁₆	⁹⁄₁₆	↓	
×14c,w	4.16	11.9	11⅞	0.200	³⁄₁₆	⅛	3.97	4	0.225	¼	0.525	¾	⁹⁄₁₆	↓	↓
W10×112	32.9	11.4	11⅜	0.755	¾	⅜	10.4	10⅜	1.25	1¼	1.75	1¹⁵⁄₁₆	1	7½	5½
× 100	29.3	11.1	11⅛	0.680	1¹⁄₁₆	⅜	10.3	10⅜	1.12	1⅛	1.62	1¹³⁄₁₆	1		
× 88	26.0	10.8	10⅞	0.605	⅝	⁵⁄₁₆	10.3	10¼	0.990	1	1.49	1¹¹⁄₁₆	¹⁵⁄₁₆		
× 77	22.7	10.6	10⅝	0.530	½	¼	10.2	10¼	0.870	⅞	1.37	1⁹⁄₁₆	⅞		
×68	19.9	10.4	10⅜	0.470	½	¼	10.1	10⅛	0.770	¾	1.27	1⁷⁄₁₆	⅞		
×60	17.7	10.2	10¼	0.420	⁷⁄₁₆	¼	10.1	10⅛	0.680	1¹⁄₁₆	1.18	1⅜	¹³⁄₁₆		
×54	15.8	10.1	10⅛	0.370	⅜	³⁄₁₆	10.0	10	0.615	⅝	1.12	1⁵⁄₁₆	¹³⁄₁₆		
×49	14.4	10.0	10	0.340	⁵⁄₁₆	³⁄₁₆	10.0	10	0.560	⁹⁄₁₆	1.06	1¼	¹³⁄₁₆	↓	↓
W10×45	13.3	10.1	10⅛	0.350	⅜	³⁄₁₆	8.02	8	0.620	⅝	1.12	1⁵⁄₁₆	¹³⁄₁₆	7½	5½
×39	11.5	9.92	9⅞	0.315	⁵⁄₁₆	³⁄₁₆	7.99	8	0.530	½	1.03	1³⁄₁₆	¹³⁄₁₆	↓	↓
×33	9.71	9.73	9¾	0.290	⁵⁄₁₆	³⁄₁₆	7.96	8	0.435	⁷⁄₁₆	0.935	1⅛	¾	↓	↓

W12-W10

Nom--inal Wt.	Compact Section Criteria		Axis X-X				Axis Y-Y				r_{ts}	h_0	Torsional Properties	
lb/ft	$\dfrac{b_f}{2t_f}$	$\dfrac{h}{t_w}$	I in.⁴	S in.³	r in.	Z in.³	I in.⁴	S in.	r in.	Z in.³	in.	in.	J in.⁴	C_w in.⁶
53	7.82	27.0	475	78.0	5.28	86.4	107	21.4	2.51	32.5	281	11.6	2.10	3570
53	8.69	28.1	425	70.6	5.23	77.9	95.8	192	2.48	29.1	279	11.5	1.58	3160
50	6.31	26.8	391	64.2	5.18	71.9	56.3	13.9	1.96	21.3	225	11.6	1.71	1880
45	7.00	29.6	348	57.7	5.15	64.2	50.0	12.4	1.95	19.0	223	11.5	1.26	1650
40	7.77	33.6	307	515	5.13	57.0	44.1	11.0	1.94	18.8	221	11.4	0.906	1440
35	6.31	362	285	45.6	525	512	24.5	7.47	1.54	11.5	1.79	12.0	0.741	879
30	7.41	41.8	238	38.6	5.21	43.1	20.3	6.24	1.52	9.56	1.77	11.9	0.457	720
26	8.54	47.2	204	33.4	5.17	37.2	17.3	5.34	1.51	8.17	1.75	11.8	0.300	607
22	4.74	41.8	156	25.4	4.91	29.3	4.66	2.31	0.848	3.66	1.04	11.9	0.293	164
19	5.72	46.2	130	213	4.82	24.7	3.76	1.88	0.822	2.98	1.02	11.9	0.180	131
16	7.53	49.4	103	17.1	4.67	20.1	2.82	1.41	0.773	2.26	0.983	11.7	0.103	96.9
14	8.82	54.3	88.6	14.9	4.62	17.4	2.36	1.19	0.753	1.90	0.961	11.7	0.0704	80.4
112	4.17	10.4	716	126	4.66	147	236	45.3	2.68	69.2	3.08	10.2	15.1	6020
100	4.62	11.6	623	112	4.60	130	207	40.0	2.65	61.0	204	10.0	10.9	5150
88	5.18	13.0	534	98.5	454	113	179	34.8	2.63	53.1	299	9.81	7.53	4330
77	5.86	14.8	455	85.9	4.49	97.6	154	30.1	2.60	45.9	295	9.73	5.11	3630
68	6.58	16.7	394	75.7	4.44	85.3	134	26.4	2.59	40.1	292	9.63	3.56	3100
60	7.41	18.7	341	66.7	4.39	74.6	116	23.0	2.57	35.0	2.88	9.52	2.48	2640
54	8.15	212	303	60.0	4.37	66.6	103	20.6	2.56	31.3	2.85	9.49	1.82	2320
49	8.93	23.1	272	54.6	4.35	60.4	93.4	18.7	2.54	28.3	2.84	9.44	1.39	2070
45	6.47	22.5	248	49.1	4.32	54.9	53.4	13.3	2.01	20.3	227	9.48	1.51	1200
39	7.53	25.0	209	42.1	4.27	46.8	45.0	11.3	1.98	17.2	224	9.39	0.976	992
33	9.15	27.1	171	35.0	4.19	38.8	36.6	9.20	1.94	14.0	220	9.30	0.583	791

8.7   DESIGN OF COMPRESSION MEMBER BY TRIAL-AND-ERROR METHOD

Example 11

A steel column of 25 ft length is supported by a fixed-hinge joint. The column is required to carry dead load of 135 kip and live load of 145 kip. Select the lightest W14 shape for the column. Assume steel is A992.

Data:

$$K = 0.80 \qquad\qquad L = 25\,\text{ft}$$
$$DL = 135\,\text{kip} \qquad\qquad LL = 145\,\text{ki}$$
$$F_y = 50\,\text{ksi}$$

Solution

$$KL = 0.8 \times 25 = 20\,\text{ft}$$
$$C_C = 4.71\sqrt{E/F_y} = 4.71\sqrt{29000/50} = 113.4$$

8.7.1   DEMAND

$$P_u = 1.2DL + 1.6LL = 1.2 \times 135 + 1.6 \times 145 = 394\,\text{kip}$$

Trial 1:

$F_{cr} = 33.3\,\text{ksi}$ (arbitrary choice of two-thirds F_y)

$$A_g = \frac{P_u}{\phi_c F_{cr}} = \frac{394}{0.9 \times 33.3} = 13.14\,\text{in}^2$$

Try $W\,14 \times 48, \left(A_g = 14.1\,\text{in}^2, r_y = 1.91\,\text{in from Table 1–1} \right).$

$$\frac{KL}{r} = \frac{20 \times 12}{1.91} = 125.6$$

Since $KL/r > C_C$, failure is by buckling.

$$F_e = \frac{\pi^2 E}{(KL/r)^2} = \frac{\pi^2 \times 29000}{125.6^2} = 18.13\,\text{ksi}$$
$$F_{Cr} = 0.877 F_e = 0.877 \times 18.13 = 15.89\,\text{ksi}$$
$$\phi_C P_n = \phi_C F_{Cr} A_g = 0.9 \times 15.89 \times 14.1$$
$$= 201.7\,\text{kip} < P_u\,(394) \therefore \text{N.G}$$

Trial 2:

Now, assuming F_{cr} to the average of previously assumed 33.3 ksi and calculated 15.89 ksi.

$$F_{cr} = (33.33 + 15.89)/2 = 25.73 \text{ ksi}$$

$$A_g = \frac{P_u}{\phi_c F_{cr}} = \frac{394}{0.9 \times 17.01} = 17.01 \text{ in}^2$$

Try W $14 \times 61, (A_g = 17.9 \text{ in}^2, r_y = 2.45 \text{ in from Table 1–1}).$

$$\frac{KL}{r} = \frac{20 \times 12}{2.45} = 97.9$$

Since $KL/r < C_C$, failure is by crushing.

$$F_e = \frac{\pi^2 E}{(KL/r)^2} = \frac{\pi^2 \times 29000}{97.9^2} = 29.83 \text{ ksi}$$

$$F_{cr} = 0.658^{F_y/F_e} F_y = 0.658^{50/29.83} \times 50 = 24.79 \text{ ksi}$$

(Since calculated F_{Cr} equals 24.79 ksi, which is quite close to the assumed value 25.73 ksi, usually this is an indication of convergence to solution.)

$$\phi_C P_n = \phi_C F_{Cr} A_g = 0.9 \times 24.79 \times 17.9$$
$$= 399.3 \text{kip} > P_u (394) \therefore \text{ O.K}$$

Answer: Select W 14×61.

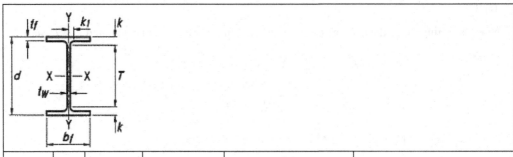

Shape	Area, A	Depth, d		Web Thickness, t_w		$\frac{t_w}{2}$	Flange Width, b_f		Thickness, t_f		Distance k k_{dos}	k_{det}	k_1	T	Workable Gage
	in.²	in.		in.		in.	in.		in.		in.	in.	in.	in.	in.
W14×132	38.8	14.7	14⅝	0.645	⅝	5/16	14.7	14¾	1.03	1	1.63	2 5/16	5½	10	5.
×120	35.3	14.5	14½	0.590	9/16	5/16	14.7	14⅝	0.940	15/16	1.54	2¼	1½		
×109	32.0	14.3	14⅜	0.525	½	¼	14.6	14⅝	0.860	⅞	1.46	2 3/16	1½		
×99f	29.1	14.2	14⅛	0.485	½	¼	14.6	14⅝	0.780	¾	1.38	2 1/16	1 7/16		
×90f	26.5	14.0	14	0.440	7/16	¼	14.5	14½	0.710	11/16	1.31	2	1 7/16	↓	↓
W14×82	24.0	14.3	14¼	0.510	½	¼	10.1	10⅛	0.855	⅞	1.45	1 11/16	1 1/16	10.	5.
×74	21.8	14.2	14⅛	0.450	7/16	¼	10.1	10⅛	0.785	13/16	1.38	1⅝	1 1/16		
×68	20.0	14.0	14	0.415	7/16	¼	10.0	10	0.720	¾	1.31	1 9/16	1 1/16		
×61	17.9	13.9	13⅞	0.375	⅜	3/16	10.0	10	0.645	⅝	1.24	1½	1	↓	↓
W14×53	15.6	13.9	13⅞	0.370	⅜	3/16	8.06	8	0.660	11/16	1.25	1½	1	10.	5.

W12-W10

Nominal Wt.	Compact Section Criteria $\frac{b_f}{2t_f}$	$\frac{h}{t_w}$	Axis X-X I	S	r	Z	Axis Y-Y I	S	r	Z	r_{ts}	h_0	Torsional Properties J	C_w
lb/ft			in.⁴	in.³	in.	in.³	in.⁴	in.	in.	in.³	in.	in.	in.⁴	in.⁶
132	7.15	17.7	1530	209	6.28	234	548	74.5	3.76	113	4.23	13.7	12.3	25500
120	7.80	19.3	1380	190	6.24	212	495	67.5	3.74	102	4.20	13.6	9.37	22700
109	8.49	21.7	1240	173	6.22	192	447	61.2	3.73	92.7	4.17	13.4	7.12	20200
99	9.34	23.5	1110	157	6.17	173	402	55.2	3.71	83.6	4.14	13.4	5.37	18000
90	10.2	25.9	999	143	6.14	157	362	49.9	3.70	75.6	4.10	13.3	4.06	16000
82	5.92	22.4	881	123	6.05	139	148	29.3	2.48	44.8	2.85	13.4	5.07	6710
74	6.41	25.4	795	112	6.04	126	134	26.6	2.48	40.5	2.83	13.4	3.87	5990
68	6.97	27.5	722	103	6.01	115	121	24.2	2.46	36.9	2.80	13.3	3.01	5380
61	7.75	30.4	640	92.1	5.98	102	107	21.5	2.45	32.8	2.78	13.3	2.19	4710
53	6.11	30.9	541	77.8	5.89	87.1	57.7	14.3	1.92	22.0	2.22	13.2	1.94	2540

8.8   DESIGN OF COMPRESSION MEMBER

8.8.1   Using AISC Column Load Table

Example 12

A steel column of 15ft length is connected by a fixed-hinge joint. Select the lightest W8 section for the column to carry a dead load of 100 kip and a live load of 120 kip. Assume the steel is A992. Use the column load table of the AISC manual.

Data:

$$K = 0.80 \qquad L = 15 \text{ ft}$$
$$DL = 100 \text{ kip} \quad LL = 120 \text{ kip}$$
$$F_y = 50 \text{ ksi}$$

Solution

$$KL = 0.8 \times 15 = 12 \text{ ft}$$
$$P_U = 1.2DL + 1.6\,LL$$
$$= 1.2 \times 100 + 1.6 \times 120 = 312 \text{ kip}$$

From Table 4–1, $\phi_C P_n$ values are listed for $KL = 12$.

Section	$\phi_C P_n$
W8×67	632
W8×58	545
W8×48	447
W8×40	367
W8×35	320(←)
W8×31	283

The section W 8×35 has the capacity of 320 kip, which is adequate for the column to carry 312 kip.

Answer: Select W 8×35.

W8

Shape		W8x											
lb/ft		67		58		48		40		35		31	
Design		Pn/Ωc	φcPn	Pn/Ωc	φcPn	Pn/Ωc	φcPn	Pn/Ωc	φcPn	Pn/Ωc	φcPn	Pn/Ωc	φcPn
		ASD	LRFD	ASD	LRFD	ASD	LRFD	ASD	LRFD	ASD	LRFD	ASD	LRFD
Effective length, KL (ft), With respect to least radius of gyration, ry	0	590	886	512	769	422	634	350	526	308	463	273	411
	6	542	815	470	706	387	581	320	481	281	423	249	374
	7	526	790	455	685	375	563	309	465	272	409	241	362
	8	508	763	439	660	361	543	298	448	262	394	232	348
	9	488	733	422	634	347	521	285	429	251	377	222	333
	10	467	701	403	606	331	497	272	409	239	359	211	317
	11	444	668	384	576	314	473	258	388	226	340	200	301
	12	421	633	363	546	297	447	243	366	213	321	189	283
	13	397	597	342	514	280	421	228	343	200	301	177	266
	14	373	560	321	482	262	394	213	321	187	281	165	248
	15	348	523	299	450	244	367	198	298	174	261	153	230
	16	324	487	278	418	226	340	183	275	160	241	141	212
	17	309	450	257	386	2C9	314	169	253	147	221	130	195
	18	276	415	236	355	192	288	154	232	135	203	118	178
	19	253	381	216	325	175	264	141	211	123	184	108	162
	20	231	347	197	296	159	239	127	191	111	166	97.2	146
	22	191	287	163	244	132	198	105	158	91.5	133	80'.3	121
	24	160	241	137	205	111	166	88.2	133	76.9	116	67.5	101
	26	137	205	116	175	94.2	142	75.2	113	65.5	98.5	57.5	86.5
	28	118	177	100	151	81.2	122	64.8	97.4	56.5	84.9	49.6	74.5
	30	103	154	87.5	131	70.7	106	56.5	84.9	49.2	74.0	43.2	64.9
	32	90.3	136	76.9	116	62.2	93.5	49.6	74.6	43.3	65.0	38.0	57.1
	34	79.9	120	68.1	102	55.1	82.8	44.0	66.1				

Example 13

A steel column of 12ft length is connected by a fixed-hinge joint. Select the lightest W8 section for the column to carry a dead load of 100 kip and a live load of 120 kip. Assume the steel is A992. Use the column load table of the AISC manual.

   Data:

$$K = 0.80 \qquad L = 15 \, \text{ft}$$
$$DL = 100 \, \text{ksi} \qquad LL = 120 \, \text{ksi}$$
$$F_y = 50 \, \text{ksi}$$

Solution

$$KL = 0.8 \times 15 = 9.6 \, \text{ft}$$
$$P_u = 1.2 DL + 1.6 \, LL$$
$$= 1.2 \times 100 + 1.6 \times 120 = 312 \, \text{kip}$$

Since there is no row for $KL = 9.6$, an interpolation is required between $KL = 9$ and $KL = 10$.

From Table 4–1, for W8 × 31, the KL are listed below.

KL	$\phi_c P_n$
9	333
10	317

$$\phi_c P_n = 333 - \frac{333 - 317}{10 - 9} \times (9.6 - 9)$$
$$= 323.4 \, \text{kip} > P_u (312) \quad \text{O.K}$$

Answer: Select W8 × 35.

W8

Shape						W8x							
lb/ft		67		58		48		40		35		31	
Design		P_n/Ω_c	$\phi_c P_n$	P_n/Ω_c	$\phi_c P_n$	P_n/Ω_c	$\phi_c P_n$	P_n/Ω_c	$\phi_c P_n$ li	P_n/Ω_c	$\phi_c P_n$	P_n/Ω_c	$\phi_c P_n$
		ASD	LRFO	ASD	LRFD	ASD	LRFD	ASD	LRFD	ASD	LRFD	ASD	LRFD
	0	590	835	512	769	422	634	350	526	303	463	273	411
	6	542	815	470	706	387	531	320	481	281	423	249	374
	7	526	790	455	635	375	563	309	465	272	409	241	362
	8	508	763	439	660	361	543	293	448	262	394	232	348
	9	438	733	422	634	347	521	285	429	251	377	222	333
	10	457	701	403	606	331	497	272	409	239	359	211	317
	11	444	663	384	576	314	473	258	388	226	340	200	301
	12	421	633	363	546	297	447	243	366	213	321	189	283
	13	397	597	342	514	280	421	223	343	200	301	177	266
	14	373	560	321	432	262	394	213	321	187	281	165	248
	15	348	523	299	450	244	367	198	298	174	261	153	230
	16	324	437	278	413	226	340	183	275	160	241	141	212
	17	300	450	257	336	209	314	169	253	147	221	130	195
	18	276	415	236	355	192	288	154	232	135	203	118	178
	19	253	331	216	325	175	264	141	211	123	184	108	162
	20	231	347	197	296	159	239	127	191	111	166	97.2	146
	22	191	287	163	244	132	198	105	158	91.5	138	80.3	121
	24	160	241	137	205	111	166	88.2	133	76.9	116	67.5	101
	26	137	205	116	175	94.2	142	75.2	113	65.5	98.5	57.5	86.5
	28	118	177	100	151	81.2	122	64.8	97.4	56.5	84.9	49.6	74.5
	30	103	154	87.5	131	70.7	106	56.5	34.9	49.2	74.0	43.2	64.9
	32	90.3	136	76.9	116	62.2	93.5	49.6	74.6	43.3	65.0	38,0	57.1
	34	79.9	120	68.1	102	55.1	82.3	44.0	66.1				

Effective length, KL (ft), With respect to least radius of gyration.ry

8.9 PLASTIC MOMENT AND SECTION MODULUS

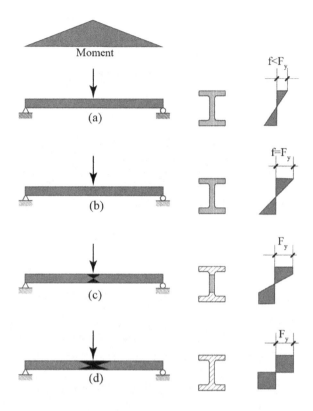

- Figure 8.20(a): At lower load, the section stress is far below the yield strength (F_y).
- Figure 8.20(b): As load is increased, the maximum stress f reaches yield strength (F_y); the corresponding moment is the yield moment (M_y), and the section modulus is denoted by S.
- Figure 8.20(c): If load is further increased, several portions of the section become plastic (shaded zone in the figure), but the region around the center of the web remains elastic.
- Figure 8.20(d): Even more increase of load causes the section to become fully plastic; the corresponding moment is the plastic moment (M_p).

Yield moment is the moment at which the furthest fiber of the beam reaches the yield point of the material.

Plastic moment is the moment at which the whole section fully becomes plastic.

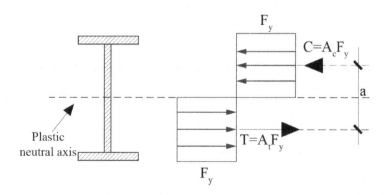

$$C = T$$
$$A_C F_y = A_t F_y$$
$$A_C = A_t$$

C	**Compressive Force**
T	Tensile force
F_y	Yield strength
A_c	Compressive area
A_t	Tensile area

$$M_P = Ca = Ta$$
$$= \left(A_C F_y\right)a = \left(A_t F_y\right)a$$
$$= F_y\left(\frac{A}{2}\right)a$$
$$= F_y Z$$

M_P	Plastic moment
a	Moment arm
Z	Plastic section modulus

Example 14

For the built-up shape shown in Figure 8.22, determine the elastic section modulus S and the yield moment M_y. Bending is about the x-axis, and the steel is A992.

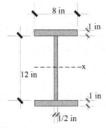

Solution

Because of symmetry, the neutral axis is at mid depth. Now, determine the moment of inertia I.

$$I = I_{web} + 2I_{flange}$$

$$= \frac{\left(\frac{1}{2}\right) \times 12^3}{12} + 2\left[\frac{8 \times 1^3}{12} + (8 \times 1) \times \left(6 + \frac{1}{2}\right)^2\right] = 749.3 \ in^4$$

Elastic section modulus:

$$S = \frac{I}{C} = \frac{749.3}{1+12/2} = 107 \text{ in}^3$$

Yield moment:

$$M_y = F_y S = 50 \text{ ksi} \times 107 \text{ in}^3 = 5352 \text{ k} - \text{in} = 446 \text{ k} - \text{ft}$$

Example 15

For the previous section, determine the plastic section modulus Z and the plastic moment M_p.

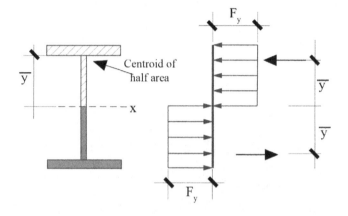

Solution

Locate the centroid of the upper T section from the neutral axis (i.e., axis x).

$$\bar{y} = \frac{\sum Ay}{\sum A} = \frac{(8 \times 6.5 + 3 \times 3)}{(8 \times 1) + (6 \times 0.5)} = 5.545 \text{ in}$$

Determine the moment arm:

$$a = 2\bar{y} = 2 \times 5.54 = 11.09 \text{ in}$$

Find the plastic section modulus:

$$Z = \left(\frac{A}{2}\right)a = \frac{2 \times (8 \times 1) + 12 \times 0.5}{2} \times 11.09 = 122 \text{ in}^3$$

Plastic moment:

$$M_p = F_y Z = 50 \times 122 = 6100 \text{ k-in} = 508.2 \text{ k} - \text{ft}$$

Example 16

Determine the plastic section modulus Z and the plastic moment M_p for the following section.

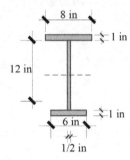

Solution

Locate the centroid of the whole I section.

$$\bar{y} = \frac{\Sigma Ay}{\Sigma A} = \frac{(8 \times 0.5) + (6 \times 7) + (6 \times 13.5)}{8 + 6 + 6}$$

$$= 6.35 \text{ in (from top surface)}$$

Locate the centroid of the upper T section.

$$y_1 = \frac{8 \times (5.35 + 0.5) + (5.35 \times 0.5) \times \dfrac{5.35}{2}}{8 + (5.35 \times 0.5)}$$

$$= 5.054 \text{ in } \left(\text{from N.A. of I section} \right)$$

Locate the centroid of the lower inverted T section.

$$y_2 = \frac{6 \times (6.65 + 0.5) + (6.65 \times 0.5) \times \dfrac{6.65}{2}}{6 + (6.65 \times 0.5)}$$

$$= 5.786 \text{ in (from N.A. of the } I \text{ section)}$$

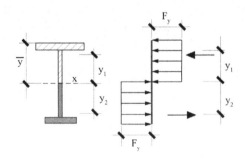

Moment arm:

$$a = y_1 + y_2 = 5.054 + 5.786 = 10.84 \, \text{in}$$

Area of compression:

$$A_C = (8 \times 1) + (5.35 \times 0.5) = 10.675 \, \text{in}$$

Plastic section modulus:

$$Z = A_C a = 10.657 \times 10.84 = 115.7 \, \text{in}^3$$

Plastic moment:

$$M_p = F_y Z = 50 \times 115.7 = 5785.8 \, \text{k} - \text{in} = 482.1 \text{k} - \text{ft}$$

Answer: $Z = 115.7 \, \text{in}^3, M_p = 482.1 k - ft$

8.10   LOCAL BUCKLING STRENGTH

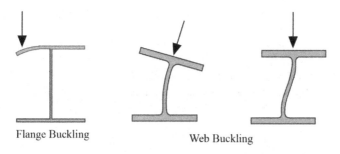

Flange Buckling   Web Buckling

- The strength corresponding to any overall buckling mode, however, such as flexural buckling, cannot be developed if the elements of the cross section are so thin that local buckling occurs.
- This type of instability is a localized buckling or wrinkling at an isolated location.
- If it occurs, the cross section is no longer fully effective and the member has failed.

8.10.1   CLASSIFICATION OF STEEL SECTION

AISC classifies cross-sectional shapes in the following three categories based on width-to-thickness ratio (λ).

- *Compact*, if $\lambda < \lambda_p$. The section can fully utilize its material strength (plastic moment), and there is no local buckling.
- *Noncompact*, if $\lambda_p < \lambda < \lambda_r$. The section cannot fully utilize its material strength. Buckling may occur inelastically or elastically before reaching the plastic moment.
- *Slender*, if $\lambda_r < \lambda$. The section definitely reaches elastic buckling prior to the plastic moment.

Element	λ	λ_p	λ_r
Flange	$\dfrac{b_f}{2t_f}$	$0.38\sqrt{\dfrac{E}{F_y}}$	$1.0\sqrt{\dfrac{E}{F_y}}$
Web	$\dfrac{h}{t_w}$	$3.76\sqrt{\dfrac{E}{F_y}}$	$5.7\sqrt{\dfrac{E}{F_y}}$

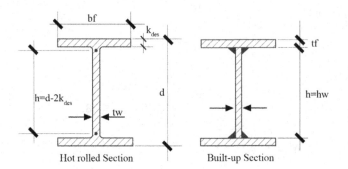

Hot rolled Section   Built-up Section

b_f	Flange width
t_f	Flange thickness
h_w	Web height
t_w	Web thickness
d	Total depth
h	Unstiffened web height
k_{des}	Distance from the outer surface of the flange to the depth of the fillet radius

Example 17

Investigate the local stability of the following section.

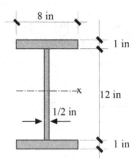

Flange buckling check:

$$\lambda = \frac{b_f}{2t_f} = \frac{8}{2 \times 1} = 4$$

$$\lambda_p = 0.38\sqrt{E/F_y} = 0.38\sqrt{29000/50} = 9.15$$

Since $\lambda(4) < \lambda_p(9.15)$, flange is compact.
 Web buckling check:

$$\lambda = \frac{h}{t_w} = \frac{12}{0.5} = 24$$

$$\lambda_p = 3.76\sqrt{E/F_y} = 3.76\sqrt{29000/50} = 90.6$$

Since $\lambda(24) < \lambda_p(90.6)$, web is also compact.

Answer: Section is compact.

Example 18

Investigate the local stability of section W14 × 90.

Solution

From Table 1–1 of the AISC manual, we find:

Section	b_f	t_f	d	k_{des}	t_w
W14×90	14.5	0.71	14	1.31	0.44

8.10.2 FLANGE BUCKLING CHECK

$$\lambda = \frac{b_f}{2t_f} = \frac{14.5}{2\times0.71} = 10.2$$

$$\lambda_p = 0.38\sqrt{E/F_y} = 0.38\sqrt{29000/50} = 9.15$$

$$\lambda = 1.00\sqrt{E/F_y} = 0.38\sqrt{29000/50} = 24.1$$

Since $\lambda_p(9.15) < \lambda(10.2) < \lambda_r(24.1)$, flange is noncompact.

8.10.3 WEB BUCKLING CHECK

$$\lambda = \frac{h}{t_w} = \frac{d - 2k_{des}}{t_w} = \frac{14 - 2\times1.31}{0.44} = 25.86$$

$$\lambda_p = 3.76\sqrt{E/F_y} = 3.76\sqrt{29000/50} = 90.6$$

Since $\lambda(25.8) < \lambda_p(90.6)$, web is compact.

Answer: Section is noncompact (flange governs).

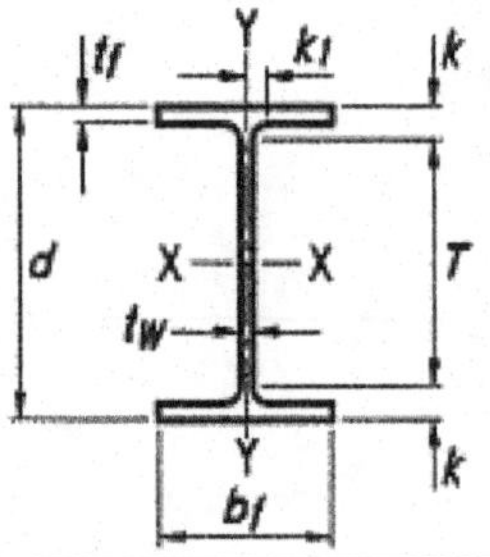
tf
Y
k1
k
d
X
X
T
tw
Y
k
bf

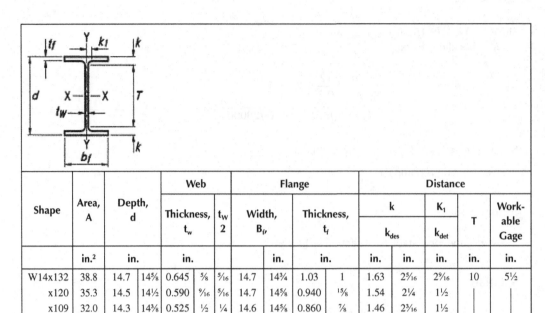

Shape	Area, A	Depth, d		Web			Flange				Distance				
				Thickness, t_w		$\frac{t_w}{2}$	Width, B_f		Thickness, t_f		k		K_1	T	Workable Gage
											k_{des}		k_{det}		
	in.²	in.		in.				in.	in.		in.	in.	in.	in.	in.
W14x132	38.8	14.7	14⅝	0.645	⅝	5/16	14.7	14¾	1.03	1	1.63	2 5/16	2 9/16	10	5½
x120	35.3	14.5	14½	0.590	9/16	5/16	14.7	14⅝	0.940	15/8	1.54	2¼	1½	↓	↓
x109	32.0	14.3	14⅜	0.525	½	¼	14.6	14⅝	0.860	⅞	1.46	2 3/16	1½	↓	↓
x99f	29.1	14.2	14⅛	0.485	½	¼	14.6	14⅝	0.780	¾	1.38	2 1/16	1 7/16	↓	↓
x90f	26.5	14.0	14.0	0.440	7/16	7/4	14.5	14½	0.710	11/16	1.31	2	1 7/16	↓	↓

Example 19

The following beam is W 16×31 of A992 steel. It supports a reinforced concrete floor slab that provides continuous lateral support of the compression flange. The service dead load is 450lb / ft, and the service live load is 550lb / ft. Does the beam have adequate moment strength?

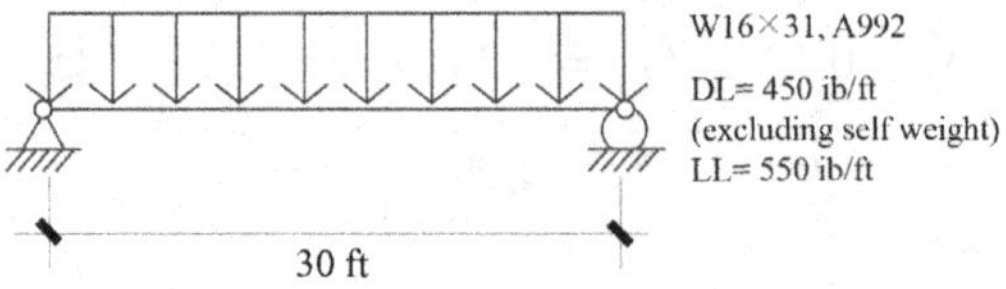
W16×31, A992
DL= 450 ib/ft
(excluding self weight)
LL= 550 ib/ft
30 ft

8.10.4 Flange Check

From Table 1–1, $b_f = 5.53$ in, $t_f = 0.44$ in.

$$\lambda = \frac{b_f}{2t_f} = \frac{5.53}{2 \times 0.44} = 6.28$$

$$\lambda_p = 0.38\sqrt{E / F_y} = 0.38\sqrt{29000 / 50} = 9.15 > 6.28$$

Since $\lambda < \lambda_p$ for flange, there is no local buckling in flange.

8.10.5   Web Check

From Table $1-1, d = 15.9$ in, $k_{des} = 0.842$ in, and $t_w = 0.275$ in.

$$\lambda = \frac{h}{t_w} = \frac{d - 2k_{des}}{t_w} = \frac{15.9 - 2 \times 0.842}{0.275} = 51.7$$

$$\lambda_p = 3.76\sqrt{E/F_y} = 3.76\sqrt{29000/50} = 90.5 > 51.7$$

Since $\lambda < \lambda_p$ for web, there is no local buckling in web either.

8.10.6   Determine Capacity

Since both flange and web have no local buckling, the section can reach up to plastic moment before failure.

From Table 1–1, $Z_x = 54^3$.

$$M_p = F_y Z_x = 50 \times 54 = 2700 \text{k-in}$$

$$\phi_b M_n = \phi_b M_p = 0.9 \times 2700 \text{k-in} = 2430 \text{k-in } 202.5 \text{k-ft}$$

8.10.7   Determine Demand

The dead load should be increased by self-weight (31 lb / ft) of the beam since the given dead load (450lb / ft) is exclusive of self-weight.

$$w_D = 450 + 31 = 481 \text{ lb / ft}$$

$$w_L = 550 \text{ lb / ft}$$

$$w_u = 1.2 w_D + 1.6 w_L$$

$$= 1.2 \times 481 + 1.6 \times 550 = 1457 \text{ lb / ft} = 1.46 \text{ k / f}$$

$$M_u = \frac{w_u L^2}{8} = \frac{1.46 \times 30^2}{8} = 164.3 \text{ k} - \text{ft} < \phi_b M_n$$

Since $\phi_b M_n \left(202.5k - ft\right) > M_u \left(164.3k - ft\right)$, the section W 16×31 has adequate moment capacity.
   Answer: Yes. The beam has adequate moment strength.

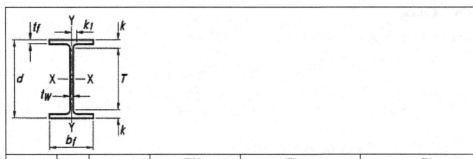

Shape	Area, A	Depth, d		Web Thickness, t_w	$\frac{t_w}{2}$	Flange Width, b_f		Flange Thickness, t_f		Distance k K_{des}	K_{det}	K_1	T	Workable Gage
	In.2	In.		In.	In.	In.		In.		In.		In.	In.	In.
W16x100	29.4	17.0	17	0.585	9/16	5/16	10.4	10⅜	0.985	1	1.39	1⅞	1⅛	131/4	51/2
x89	26.2	16.8	16¾	0.525	½	¼	10.4	10⅜	0.875	⅞	1.28	1¾	1¹⁄₁₆		
x77	22.6	16.5	16½	0.455	7/16	¼	10.3	10¼	0.760	¾	1.16	1⅝	1 ¹⁄₁₆		
x67ᶜ	19.6	16.3	16⅜	0.395	⅜	3/16	10.2	10¼	0.665	1¹⁄₁₆	1.07	19/16	1		
W16x57	16.8	16.4	16⅜	0.430	7/16	¼	7.12	7⅛	0.715	1¹⁄₁₆	1.12	1⅜	⅞	135/8	31/29
x50ᶜ	14.7	16.3	16¼	0.380	⅜	3/16	7.07	7⅛	0.630	⅝	1.03	1⁵⁄₁₆	1³⁄₁₆		
x45ᶜ	13.3	16.1	16⅛	0.345	⅜	3/16	7.04	7	0.565	⁹⁄₁₆	0.967	1¼	1³⁄₁₆		
x40ᶜ	11.8	16.0	16	0.305	5/16	3/16	7.00	7	0.505	½	0.907	1³⁄₁₆	1³⁄₁₆		
x36ᶜ	10.6	15.9	15⅞	0.295	5/16	3/16	6.99	7	0.430	⁷⁄₁₆	0.832	1⅛	¾		
W16x31ᶜ	9.13	15.9	15⅞	0.275	¼	⅛	5.53	5½	0.440	⁷⁄₁₆	0.842	1⅛	¾	135/8	31/2
x26ᶜ·ᵛ	7.68	15.7	15¾	0.250	¼	⅛	5.50	5½	0.345	⅜	0.747	1¹⁄₁₆	¾	135/8	31/2

W16-W14

Nominal Wt.	Compact Section Criteria		Axis X-X				Axis Y-Y				r_{ts}	h_o	Torsional Properties	
	bt/2ti	h/t_w	I In.⁴	S In.³	r In.	Z In.³	I In.⁴	S In.³	r In.	Z In.³	In.	In.	J In.⁴	Cw In.⁶
lb/ft														
100	5.29	24.3	1490	175	7.10	198	186	35.7	2.51	54.9	2.92	16.0	7.73	11900
89	5.92	27.0	1300	155	7.05	175	163	31.4	2.49	48.1	2.88	15.9	5.45	10200
77	6.77	31.2	1110	134	7.00	150	138	26.9	2.47	41.1	2.85	15.7	3.57	8590
67	7.70	35.9	954	117	6.96	130	119	23.2	2.46	35.5	2.82	15.6	2.39	7300
57	4.98	33.0	758	92.2	6.72	105	43.1	12.1	1.60	18.9	1.92	15.7	2.22	2660
50	5.61	37.4	659	81.0	6.68	92.0	37.2	10.5	1.59	16.3	1.89	15.7	1.52	2270
45	6.23	41.1	586	72.7	6.65	82.3	32.8	9.34	1.57	14.5	1.87	15.5	1.11	1990
40	6.93	46.5	518	64.7	6.63	73.0	28.9	8.25	1.57	12.7	1.86	15.5	0.794	1730
36	8.12	48.1	448	56.5	6.51	64.0	24.5	7.00	1.52	10.8	1.83	15.5	0.545	1460
31	6.28	51.6	375	47.2	6.41	54.0	12.4	4.49	1.17	7.03	1.42	15.5	0.461	739
26	7.97	56.8	301	38.4	6.26	44.2	9.59	3.49	1.12	5.48	1.38	15.4	0.262	565

Example 20

The beam shown in following figure supports a reinforced concrete floor slab that provides continuous lateral support of compression flange. Does the beam have adequate moment strength?

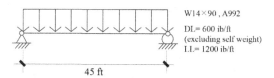

W14×90 , A992

DL= 600 ib/ft
(excluding self weight)
LL= 1200 ib/ft

45 ft

8.10.8   FLANGE CHECK

From Table $1-1, b_f = 14.5$ in, $t_f = 0.71$ in.

$$\lambda = \frac{b_f}{2t_f} = \frac{14.5}{2\times0.71} = 10.21$$

$$\lambda_p = 0.38\sqrt{E / F_y} = 0.38\sqrt{29000 / 50} = 9.15 < 10.21$$

$$\lambda_r = 1.0\sqrt{E / F_y} = 1.0\sqrt{29000 / 50} = 24.08 > 10.21$$

Since $\lambda_p < \lambda < \lambda_r$, flange is noncompact.

8.10.9   WEB CHECK

From Table 1–1, $d = 14$ in, $k_{des} = 1.31$ in and $t_w = 0.44$ in

$$\lambda = \frac{h}{t_w} = \frac{d - 2k_{des}}{t_w} = \frac{14 - 2\times1.31}{0.44} = 25.8$$

$$\lambda_p = 3.76\sqrt{E / F_y} = 3.76\sqrt{29000 / 50} = 90.5 > 25.8$$

Since $\lambda < \lambda_p$, web is compact.
The shape is therefore noncompact.

8.10.10   DETERMINE CAPACITY

Since flange has local buckling, the section cannot reach up to the plastic moment before failure.

From Table 1-1, $S_x = 143$ in³, $Z_x = 157$ in³

$$M_p = F_y Z_x = 50\times157 = 7850 \text{ k-in}$$

$$\phi_b M_n = \phi_b \left[M_p - \left(M_p - 0.7 F_y S_x \right) \left(\frac{\lambda - \lambda_p}{\lambda_r - \lambda_p} \right) \right]$$

$$= 0.9\times\left[7850 - \left(7850 - 0.7\times50\times143 \right) \left(\frac{10.2 - 9.15}{24.08 - 9.15} \right) \right]$$

$$= 7650 \text{k-in} = 637.5 \text{ k} - \text{ft}$$

8.10.11 Determine Demand

The dead load should be increased by self-weight $(90\text{lb}/\text{ft})$ of the beam since the given dead load $(600\text{lb}/\text{ft})$ is exclusive of self-weight.

$$w_D = 600 + 90 = 690\text{lb}/\text{ft}$$

$$w_L = 1200\text{lb}/\text{ft}$$

$$w_u = 1.2w_D + 1.6w_L$$

$$= 1.2 \times 690 + 1.6 \times 1200 = 2748\text{lb}/\text{ft} = 2.75\text{k}/\text{ft}$$

$$M_u = \frac{w_u L^2}{8} = \frac{2.75 \times 45^2}{8} = 696.1\text{k} - \text{ft} > \phi_b M_n$$

Since $\phi_b M_n (637.5\text{k} \text{ -ft})< M_u (696.1\text{k} \text{ -ft})$, the section $W14 \times 90$ has no adequate moment capacity.

Answer: No. The beam has no adequate moment strength.

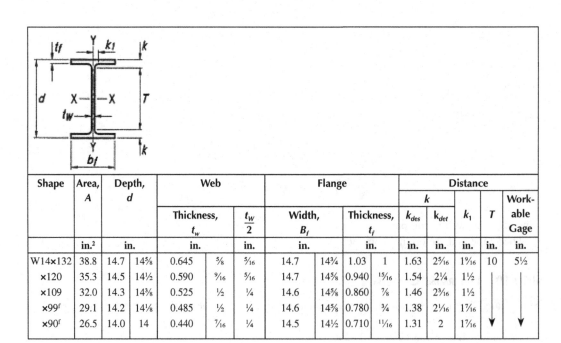

Shape	Area, A	Depth, d		Web			Flange				Distance				
				Thickness, t_w		$\dfrac{t_w}{2}$	Width, B_f		Thickness, t_f		k_{des}	k_{det}	k_1	T	Work-able Gage
												k			
	in.²	in.		in.		in.	in.		in.		in.	in.	in.	in.	in.
W14×132	38.8	14.7	14⅝	0.645	⅝	⁵⁄₁₆	14.7	14¾	1.03	1	1.63	2⁵⁄₁₆	1⁹⁄₁₆	10	5½
×120	35.3	14.5	14½	0.590	⁹⁄₁₆	⁵⁄₁₆	14.7	14⅝	0.940	¹⁵⁄₁₆	1.54	2¼	1½		
×109	32.0	14.3	14⅜	0.525	½	¼	14.6	14⅝	0.860	⅞	1.46	2³⁄₁₆	1½		
×99[f]	29.1	14.2	14⅛	0.485	½	¼	14.6	14⅝	0.780	¾	1.38	2¹⁄₁₆	1⁷⁄₁₆		
×90[f]	26.5	14.0	14	0.440	⁷⁄₁₆	¼	14.5	14½	0.710	¹¹⁄₁₆	1.31	2	1⁷⁄₁₆		

Nom inal Wt.	Compact Section Criteria		Axis X-X				Axis Y-Y				r_{ts}	h_0	Torsional Properties	
	$\dfrac{b_f}{2t_f}$	$\dfrac{h}{t_w}$	I	S	r	Z	I	S	r	Z			J	C_W
lb/ft			in.⁴	in.³	in.	in.³	in.⁴	in.³	in.	in.³	in.	in.	in.⁴	in.⁶
132	7.15	17.7	1530	209	6.28	234	548	74.5	3.76	113	4.23	13.7	12.3	25500
120	7.80	19.3	1380	190	6.24	212	495	67.5	3.74	102	4.0	13.6	9.37	22700
109	8.49	21.7	1240	173	6.22	192	447	61.2	3.73	92.7	4.17	13.4	7.12	20200
99	9.34	23.5	1110	157	6.17	173	402	55.2	3.71	83.6	4.14	13.4	5.37	18000
90	10.2	25.9	999	143	6.14	157	362	49.9	3.70	75.6	4.10	13.3	4.06	16000

W14-W12

8.11 LATERAL TORSIONAL BUCKLING STRENGTH

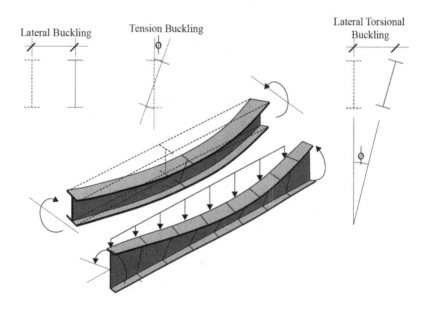

Lateral torsional buckling is a torsional deformation that is accompanied by flexural deformation

Lateral support, L_b:

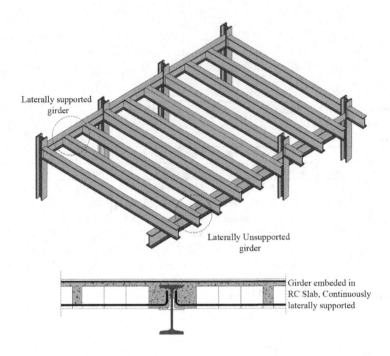

Laterally supported girder

Laterally Unsupported girder

Girder embeded in RC Slab, Continuously laterally supported

Bending strength (M_n) vs. unsupported length (L_b):

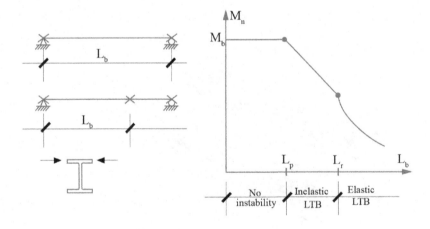

No instability

Inelastic LTB

Elastic LTB

1. No LTB.
2. Inelastic LTB.
3. Elastic LTB.

$$M_n = M_p = F_y Z_x$$

$$M_n = C_b \left[M_p - \left(M_p - 0.7 F_y S_x \right) \left(\frac{L_b - L_p}{L_r - L_p} \right) \right]$$

$$M_n = \frac{C_b \pi^2 E S_x}{\left(L_b / r_{ts} \right)^2} \sqrt{1 + 0.078 \frac{1}{S_x h_o} \left(\frac{L_b}{r_{ts}} \right)^2}$$

8.11.1 MOMENT GRADIENT, C_B

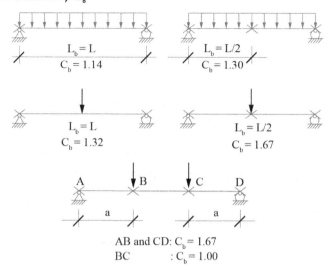

$$L_b = L$$
$$C_b = 1.14$$

$$L_b = L/2$$
$$C_b = 1.30$$

$$L_b = L$$
$$C_b = 1.32$$

$$L_b = L/2$$
$$C_b = 1.67$$

$$\text{AB and CD: } C_b = 1.67$$
$$\text{BC} \qquad : C_b = 1.00$$

Moment gradient factor (C_b) accounts for the variable moment that occurs along the span of the beam. For constant bending moment, $C_b = 1.0$; otherwise:

$$C_b = \frac{12.5M_{max}}{2.5M_{max} + 3M_A + 4M_B + 3M_C}$$

The Ms are an absolute value of the moment in the unbraced segment in the following position:

M_{max}	Entire beam
M_A	First quarter point
M_B	Midspan
M_C	Third quarter point

Example 21

Determine the moment gradient of the following beam.

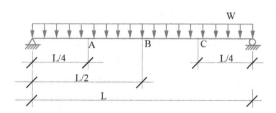

Solution

$$M_A = \frac{wL}{2}\left(\frac{L}{4}\right) - \frac{wL}{4}\left(\frac{L}{8}\right) = \frac{3}{32}wL^2$$

$$M_B = \frac{wL^2}{8} \text{ (Simply supported, midspan)}$$

$$M_C = \frac{3}{32}wL^2 \text{ (Same as } M_A \text{ due to symmetry)}$$

$$M_{max} = \frac{wL^2}{8} \text{ (At midspan)}$$

$$C_b = \frac{12.5M_{max}}{2.5M_{max} + 3M_A + 4M_B + 3M_C}$$

$$= \frac{12.5\left(\frac{1}{8}\right)wL^2}{\left[2.5\left(\frac{1}{8}\right) + 3\left(\frac{3}{32}\right) + 4\left(\frac{1}{8}\right) + 3\left(\frac{3}{32}\right)\right]wL^2}$$

$$= 1.14$$

Example 22

Determine the moment gradient of the following beam, which is laterally supported at midspan.

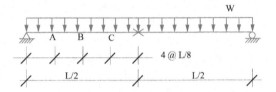

Solution

$$M_A = \frac{wL}{2}\left(\frac{L}{8}\right) - \frac{wL}{8}\left(\frac{L}{16}\right) = \frac{7}{128}wL^2$$

$$M_B = \frac{wL}{2}\left(\frac{L}{4}\right) - \frac{wL}{4}\left(\frac{L}{8}\right) = \frac{3}{32}wL^2$$

$$M_C = \frac{wL}{2}\left(\frac{3L}{8}\right) - \frac{3wL}{8}\left(\frac{3L}{16}\right) = \frac{15}{128}wL^2$$

$$M_{max} = \frac{wL^2}{8} \text{ (At midspan)}$$

$$C_b = \frac{12.5M_{max}}{2.5M_{max} + 3M_A + 4M_B + 3M_C}$$

$$= \frac{12.5\left(\frac{1}{8}\right)wL^2}{\left[2.5\left(\frac{1}{8}\right) + 3\left(\frac{7}{128}\right) + 4\left(\frac{3}{32}\right) + 3\left(\frac{15}{128}\right)\right]wL^2}$$

$$= 1.30$$

Example 23

Does the following beam have adequate moment strength if:

1. $L = 8$ ft
2. $L = 20$ ft
3. $L = 44$ ft
4. $L = 44$ ft and if there is a lateral support at midspan?

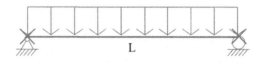

W14×68 , A992
DL= 400ib/ft(excluding self weight)
LL=700ib/ft

Figure 8.38	C_b	1.14
Tab. 3–2	L_p	8.69 ft
	L_r	29.3 ft
Tab. 1–1	Z_x	115 in³
	S_X	103 in³
	J	3.01 in⁴
	r_{ts}	2.80 in
	h_O	13.3 in

$$w_u = 1.2 \times (400 + 68) + 1.6 \times 700$$
$$= 1.68 \text{k / ft}$$
$$M_p = F_y Z_x = 50 \times 115 = 5750 \text{k -in}$$

Case 1: $L_b = 8$ft. Since $L_b < L_p$, no LTB occurs.

$$\phi_b M_n = \phi_b M_p = 0.9 \times 5750 \text{k-in} = 5175 \text{k} - \text{in} = 431.2 \text{k} - \text{ft}$$
$$M_u = w_u l^2 / 8 = 1.68 \times 8^2 / 8 = 13.4 \text{k} - \text{ft} < \phi_b M_n (431.2 \text{k} - \text{ft}) \therefore \text{O.K}$$

Case 2: $L_b = 20$ft. Since $L_p < L_b < L_r$, inelastic LTB occurs.

$$\phi_b M_n = \phi_b C_b \left[M_p - (M_p - 0.7 F_y S_x) \left(\frac{L_b - L_p}{L_r - L_p} \right) \right]$$
$$= 0.90 \times 1.14 \left[5750 - (5750 - 0.7 \times 50 \times 103) \left(\frac{20 - 8.69}{29.3 - 8.69} \right) \right] = 4691 = 390.9 \text{k} - \text{f}$$

$$M_u = w_u l^2 / 8 = 1.68 \times 20^2 / 8 = 84 \text{k} - \text{ft} < \phi_b M_n (390.9 \text{ k} - \text{ft}) \therefore \text{O.K}$$

Case 3: $L_b = 44$ ft Since $L_r < L_b$, elastic LTB occurs.

$$\phi_b M_n = \frac{\phi_b C_b \pi^2 S_x E}{\left(L_b / r_{ts}\right)^2} \sqrt{1 + 0.078 \frac{J}{S_x h_O} \left(\frac{L_b}{r_{ts}}\right)^2}$$

$$= \frac{0.9 \times 1.14 \pi^2 \times 103 \times 29000}{\left(44 \times 12 / 2.80\right)^2} \sqrt{1 + 0.078 \frac{3.01}{103 \times 13.3} \left(\frac{44 \times 12}{2.80}\right)^2} = 2265 = 188.7 \, k - ft$$

$$M_u = w_u L^2 / 8 = 1.68 \times 44^2 / 8 = 406.5 k - ft > \phi_b M_n \left(188.7 k - ft\right) \therefore \text{ N.G}$$

Case 4: $L_b = 44 / 2 = 22$ ft. Since $L_p < L_b < L_r$, inelastic LTB occurs.

$$\phi_b M_n = 0.90 \times 1.30 \left[5750 - \left(5750 - 0.7 \times 50 \times 103\right) \left(\frac{22 - 8.69}{29.3 - 8.69}\right)\right] = 425.5 k - ft$$

$$M_u = w_u L^2 / 8 = 1.68 \times 44^2 / 8 = 406.5 k - ft < \phi_b M_n \left(425.5 k - ft\right) \therefore \text{ OK}$$

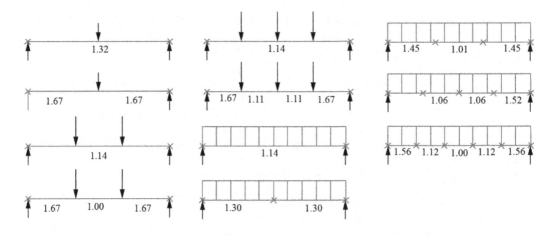

Shape	z_x	M_{px}/Ω_b	$\phi_b M_{px}$	M_{rx}/Ω_b	$\phi_b M_{rx}$	BF/Ω_b	$\phi_b BF$	L_p	L_r	I_x	V_{nx}/Ω_v	$\phi_v V_{nx}$
		kip-ft	kip-ft	kip-ft	kip-ft	kips	kips				kips	kips
	in.³	ASD	LRFD	ASD	LRFD	ASD	LRFD	ft	ft	in.⁴	ASD	LRFD
W21×55	126	314	473	192	289	10.8	16.3	6.11	17.4	1140	156	234
W14×74	126	314	473	196	294	5.31	8.05	8.76	31.0	795	128	192
W18×60	123	307	461	189	284	9.62	14.4	5.93	18.2	984	151	227
W12×79	119	297	446	187	281	3.78	5.67	10.8	39.9	662	117	175
W14×68	115	287	431	180	270	5.19	7.81	8.69	29.3	722	116	174
W10×88	113	282	424	172	259	2.62	3.94	9.29	51.2	534	131	196

$F_y = 50$ ksi

Z_x

											W14-W12		
Nom-inal Wt.	Compact Section Criteria		Axis X-X				AxisY-Y				r_{ts}	h_0	Torsional Properties	
lb/ft	$\dfrac{bf}{2tf}$	$\dfrac{h}{tw}$	I in.⁴	S in.³	r in.	Z in.³	I in.⁴	Z in.³	r in.	Z in.³	in.	in.	J in.⁴	C_w in.⁶
132	7.15	17.7	1530	209	6.28	234	548	74.5	3.76	113	4.23	13.7	12.3	25500
120	7.80	19.3	1380	190	6.24	212	495	67.5	3.74	102	4.20	13.6	9.37	22700
109	8.49	21.7	1240	173	6.22	192	447	61.2	3.73	92.7	4.17	13.4	7.12	20200
99	9.34	23.5	1110	157	6.17	173	402	55.2	3.71	83.6	4.14	13.4	5.37	18000
90	10.2	25.9	999	143	6.14	157	362	49.9	3.70	75.6	4.10	13.3	4.06	16000
82	5.92	22.4	881	123	6.05	139	148	29.3	2.48	44.8	2.85	13.4	5.07	6710
74	6.41	25.4	795	112	6.04	126	134	26.6	2.48	40.5	2.83	13.4	3.87	5990
68	6.97	27.5	722	103	6.01	115	121	24.2	2.46	36.9	2.80	13.3	3.01	5380
61	7.75	30.4	640	92.1	5.98	102	107	21.5	2.45	32.8	2.78	13.3	2.19	4710

8.12   BEAM DESIGN

Example 24

Select the lightest section for the following beam, which has continuous lateral support. Steel is A992.

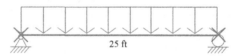

25 ft

DL=1.2 k/ft, LL= 2.0 k/ft

8.12.1   DETERMINE DESIGN MOMENT

Ignore self-weight since the self-weight of the beam is unknown.

$$w_u = 1.2w_D + 1.6w_L = 1.2(1.2+0) + 1.6 \times 2 = 4.64 \text{k} / \text{ft}$$
$$M_u = w_u L^2 / 8 = 4.64 \times 25^2 / 8 = 362.5 \text{k} - \text{ft}$$

Trial 1:

From Table 3–2, try $W\,10\times77$ since its $\phi_b M_p = 366\text{k -ft.}$

$$w_u = 1.2(1.2+0.077)+1.6\times2 = 4.73\text{k / ft}$$
$$M_u = w_u L^2 / 8 = 4.73\times25^2 / 8 = 369.5 > \phi_b M_p\,(366)\ \ \text{N.G}$$

Trial 2:

From Table 3–2, try $W\,18\times50, \phi_b M_p = 379$ k-ft.

$$w_u = 1.2(1.2+0.050)+1.6\times2 = 4.70\text{k / ft}$$
$$M_u = w_u L^2 / 8 = 4.70\times25^2 / 8 = 367.2 < \phi_b M_p\,(379)\ \ \text{O.K}$$

8.12.2   CHECK LIVE LOAD DEFLECTION

From Table 3–2, $I_x = 800\ \text{in}^4$:

$$\Delta_{max} = \frac{L}{360} = \frac{25\times12}{360} = 0.83\ \text{in}$$
$$\Delta = \frac{5}{384}\frac{wL^4}{EI} = \frac{5}{384}\frac{(2/12)\times(25\times12)^4}{29000\times800} = 0.75\ \text{in}$$

Since $\Delta(0.75) < \Delta_{max}(0.83)$, live load deflection is okay.

8.12.3   CHECK DEAD LOAD PLUS LIVE LOAD DEFLECTION

$$w_s = w_D + w_L = (1.2+0.050)+2 = 3.25\text{k / ft}$$
$$\Delta_{max} = \frac{L}{240} = \frac{25\times12}{240} = 1.25\ \text{in}$$
$$\Delta = \frac{5}{384}\frac{wL^4}{EI} = \frac{5}{384}\frac{(3.25/12)\times(25\times12)^4}{29000\times800} = 1.23\ \text{in}$$

Since $\Delta(1.23) < \Delta_{max}(1.25)$, dead load plus live load deflection is also okay.

8.12.4   CHECK SHEAR

$$V_u = \frac{w_u L}{2} = \frac{4.70\times25}{2} = 58.75\ \text{kip}$$

From Table 3–2, $\phi_v V_n = 192$ kip.
  Since $V_u\,(58.75) < \phi_v V_n\,(192)$, shear is also checked.
  Answer: Select $W\,18\times50$ section.

		M_{px}/Ω_b	$\phi_b M_{px}$	M_{rx}/Ω_b	$\phi_b M_{rx}$	BF/Ω_b	$\phi_b BF$				V_{nx}/Ω_v	$\phi_v v_{nx}$
Shape	Z_x	kip-ft	kip-ft	kip-ft	kip-ft	kips	kips	L_p	Lr	L_x	kips	kips
	in.³	ASD	LRFD	ASD	LRFD	ASD	LRFD	ft	ft	in.⁴	ASD	LRFD
W21x55	126	314	473	192	289	10.8	16.3	6.11	17.4	1140	156	234
W14x74	126	314	473	196	294	5.31	8.05	8.76	31.0	795	128	192
W18x60	123	307	461	189	284	9.62	14.4	5.93	18.2	984	151	227
W12x79	119	297	446	187	281	3.78	5.67	10.8	39.9	662	117	175
W14x68	115	287	431	180	270	5.19	7.81	8.69	29.3	722	116	174
W10x88	113	282	424	172	259	2.62	3.94	9.29	51.2	534	131	196
W18x55	112	279	420	172	258	9.15	13.8	5.90	17.6	890	141	212
W21x50	110	274	413	165	248	12.1	18.3	4.59	13.6	984	158	237
W12x72	108	269	405	170	256	3.69	5.56	10.7	37.5	597	106	159
W21x48[f]	107	265	398	162	244	9.89	14.8	5.86	16.5	959	144	216
W16x57	105	262	394	161	242	7.98	12.0	5.65	18.3	758	141	212
W14x61	102	254	383	161	242	4.93	7.48	8.65	27.5	640	104	156
W18x50	101	252	379	155	233	8.76	13.2	5.83	16.9	800	128	192
W10x77	97.6	244	366	150	225	2.60	3.90	9.18	45.3	455	112	169
W12x65[f]	96.8	237	356	154	231	3.58	5.39	10.7	35.1	533	94.4	142

F_y = 50 ksi

$$Z_x$$

8.13   BOLT CONNECTION

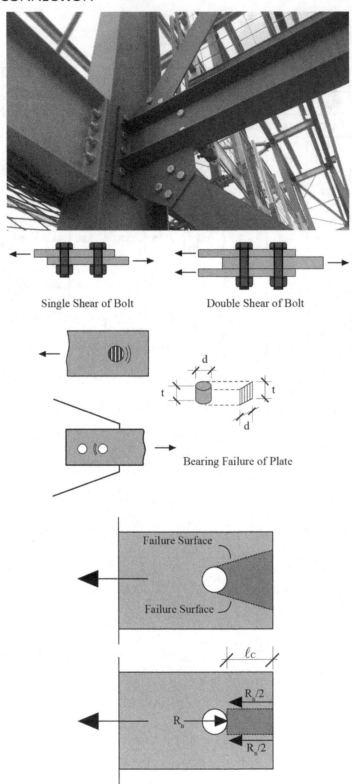

Single Shear of Bolt   Double Shear of Bolt

Bearing Failure of Plate

Failure Surface

Failure Surface

8.13.1 Bearing Strength Criteria

A possible failure mode resulting from excessive bearing is shear tear-out at the end of a connected element.

$$\frac{R_n}{2} = 0.6F_u \ell_c t; \; R_n = 1.2F_u \ell_c t$$

$0.6F_u$: Shear fracture stress of the connected part
ℓ_C : Distance from the edge of the hole to the edge of the connected part
d : Bolt diameter
t : Thickness of the connected part

8.13.2 Excessive Elongation of Hole Criteria

To prevent excessive elongation of the hole, an upper limit is placed on the bearing load, given by the previous equation.

$$R_n = 2.4dtF_u$$

So in combination of both the bearing criteria and the excessive elongation of hole criteria:

$$R_n = 1.2F_u \ell_c t \leq 2.4dtF_u$$

AISC recommends to use:

$$\ell_C \geq 2.67d \qquad \text{preferebly} \; 3d$$
$$\ell_e \geq (d+0.25) \quad \text{upto1 inch dia bolt}$$

For determining ℓ_c :

$$\ell_C = \ell_e - \frac{h}{2} \quad \text{For edge bolts}$$
$$\ell_C = s - h \qquad \text{For other bolts}$$

Use actual hole diameter for h :

$$h = d + \frac{1}{16}$$

ℓ_c : Clear distance to the perimeter of the hole
ℓ_e : Edge distance to the center of the hole
s : Center-to-center spacing of the holes
h : Hole diameter

Example 25

Check bolt spacing, edge distances, and bearing on tension member and gusset plate of the connection. Bolt is A325.

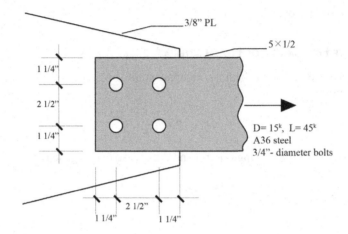

Check spacing.
   Minimum spacing in any direction:

$$2.67d = 2.67 \times (3/4) = 2 \text{ in} < 2.50 \text{ in}   \text{O.K}$$

Minimum edge distance in any direction (for up to $d = 1.0$ in):

$$d + 0.25 = 3/4 + 0.25 = 1.00 < 1.25 \text{ in}   \text{O.K}$$

8.13.3 HOLE DIAMETER

$$h = d + (1/16) = (3/4) + (1/16) = (13/16) \text{ in}$$

8.13.4 BEARING OF EDGE BOLT ON TENSION MEMBER

$$\ell_c = \ell_e - h/2 = 1.25 - (13/16)/2 = 0.843 \text{ in}$$
$$R_n = 1.2\ell_c t F_u \leq 2.4 dt F_u$$
$$= 1.2 \times \left(0.843 \times \frac{1}{2}\right) \times 58 \leq 2.4 \times \left(\frac{3}{4} \times \frac{1}{2}\right) \times 58$$
$$= 29.36 \text{ kip} \leq 52.2 \text{ kip}; \quad \text{Use } R_n = 29.36 \text{ kip / bolt}$$

8.13.5   Bearing of Inner Bolt on Tension Member

$$\ell_c = s - h = 2.5 - (13/16) = 1.688 \text{ in}$$
$$R_n = 1.2\ell_c t F_u \le 2.4 dt F_u$$
$$= 1.2 \times \left(1.688 \times \frac{1}{2}\right) \times 58 \le 52.20 \text{ kip}$$
$$= 58.74 \text{ kip} \not\le 52.2 \text{ kip} \quad \text{Use } R_n = 52.2 \text{ kip} / \text{bolt}$$

8.13.6   Bearing on Tension Member

$$R_n = 2 \times 29.36 + 2 \times 52.20 = 163.1 \text{ kip}$$

8.13.7   Bearing on Gusset Plate

Repeating the same process for edge and inner bolt, same as bearing on the tension member, but taking $t = 3/8$ in, we find:

$$R_n = 2 \times 22.02 + 2 \times 39.15 = 122.3 \text{ kip}$$

8.13.8   Shearing on Bolt

$$R_n = nF_n A_b = 4 \times 48 \times \frac{\pi}{4} \times \left(\frac{3}{4}\right)^2 = 84.8 \text{ kip}$$

8.13.9   Design Strength

$$\phi R_n = 0.75 \times 84.8 = 63.6 \text{ kip}$$

8.13.10   Demand

$$P_u = 1.2DL + 1.6LL = 1.2 \times 15 + 1.6 \times 45 = 90 \text{kip} > \phi R_n$$

Answer: Connection is not adequate.

Example 26

The C8×18.75 shown has been selected to resist a service dead load of 18 kips and a service live load of 54 kips. It is to be attached to a 3/8 in gusset plate with 7 / 8 in diameter. Determine the number and required layout of bolts such that the length of connection L is reasonably small. Assume A36 steel.

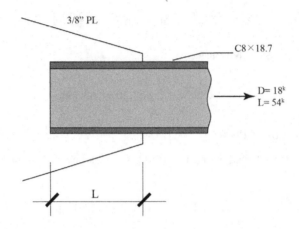

8.13.11   SHEARING ON EACH BOLT

$$R_n = F_n A_b = 48 \times \frac{\pi}{4} \times \left(\frac{7}{8}\right)^2 = 28.86 \text{ kip / bolt}$$

8.13.12   DEMAND

$$P_u = 1.2DL + 1.6LL = 1.2 \times 18 + 1.6 \times 54 = 108 \text{ kip}$$

8.13.13   BOLTS REQUIRED

$$n = \frac{P_u}{\phi R_n} = \frac{108}{0.75 \times 28.86} = 4.99 \text{ bolts}$$

Try using six bolts for a symmetrical layout with two gage lines of three bolts.

8.13.14 DETERMINE SPACING

Minimum spacing in any direction:

$$2.67d = 2.67 \times (7/8) = 2.33 \text{ in} \quad \text{Try} 2.50 \text{ in}$$

Minimum edge distance in any direction (for up to $d = 1.0$ in):

$$d + 0.25 = 7/8 + 0.25 = 1.125 \text{ in} \quad \text{Try} 1.125 \text{ in}$$

8.13.15 HOLE DIAMETER

$$h = d + (1/16) = (7/8) + (1/16) = (15/16) \text{ in}$$

Only the bearing of the gusset plate should be checked since it is thinner than the channel.

8.13.16 BEARING OF EDGE BOLT ON GUSSET PLATE

$$\ell_C = \ell_e - h/2 = 1.125 - (15/16)/2 = 0.656 \text{ in}$$
$$R_n = 1.2\ell_c tF_u \leq 2.4dtF_u$$
$$= 1.2 \times \left(0.656 \times \frac{3}{8}\right) \times 58 \leq 2.4 \times \left(\frac{7}{8} \times \frac{3}{8}\right) \times 58$$
$$= 17.13 \text{ kip} \leq 45.68 \text{ kip}; \quad \text{Use } R_n = 17.13 \text{ kip / bolt}$$

8.13.17 BEARING OF INNER BOLT ON GUSSET PLATE

$$\ell_C = s - h = 2.5 - (15/16) = 1.563 \text{ in}$$
$$R_n = 1.2\ell_c tF_u \leq 2.4dtF_u$$
$$= 1.2 \times \left(1.563 \times \frac{3}{8}\right) \times 58 \leq 45.68 \text{ kip}$$
$$= 40.79 \text{kip} \leq 45.68 \text{kip}; \quad \text{Use } R_n = 40.79 \text{ kip / bolt}$$

8.13.18 BEARING ON GUSSET PLATE

$$R_n = 2 \times 17.13 + 4 \times 40.79 = 197.1 \text{kip} > P_u \quad \text{O.K}$$

8.14 WELD JOINTS

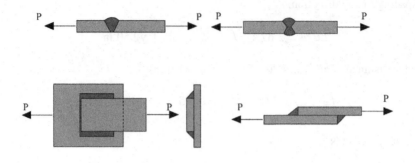

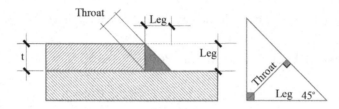

$$\text{Throat} = \text{Leg} \times \sin 45 = \text{Leg} \times \frac{1}{\sqrt{2}} = 0.707 \times \text{Leg}$$

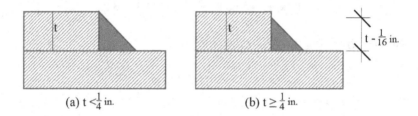

(a) $t < \frac{1}{4}$ in.   (b) $t \geq \frac{1}{4}$ in.

- If plate thickness $t < \dfrac{1}{4}$ in, then the maximum weld thickness $= t$.

- If plate thickness $t >= \dfrac{1}{4}$ in, then the maximum weld thickness $= t - \dfrac{1}{16}$.

Example 27

A steel plate (PL $10 \times 3/4$) and an angle member (L $6 \times 5 \times 1/4$) will be joined together by E60 weld. Determine the length of the welds if the angle member carries a centric force $P = 60$ kip, given that $a = 4.5$ in.

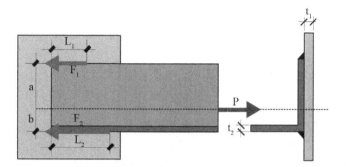

Solution

8.14.1 ALLOWABLE WELD SHEAR STRENGTH

$$\sigma_S = 0.30 \times \sigma_T = 0.30 \times 60 = 18 \text{ ksi}$$

8.14.2 MAXIMUM WELD SIZE

$$L = \frac{1}{4} - \frac{1}{16} = \frac{3}{16} \text{ in}$$

8.14.2 WELD CAPACITY PER UNIT LENGTH

$$f_s = \sigma_s \times 0.707 \times L = 18 \times 0.707 \times \frac{3}{16} = 2.38 \text{ k / in}$$

8.14.3 DETERMINE F_1 AND F_2

$$\sum M_{\text{bottom weld}} = 00^+ - F_1 \times 6 + 60 \times 1.5 = 0 \ F_1 = 15 \text{ kip}$$
$$\sum F_x = 0 \rightarrow^+ - F_1 - F_2 + 60 = 0 \quad F_2 = 45 \text{ kip}$$

Determine L_1 and L_2 :

$$L_1 = 15 / 2.38 = 6.30 \text{ in}$$
$$L_2 = 45 / 2.38 = 18.9 \text{ in}$$

9 Design for Lateral Load and Building Systems

Topics to be covered in this chapter are as follows:

- Concrete column.
- Approximate analysis.
- Shear walls.
- Prestressed concrete.
- Structural forms.

9.1 SHORT COLUMN

9.1.1 CLASSIFICATION OF COLUMNS

9.1.1.1 Pedestals

If the height of an upright compression member is less than three times its least lateral dimensions, it may be considered to be a pedestal. The ACI318 states that a pedestal may be designed with unreinforced or plain concrete.

9.1.2 SHORT COLUMNS

Should a reinforced concrete column fail due to initial material failure, such as by crushing of concrete, it is classified as a short column. We think of a short column as being a rather stocky member with little flexibility.

DOI: 10.1201/9781032638072-9

9.1.3 Long Columns

As columns become more slender, secondary moments $(P\Delta)$ are generated due to bending deformations (Δ). These secondary moments add up with existing moments in joints (M).

Therefore, the column needs to carry the total moment of $M + P\Delta$. If these secondary moments are of such magnitude as to significantly reduce the axial load capacities of columns, those columns are referred to as being long or slender.

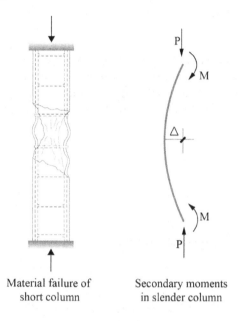

Material failure of   Secondary moments
short column   in slender column

9.1.4 Type of Columns

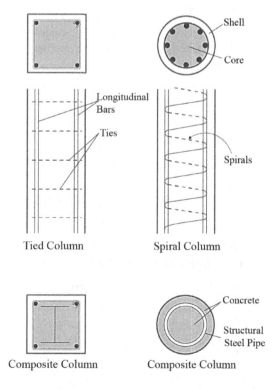

Tied Column   Spiral Column

Composite Column   Composite Column

- A *plain concrete column* can support very little load, but its load-carrying capacity will be greatly increased if longitudinal bars are added.
- Under compressive loads, columns tend not only to shorten lengthwise but also to expand laterally due to the Poisson effect. The capacity of such members can be greatly increased by providing lateral restraint.
- If the column has a series of closed ties, it is referred to as a tied column. They prevent the longitudinal bars from being displaced during construction, and they resist the tendency of the same bars to buckle outward under load.
- Sometimes, when they are used in open spaces, *circular columns* are very attractive. If a continuous helical spiral made from bars or heavy wire is wrapped around the longitudinal bars, the column is referred to as a *spiral column*.
- Spirals are even more effective than ties in increasing a column's strength. The closely spaced spirals do a better job of holding the longitudinal bars in place, and they also confine the concrete inside and greatly increase its resistance to axial compression.
- *Composite columns* are concrete columns that are reinforced longitudinally by structural steel shapes, which may or may not be surrounded by structural steel bars, or they may consist of structural steel tubing filled with concrete.

9.1.5 COLUMN SUBJECTED TO AXIAL LOAD ONLY

9.1.5.1 Nominal Strength of Column

The nominal strength of a purely axially loaded column is given by the following expression:

$$P_n = 0.85 f_C' \left(A_g - A_s \right) + f_y A_s$$

The part $0.85 f_C'(A_g - A_s)$ is the compressive force that is carried by concrete, and the latter part, $f_y A_s$, is the compressive force that is carried by steel.

9.1.6 DESIGN STRENGTH OF COLUMN

For design strength, an additional ϕ factor is applied to the nominal strength expression. In actual practice, there are no perfect axially loaded column; therefore, another factor α is applied to account for the eccentricity of the axial load.

$$\phi P_n = \phi \alpha \left[0.85 f_C' \left(A_g - A_s \right) + f_y A_s \right]$$

9.1.7 DESIGN STRENGTH OF TIED COLUMN

ACI states that, for tied column, $\phi = 0.65$ and $\alpha = 0.80$. Therefore:

$$\phi P_n = 0.65 \times 0.80 \left[0.85 f_C' \left(A_g - A_s \right) + f_y A_s \right]$$

TABLE 9.1

Description of the Symbols

Symbol	Description
f_C'	Compressive stress of concrete
f_y	Yield strength of steel
A_g	Gross area of concrete
A_s	Steel area

9.1.8 Design Strength of Spiral Column

For spiral column, $\phi = 0.75$ and $\alpha = 0.85$. Therefore:

$$\phi P_n = 0.75 \times 0.85 \left[0.85 f_C' \left(A_g - A_s \right) + f_y A_s \right]$$

Example 1

Design a square tied column to support an axial dead load of 130 k and an axial live load of 180 k. Assume that 2% longitudinal steel is desired, $f_C' = 4000 \, \text{psi}$, and $f_y = 60,000 \, \text{psi}$.

Solution

9.1.9 Calculate Design Load

$$P_u = 1.2 \, DL + 1.6 \, LL = 1.2 \times 130 + 1.6 \times 180 = 444 \, \text{kip}$$

9.1.10 Determine Required Gross Area of the Column

$$\phi P_n = \phi \alpha \left[0.85 f_C' \left(A_g - A_s \right) + f_y A_s \right]$$
$$444 = 0.65 \times 0.80 \left[0.85 \times 4 \left(A_g - 0.02 A_g \right) + 60 \times 0.02 A_g \right]$$
$$A_g = 188.40 \, \text{in}^2$$

Use $14 \, \text{in} \times 14 \, \text{in} \left(A_g = 196 \, \text{in}^2 \right)$ for column size.

9.1.11 Select Longitudinal Bars

$$\phi P_n = \phi \alpha \left[0.85 f_C' \left(A_g - A_s \right) + f_y A_s \right]$$
$$444 = 0.65 \times 0.80 \left[0.85 \times 4 \left(196 - A_s \right) + 60 \times A_s \right]$$
$$A_s = 3.31 \, \text{in}^2$$

Use 6 #7 bars $\left(A_s = 3.60 \, \text{in}^2 \right)$.

Design of ties (assuming #3 bars):

$$\text{Spacing} = \text{Min. of} \begin{cases} 48 d_b \text{ of tie} & = 48 \times (3/8) = 18 \, \text{in} \\ 16 d_b \text{ of main bar} & = 16 \times (7/8) = 14 \, \text{in} (\leftarrow) \\ \text{Least col. dim.} & = 14 \, \text{in} (\leftarrow) \end{cases}$$

Use #3 @ 14 in c/c.

9.1.12   DESIGNED COLUMN SECTION

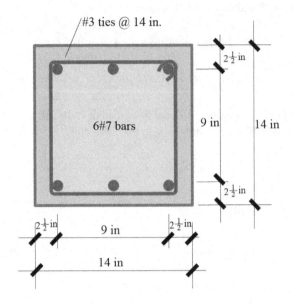

Example 2

Design a round spiral column to support an axial dead load of 130 kip and an axial live load of 180 kip. Assume that 2% longitudinal steel is desired, $f_c' = 4000$ psi, and $f_y = 60,000$ psi.

Solution

9.1.13   CALCULATE DESIGN LOAD

$$P_u = 1.2\,DL + 1.6\,LL = 1.2 \times 130 + 1.6 \times 180 = 444\,\text{kip}$$

Determine the required gross area of the column.

$$\phi P_n = \phi\alpha\left[0.85 f_c'\left(A_g - A_s\right) + f_y A_s\right]$$
$$444 = 0.75 \times 0.85\left[0.85 \times 4\left(A_g - 0.02 A_g\right) + 60 \times 0.02 A_g\right]$$
$$A_g = 153.67\,\text{in}^2$$

Use 14 in diameter column $\left(A_g = 153.93\,\text{in}^2\right)$.

9.1.14 SELECT LONGITUDINAL BARS

$$\phi P_n = \phi\alpha\left[0.85f_C'\left(A_g - A_s\right) + f_y A_s\right]$$
$$444 = 0.75\times0.85\left[0.85\times4\left(153.67 - A_s\right) + 60\times A_s\right]$$
$$A_s = 3.08\,\text{in}^2$$

Use 6#7 bars $\left(A_s = 3.60\,\text{in}^2\right)$.

9.1.15 REINFORCEMENT RATIO OF SPIRAL

Assuming a concrete core diameter $D_C = 11\text{in}$:

$$A_C = \frac{\pi D_C^2}{4} = \frac{\pi\times11^2}{4} = 95.03\,\text{in}^2$$

Reinforcement ratio of the spiral:

$$\rho_s = 0.45\left(\frac{A_g}{A_c} - 1\right)\frac{f_C'}{f_y} = 0.45\left(\frac{153.93}{95.03} - 1\right)\frac{4}{60} = 0.0186$$

9.1.16 SPACING OF SPIRAL

Assuming #3 bars for spiral, therefore, $a_s = 0.11\text{in}^2$ and $d_b = 3/8$ in.

$$s = \frac{4a_s\left(D_c - d_b\right)}{\rho_s D_c^2} = \frac{4\times0.11\times\left(11-3/8\right)}{0.0186\times11^2} = 2.07\,\text{in}$$

Use #3 @ 2 in c/c.

9.1.17 DESIGNED COLUMN SECTION

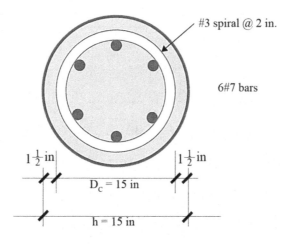

9.1.18 COLUMN SUBJECTED TO AXIAL LOAD AND BENDING

Consider the following frame. Due to symmetry, the column *BD* is purely axially loaded. But columns *AE* and *CF* experience both axial load *P* and bending moment *M*.

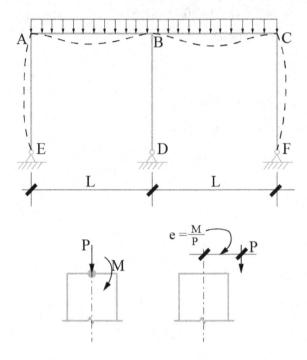

When both *P* and *M* are present, it is convenient to replace the axial load and moment with an equal load *P* applied at an eccentricity e = *M*/*P*. The two loadings are statically equivalent.

9.1.19 INTERACTION DIAGRAM

The following diagram is known as an *interaction diagram*, used to analyze a column that is subjected to both *P* and *M*.

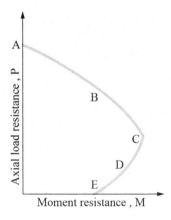

The point A on the interaction diagram represents $M = 0$, thus a purely axially loaded column. The point E represents $P = 0$; therefore, it is essentially a beam. The points in between A and E represent a member that acts like both a beam and a column. In actual practice, a modified version of this interaction diagram is used. This is just the basic.

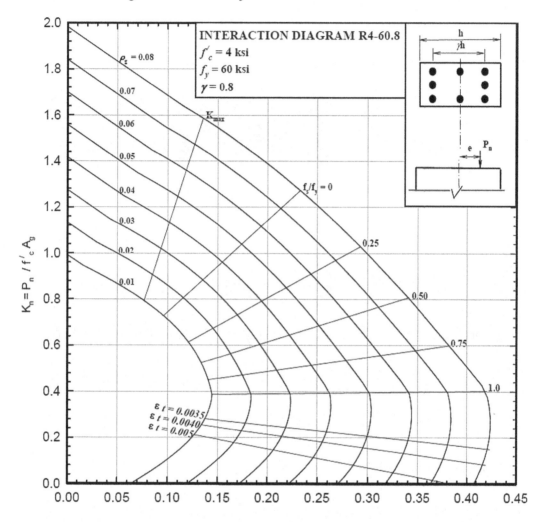

- This figure is the actual interaction diagram used in practice.
- In fact, there are eight interaction diagrams in this figure, and more can be approximated from this.
- Here, the P and M are replaced by K_n and R_n, respectively, where:

$$K_n = \frac{P_n}{f'_c A_g}; \quad R_n = \frac{P_n e}{f'_c A_g h}$$

- The interaction line represents reinforcement ratio ρ, which is defined as:

$$\rho = \frac{A_s}{A_g}$$

- The gamma (γ) is defined as $\gamma = \dfrac{\text{Distance between reinforcement layers}}{\text{Width of column in the same direction}}$.
- Use of the interaction diagram is illustrated in the following examples.

Example 3

The short 14 in ×20 in column shown in the figure is to be used to support the following loads and moments:

$$P_D = 125 \text{ kip} \qquad P_L = 140 \text{ kip}$$
$$M_D = 75 \text{ k–ft} \qquad M_L = 90 \text{ k–ft}$$

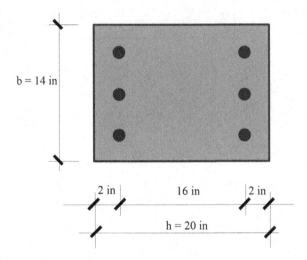

b = 14 in

2 in   16 in   2 in

h = 20 in

If $f_c' = 4000\text{psi}$ and $f_y = 60,000\text{psi}$, select reinforcing bars using the appropriate column interaction diagrams.

Solution

$$A_g = 14 \times 20 = 280 \text{in}^2$$

Calculate the design load and moments.

$$P_u = 1.2P_D + 1.6P_L = 1.2 \times 125 + 1.6 \times 140 = 374 \text{kip}$$
$$M_u = 1.2M_D + 1.6M_L = 1.2 \times 75 + 1.6 \times 90 = 234 \text{k–ft}$$

9.1.20   Determine Eccentricity and γ

$$e = \frac{M_u}{P_u} = \frac{234 \times 12}{374} = 7.51 \text{in}$$

$$\gamma = \frac{16}{20} = 0.8$$

Compute K_n and R_n.

$$K_n = \frac{P_u}{\phi f_c' A_g} = \frac{374}{0.65 \times 4 \times 280} = 0.513$$

$$R_n = \frac{P_u e}{\phi f_c' A_g h} = \frac{374 \times 7.51}{0.65 \times 4 \times 280 \times 20} = 0.193$$

9.1.21 FIND ρ FROM THE INTERACTION DIAGRAM

For $K_n = 0.513$ and $R_n = 0.193$, we find the intersecting point to be very close to the $\rho = 0.02$ line. Therefore, by visual judgment, we take $\rho = 0.19$. Now:

$$A_s = \rho A_g = 0.029 \times 280 = 8.12 \text{in}^2$$

Answer: Use 6#9 bars $\left(A_s = 6.00 \text{in}^2\right)$.

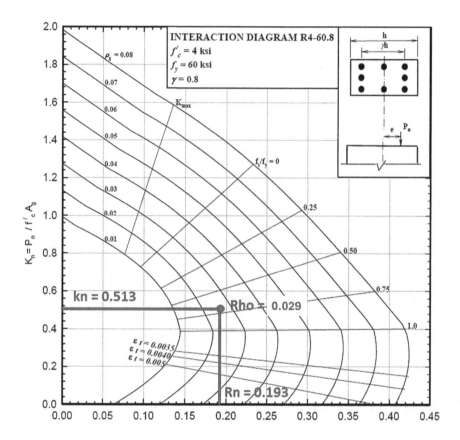

Example 4

The short column of 20 in diameter shown in the figure is to be used to support the following loads and moments:

$$P_D = 125 \text{kip} \qquad P_L = 140 \text{kip}$$
$$M_D = 75 \text{k} - \text{ft} \qquad M_L = 90 \text{k} - \text{ft}$$

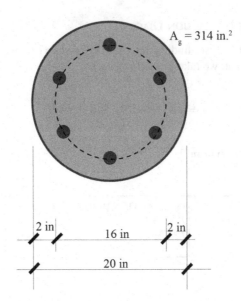

If $f'_c = 4000\,\text{psi}$ and $f_y = 60,000\,\text{psi}$, select reinforcing bars using the appropriate column interaction diagrams.

Solution

$$A_g = \frac{\pi \times 20^2}{4} = 314\,\text{in}^2$$

9.1.22 Calculate Design Load and Moments

$$P_u = 1.2P_D + 1.6P_L = 1.2 \times 125 + 1.6 \times 140 = 374\,\text{kip}$$
$$M_u = 1.2M_D + 1.6M_L = 1.2 \times 75 + 1.6 \times 90 = 234\,\text{k} - \text{ft}$$

9.1.23 Determine Eccentricity and γ

$$\text{e} = \frac{M_u}{P_u} = \frac{234 \times 12}{374} = 7.51\,\text{in}$$

$$\gamma = \frac{16}{20} = 0.8$$

Compute K_n and R_n.

$$K_n = \frac{P_u}{\phi f'_C A_g} = \frac{374}{0.75 \times 4 \times 314} = 0.397$$

$$R_n = \frac{P_u e}{\phi f'_C A_g h} = \frac{374 \times 7.51}{0.75 \times 4 \times 314 \times 20} = 0.149$$

9.1.24 FIND ρ FROM THE INTERACTION DIAGRAM

For $K_n = 0.397$ and $R_n = 0.149$, we find the intersecting point to be on the $\rho = 0.02$ line. Therefore, $\rho = 0.02$. Now:

$$A_s = \rho A_g = 0.02 \times 314 = 6.28 \text{in}^2$$

Answer: Use 8#8 bars $\left(A_s = 6.32 \text{in}^2\right)$.

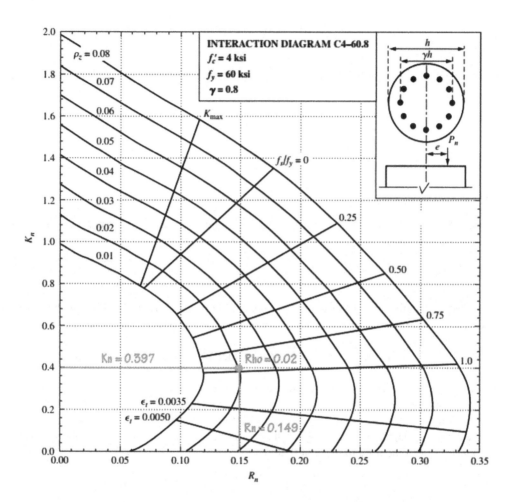

9.2 SLENDER COLUMN

When a column bends or deflects laterally an amount Δ, its axial load will cause an increased column moment equal to $P\Delta$. This moment will be superimposed onto any moments already in the column. Should this $P\Delta$ moment be of such magnitude as to reduce the axial load capacity of the column significantly, the column will be referred to as a slender column.

9.2.1 SLENDERNESS RATIO

The *slenderness ratio* of a column is a parameter that indicates how much slender a column is and is defined by the following expression:

$$\text{Slenderness Ratio} = \frac{kL_u}{r}$$

The effective length factor depends on the support condition. For a single column with both ends pinned, $k = 1.0$; for both ends fixed, $k = 0.5$; for one end pinned and the other fixed, $k = 0.707$.

- But for a column in building frame, k is determined by using Jackson and Moreland alignment charts.
- The unsupported length is simply the distance that is visible to an observer, that is, the clear distance between floors.
- The radius of gyration is equal to 0.289 times the dimension of a rectangular column in the direction that stability is being considered. But the ACI code permits the approximate value of 0.30 to be used in place of 0.289

TABLE 9.2
Description of the Symbols

Symbol	Description
k	Effective length factor
L_u	Unsupported length of column
r	Radius of gyration of column section

9.2.2 CRITERIA FOR SLENDER COLUMN

A column is a slender one if the following criteria is satisfied:

$$\left(\frac{kL_u}{r}\right) > 36 - 12\left(\frac{M_A}{M_B}\right)$$

Where M_A and M_B are the end moments of the column and M_B is the larger one.

9.2.3 UNBRACED FRAME VS. BRACED FRAME

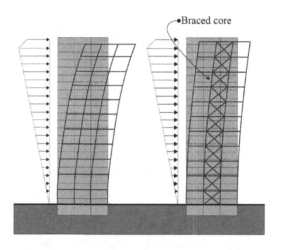

9.2.4 JACKSON AND MORELAND ALIGNMENT CHARTS FOR K

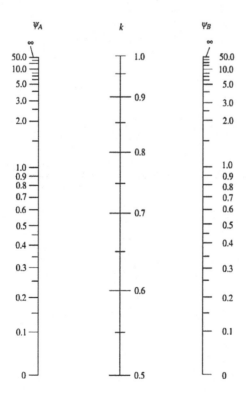

- To use the alignment charts for a particular column, ψ factors are computed at each end of the column.
- The ψ factor at one end of the column equals the sum of the stiffness $\sum (EI/I)$ of the columns meeting at that joint, divided by the sum of all the stiffnesses of the beams meeting at the joint.

$$\psi = \frac{\sum (EI/L)_{\text{columns}}}{\sum (EI/L)_{\text{beams}}}$$

- If the structure is entirely steel or concrete, that is, E is same for both beams and columns, the relative stiffness $(K = I / L)$ could be used instead. (Note that it is uppercase K.)

$$\psi = \frac{\sum (K)_{\text{columns}}}{\sum (K)_{\text{beams}}}$$

- One of the two ψ values is called ψ_A, and the other is called ψ_B.
- After these values are computed, the effective length factor, k, is obtained by placing a straightedge between ψ_A and ψ_B.
- The point where the straightedge crosses the middle nomograph is k.
- In ACl Section 10.10.4.1, it is stated that for determining ψ values for use in evaluating k factors, the rigidity of the beams may be calculated on the basis of 0.35 g to account for cracking and reinforcement, while $0.70I_g$ may be used for columns.

Example 5

Determine whether the column AB is a slender column. End moments are $M_A = 115\text{k-ft}$ and $M_B = 120\text{k-ft}$. Consider the frame as a braced frame.

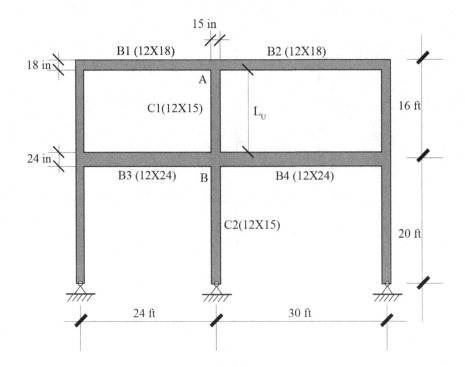

9.2.5 MOMENT OF INERTIA

Applying $I_g = bh^3/12$ for each member of the frame and multiplying it with 0.70 for columns and 0.35 for beams, we determine:

$$I_{C1} = 0.70 \times \left(12 \times 15^3/12\right) = 2362\,\mathrm{in}^4 \quad I_{C2} = 2362\,\mathrm{in}^4$$

$$I_{B1} = 0.35 \times \left(12 \times 18^3/12\right) = 2041\,\mathrm{in}^4 \quad I_{B2} = 2041\,\mathrm{in}^4$$

$$I_{B3} = 0.35 \times \left(12 \times 24^3/12\right) = 4838\,\mathrm{in}^4 \quad I_{B4} = 4838\,\mathrm{in}^4$$

9.2.6 RELATIVE STIFFNESS

Applying I/L for each member, we write:

$$K_{C1} = 2362/16 = 147 \quad K_{C2} = 2362/20 = 118$$
$$K_{B1} = 2041/24 = 85 \quad K_{B2} = 2041/30 = 68$$
$$K_{B3} = 2041/24 = 202 \quad K_{B4} = 2041/30 = 161$$

The ψ factors:

$$\psi_A = \frac{K_{C1}}{K_{B1} + K_{B2}} = \frac{147}{85 + 68} = 0.96$$

$$\psi_B = \frac{K_{C1} + K_{C2}}{K_{B3} + K_{B4}} = \frac{147 + 118}{202 + 161} = 0.73$$

From the alignment chart of the braced frame, for $\psi_A = 0.96$ and $\psi_B = 0.73$, we find effective length factor $k_{AB} = 0.76$.

9.2.7 UNSUPPORTED LENGTH OF *AB* COLUMN

$$L_u = 16 - \left(\frac{18}{2} + \frac{24}{2}\right)/12 = 14.25\mathrm{ft}$$

9.2.8 SLENDERNESS RATIO OF *AB* COLUMN

$$\left(\frac{kL_u}{r}\right)_{AB} = \frac{0.76 \times 14.25 \times 12}{0.3 \times 15} = 28.9$$

9.2.9 MAXIMUM SLENDERNESS RATIO ALLOWED FOR NON-SWAY FRAME

$$34 - 12\left(\frac{M_A}{M_B}\right) = 34 - 12\left(\frac{115}{120}\right) = 22.5 < \left(\frac{kL_u}{r}\right)_{AB}$$

Therefore, *AB* is a slender column.

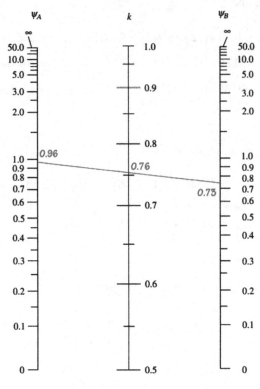

Ψ_A k Ψ_B

Alignment Chart for Braced Frame

Example 6

Determine the magnified moment for the slender column *AB* of example 5, given that:

$$M_A = 115\,\text{k-ft} \quad M_B = 120\,\text{k-ft} \quad f'_c = 4000\,\text{psi}$$
$$P_D = 60\,\text{kip} \qquad\quad P_u = 180\,\text{kip}$$

Also, from example 1, it was determined that:

$$k = 0.76, \quad L_u = 14.25\text{ft}$$

Solution

9.2.10 DETERMINE FLEXURAL RIGIDITY (EI) OF COLUMN AB

1. Elastic modulus of concrete:

$$E_C = 57000\sqrt{f'_c} = 57000\sqrt{4000} = 3{,}604{,}996\,\text{psi} = 3605\,\text{ksi}$$

2. Gross moment of inertia:

$$I_g = \frac{bh^3}{12} = \frac{12 \times 15^3}{12} = 3375 \, in^4$$

3. Stiffness reduction factor due to axial load:

$$\beta_d = \frac{1.2P_D}{P_u} = \frac{1.2 \times 60}{180} = 0.4$$

4. Flexural rigidity of column AB:

$$EI = \frac{0.4E_c l_g}{1+\beta_d} = \frac{0.4 \times 3605 \times 3375}{1+0.4} = 3476 \times 10^3 \, k - in^2$$

9.2.11 FIND MOMENT MAGNIFIER, δ

5. Effective length:

$$kL_u = 0.76 \times (14.25 \times 12) = 129.96 \, in$$

6. Euler critical load:

$$P_e = \frac{\pi^2 EI}{(kL_u)2} = \frac{\pi^2 \times (3476 \times 10^3)}{129.96^2} = 2031 \, kip$$

7. Moment gradient:

$$C_m = 0.6 + 0.4 \left(\frac{M_A}{M_B} \right) = 0.6 + 0.4 \left(\frac{115}{120} \right) = 0.983$$

8. Moment magnifier:

$$\delta = \frac{C_m}{1 - \dfrac{P_u}{0.75P_e}} = \frac{0.983}{1 - \dfrac{180}{0.75 \times 2031}} = 1.115$$

9.2.12 CALCULATE MAGNIFIED MOMENT

$$M = \delta M_B = 1.115 \times 120 = 133.8 \, k - ft$$

Answer: 133.8 kft

Note that though 120 k-ft moment was applied on the column, an excess of 13.8 k-ft secondary moment was generated due to its slenderness.

Example 7

Select reinforcing for the column AB of example 2 using the appropriate column interaction diagrams, given that:

$$P_u = 180 \text{ kip} \quad f_C' = 4000 \text{ psi} \quad f_y = 60000 \text{ psi}$$

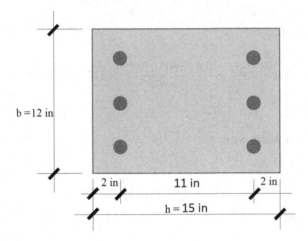

In example 1, it was determined that the magnified moment is 133.8 kft; therefore, take the design moment $M_u = 133.8$ k-ft.

Solution

$$A_g = 12 \times 15 = 180 \text{ in}^2$$

9.2.13   DETERMINE ECCENTRICITY AND γ

$$e = \frac{M_u}{P_u} = \frac{133.8 \times 12}{180} = 8.92 \text{ in}$$

$$\gamma = \frac{12}{15} = 0.80$$

Compute K_n and R_n.

$$K_n = \frac{P_u}{\phi f_c' A_g} = \frac{180}{0.65 \times 4 \times 180} = 0.39$$

$$R_n = \frac{P_u e}{\phi f_c' A_g h} = \frac{180 \times 8.92}{0.65 \times 4 \times 180 \times 15} = 0.23$$

9.2.14 FIND ρ FROM INTERACTION DIAGRAM

For $K_n = 0.39$ and $R_n = 0.23$, we find the intersecting point to be just above the $\rho = 0.02$ line. Therefore, by visual judgment, we take $\rho = 0.022$. Now:

$$A_s = \rho A_g = 0.022 \times 180 = 3.96\, \text{in}^2$$

Answer: Use 4#9 bars $\left(A_s = 4.00\, \text{in}^2\right)$.

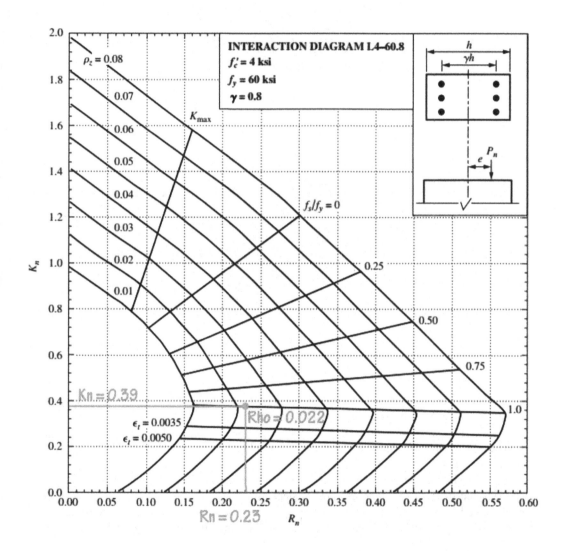

9.3 APPROXIMATE ANALYSIS OF GRIDS

9.3.1 GRID STRUCTURE

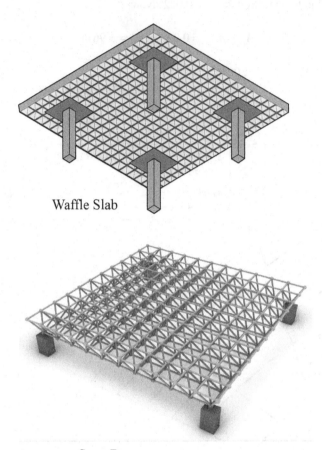

Waffle Slab

Space Frame

Plane grid structures are typically made of a series of intersecting long rigid linear elements, such as beams or trusses, with parallel upper and lower chords.

- These structures are essentially planar and are relatively thin. Joints at points of intersection are rigid.
- Waffle slabs are common example of grid structure. These slabs are able to carry heavier loads and span longer distances than flat slabs.
- Suitable for spans of 24 ft to 54 ft; longer spans may be possible with posttensioning.
- For maximum efficiency, bays should be square or as nearly square as possible.

- Another example of grid structure is space frame, which is a long-spanning three-dimensional plate structure based on the rigidity of the triangle and composed of linear elements subject only to axial tension or compression.
- Span range for column-supported space frames is 30 ft to 80 ft.
- Mechanical services such as piping, conduit, and duct may pass through the web spaces.

9.3.2 DEFLECTION

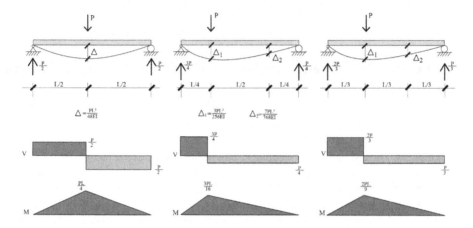

Expressions of deflection (Δ) for various loading cases are required to analyze grids.

9.3.3 SIMPLE GRID ANALYSIS

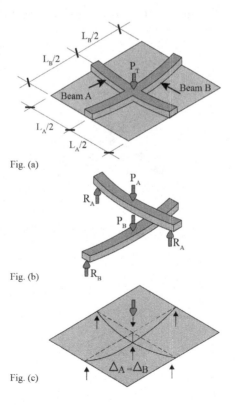

Fig. (a)

Fig. (b)

Fig. (c)

- Consider the simple crossed-beam system, which is the simplest grid structure, supported on four sides, as shown in Figure 9.22(a).
- The key to analyzing a grid structure of this type is to recognize that a state of deflection compatibility must exist at each point of connection in a crossed-beam system.
- This compatibility requirement assumes that the beams are rigidly connected such that both undergo an identical deflection due to a load. By equating deflection expressions appropriate for each beam, it is possible to determine the load carried by each element.
- Let P_A be defined as the load carried by member A, and P_B the load carried by member B. Since the total applied load is P_T, from statics, we write:

$$P_T = P_A + P_B$$

- Now, by equating deflection expressions for each member, we obtain (assuming EI is same for both beams):

$$\Delta_A = \Delta_B; \frac{P_A L_A^3}{48EI} = \frac{P_B L_B^3}{48EI}; \frac{P_A}{P_B} = \left(\frac{L_B}{L_A}\right)^3$$

- From these two boxed equations, P_A and P_B can be solved simultaneously.
- Notice that for each crossed joint, there is one equation of statics and one equation of deflection.
- This procedure of analysis is approximate because torsional deformations of beams are ignored.

Example 8

Determine the support reactions of the following grid and maximum moments in the members. Assume the members are simply supported. EI is constant as well as same.

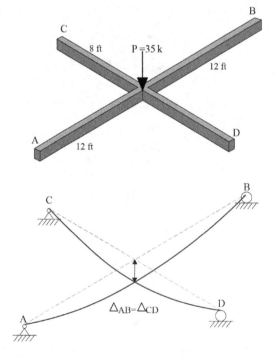

Solution

Assume P_{AB} is the portion of P that is carried by AB, and P_{CD} is the portion that is carried by CD. Therefore:

$$P_{AB} + P_{CD} = 35$$

Now, from deflection compatibility, we write:

$$\Delta_{AB} = \Delta_{CD}; \left(\frac{PL^3}{48EI}\right) = \left(\frac{PL^3}{48EI}\right)$$

$$\frac{P_{AB} \times 24^3}{48EI} = \frac{P_{CD} \times 16^3}{48EI}; P_{AB} = 8/27 P_{CD}$$

Solving these equations simultaneously, we find:

$$P_{AB} = 8\,\text{kip}, P_{CD} = 27\,\text{kip}$$

9.3.4 Support Reactions

Member AB: Due to symmetry, $P_{AB} = 8$ kip force on beam AB will be equally carried by both supports. Therefore:

$$R_A = 4\,\text{kip}, R_B = 4\,\text{kip}$$

Member CD: Similarly, $P_{CD} = 27$ kip will be equally carried by support C and D.

$$R_C = 13.5\,\text{kip}, R_D = 13.5\,\text{kip}$$

9.3.5 Maximum Moments

Member AB:

$$M = (PL/4)_{AB} = (8 \times 24)/4 = 48\,\text{k} - \text{ft}$$

Member CD:

$$M = (PL/4)_{CD} = (27 \times 16)/4 = 108\,\text{k} - \text{ft}$$

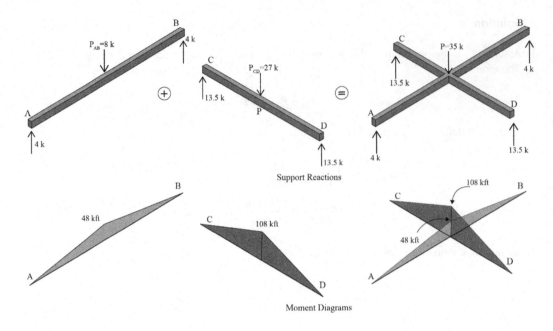

Moment Diagrams

Example 9

Determine the support reactions of the following grid and maximum moments in the members. Assume the members are simply supported; *EI* is constant as well as same.

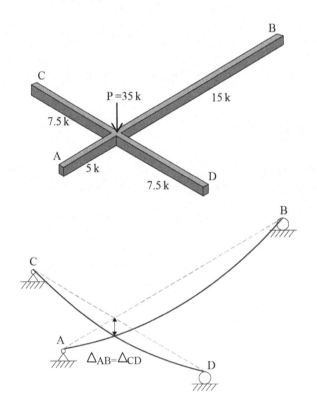

Solution

Assume P_{AB} is the portion of P that is carried by AB, and P_{CD} is the portion that is carried by CD. Therefore:

$$P_{AB} + P_{CD} = 35$$

Now, from deflection compatibility, we write:

$$\Delta_{AB} = \Delta_{CD}; \left(\frac{3PL^3}{256EI}\right)_{AB} = \left(\frac{PL^3}{48EI}\right)_{CD}$$

$$\frac{3P_{AB} \times 20^3}{256EI} = \frac{P_{CD} \times 15^3}{48EI}; P_{AB} = 3/4\, P_{CD}$$

Solving these equations simultaneously, we find:

$$P_{AB} = 15\,\text{kip}, P_{CD} = 20\,\text{kip}$$

9.3.6 SUPPORT REACTIONS

Member AB:

$$R_A = (15/20) \times 15 = 11.25\,\text{kip}, R_B = (5/20) \times 15 = 3.25\,\text{kip}$$

Member CD: Due to symmetry, $P_{CD} = 20$ kip force on beam CD will be equally carried by both supports. Therefore:

$$R_C = 10\,\text{kip}, R_D = 10\,\text{kip}$$

9.3.7 MAXIMUM MOMENTS

Member AB:

$$M = (3PL/16)_{AB} = (3 \times 15 \times 20)/16 = 56.25\,\text{k} - \text{ft}$$

Member CD:

$$M = (PL/4)_{CD} = (10 \times 1.5)/4 = 75\,\text{k} - \text{ft}$$

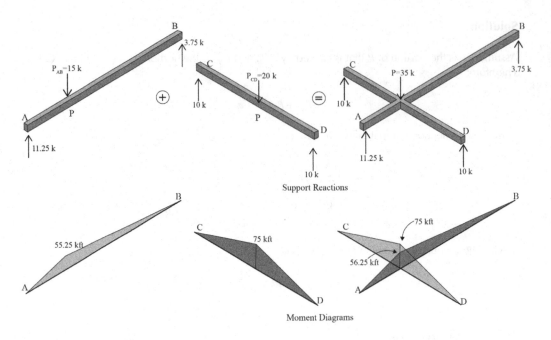

Support Reactions

Moment Diagrams

9.3.8 COMPLEX GRID ANALYSIS

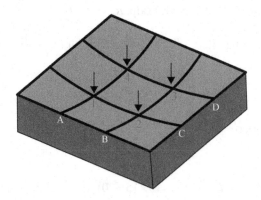

- The analysis of more complex grids with multiple crossed beams proceeds in a fashion similar to a simple grid.
- Deflections at the intersections of the various beams can be equated.
- In this figure, deflections at point 1 of both A and C beams must be equal. For point 2, deflections at both B and C beams are equal. And so on.
- An analytical difficulty arises, however, because deflections must be compatible at multiple points and the interactive reaction forces between members at one point contribute to the deflections at another.
- Invariably, for a complex grid, several equations are generated that must be solved simultaneously – a task for computer-based analysis systems.
- In this grid, there will be four equations of statics and four equations of deflections. Thus, a total of eight equations must be simultaneously solved to analyze this grid.

Example 10

Determine the support reactions of the following grid and maximum moments in the members. Assume the members are simply supported; EI is constant as well as same.

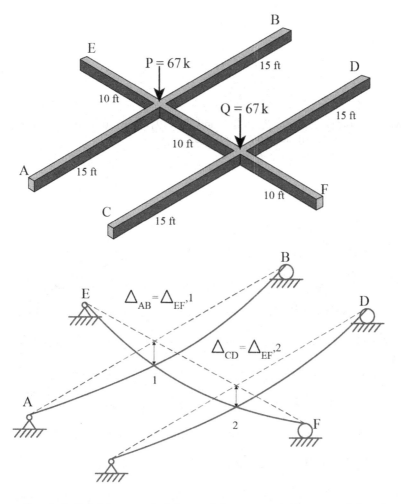

Note: The author feels that this example is beyond the scope of this course. It is only added to illustrate that the concept learned in the previous examples can be applied to solve more complex grid structures.

Solution

Assume P_{AB} is the portion of P that is carried by AB, and P_{EF} is the portion that is carried by EF. Therefore:

$$P_{AB} + P_{EF} = 67$$

Now, at point 1, deflection on beam AB caused by load P_{AB} is equal to deflection on beam EF caused by loads P_{EF} and Q_{EF}.

$$\frac{P_{AB} \times 30^3}{48EI} = \frac{4P_{EF} \times 30^3}{243EI} + \frac{7Q_{EF} \times 30^3}{486EI}$$

Assume Q_{CD} is the portion of Q that is carried by CD, and Q_{EF} is the portion that is carried by EF. Therefore:

$$Q_{CD} + Q_{EF} = 67$$

Now, at point 2, deflection on beam CD caused by load Q_{CD} is equal to deflection on beam EF caused by loads P_{EF} and Q_{EF}.

$$\frac{Q_{CD} \times 30^3}{48\,EI} = \frac{7P_{EF} \times 30^3}{486\,EI} + \frac{4Q_{EF} \times 30^3}{243\,EI}$$

Due to symmetry of the problem, $P_{AB} = Q_{CD}$ and $P_{EF} = Q_{EF}$. Now, substituting these relations into the equations and solving simultaneously, we find:

$$P_{AB} = Q_{CD} = 40\,\text{kip}, \; P_{EF} = Q_{EF} = 27\,\text{kip},$$

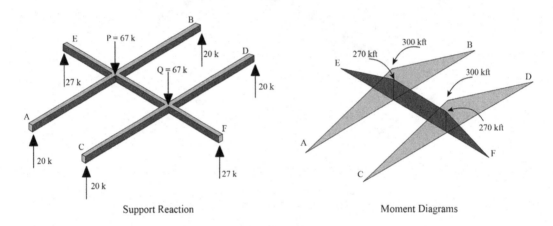

Support Reaction   Moment Diagrams

9.4   APPROXIMATE ANALYSIS FOR LATERAL LOAD

9.4.1   PORTAL METHOD

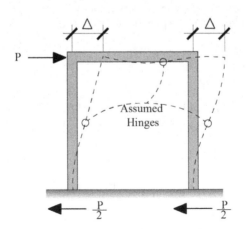

Portal frame:

- The following figure is a portal frame laterally loaded by force *P*.
- Each column equally carries a base shear of *P* = 2 in response of *P*.
- By exact analysis, it is found that the bending moment is zero near the center of both beams and columns.
- Thus, we could assume, for analysis purpose, that there are imaginary hinges, because hinge carries no moment.

Assumption of portal method of analysis:

- An inflection point is located at the middle of each member of the frame.
- On each story of the frame, interior columns carry twice as much shear as exterior columns.

The following frame could be thought of as a combination of two portal frames.

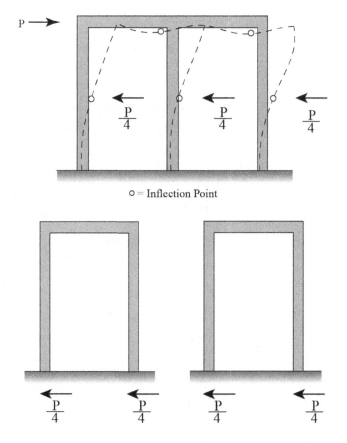

9.4.2 COLUMN SHEAR

Example 11

Determine the shear forces in columns of the following building frame, and draw the corresponding diagrams.

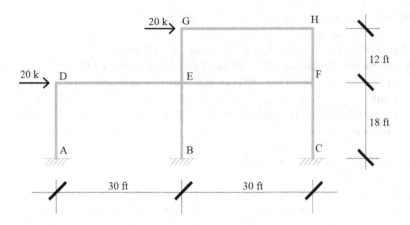

Solution

Figure 9.32(a): Imagine a horizontal section passing through the centers of all columns on the second floor, that is, *EG* and *FH*. Assume each column carries shear force of amount V. Therefore:

$$V + V = 20; \ 2V = 20; \ V = 10\text{kip}$$

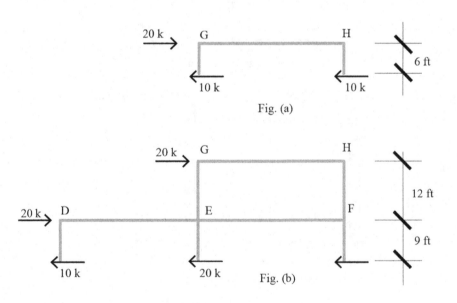

Fig. (a)

Fig. (b)

Figure 9.33(b): Now, imagine a horizontal section passing through the centers of all columns on first floor, that is, *AD,BE,* and *CF*. Assume the two external columns *AD* and *CF* carry shear force of amount V, and the internal column *BE* carries shear force of $2V$. Thus:

$$V + 2V + V = 20 + 20; \ 4V = 40; \ V = 10 \text{ kip}$$

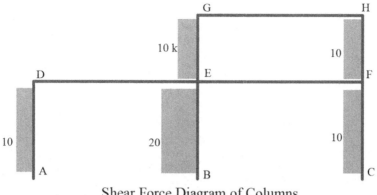

Shear Force Diagram of Columns

Example 12

9.4.3 COLUMN MOMENTS

Determine the bending moments in columns of the building frame of example 11, and draw the corresponding diagrams.

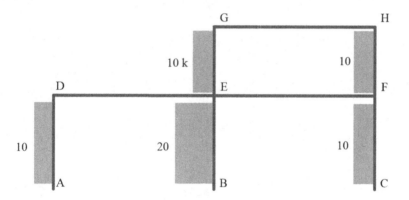

Shear Force Diagram of Columns

Solution

Figure 9.35(c): The bending moment in the columns *EG* and *FH* are found by multiplying the corresponding shear forces and their half-lengths.

$$M_{GE} = M_{HF} = \text{ Shear Force} \times \text{Half Col. Length}$$
$$= 10 \times 6 = 60 \text{k} - \text{ft}$$

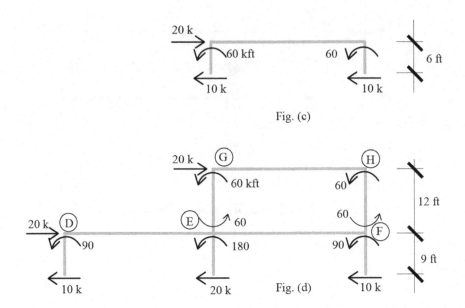

Fig. (c)

Fig. (d)

Figure 9.36(d): Similar procedure applies for columns on the first floor.

$$M_{DA} = M_{FC} = 10 \times 9 = 90\,k - ft$$
$$M_{EB} = 20 \times 9 = 180\,k - ft$$

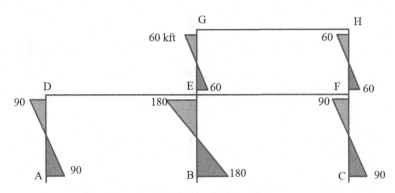

Bending Moment Diagram of Columns

Note that new information (bending moments for this example) is the upper triangle if positive or lower triangle if negative. This convention is followed in the subsequent examples.

Example 13

9.4.4 BEAM MOMENTS

Determine the bending moments in beams of the building frame of example 11 and draw the corresponding diagrams.

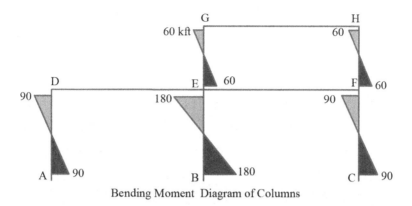

Bending Moment Diagram of Columns

Solution

At each joint, the total sum of moments must be zero, that is, $\sum M = 0$.

Beam GH: At joint G, a counterclockwise moment of 60kft exists due to column EG; therefore, to counterbalance it, another clockwise moment must exist in beam GH. Alternatively:

$$M_{GE} + M_{GH} = 0; \; -60 + M_{GH} = 0; \; M_{GH} = 60\text{k} - \text{ft}$$

Beam DE: At joint D, we write:

$$M_{DA} + M_{DE} = 0; \; -90 + M_{DE} = 0; \; M_{DE} = 90\text{k} - \text{ft}$$

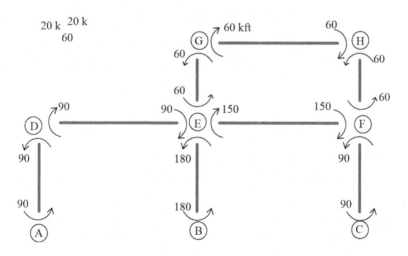

Beam EF: At joint G, we write:

$$M_{ED} + M_{EF} + M_{EG} + M_{EB} = 0$$
$$90 + M_{EF} - 60 - 180 = 0; \; M_{EF} = 150\text{k} - \text{ft}$$

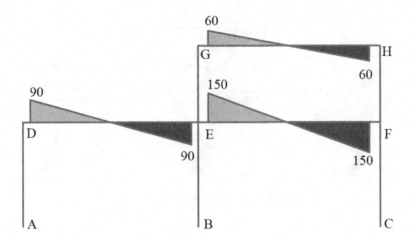

Example 14

9.4.5 BEAM SHEARS

Determine the shears in beams of the building frame of example 11, and draw the corresponding diagrams.

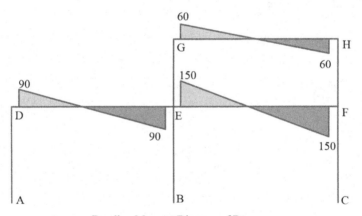

Bending Moment Diagram of Beams

Solution

The shear force in each beam is equal to the sum of the moments at its ends, divided by its length.

$$V = \frac{M+M}{L} = \frac{2M}{L}$$

Beam *GH*:

$$V_{GH} = \frac{2\times 60}{30} = 4\text{kip}$$

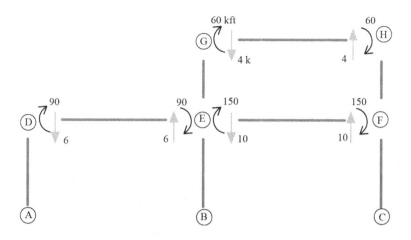

Beam *DE*:

$$V_{DE} = \frac{2 \times 90}{30} = 6\,\text{kip}$$

Beam *EF*:

$$V_{EF} = \frac{2 \times 150}{30} = 10\,\text{kip}$$

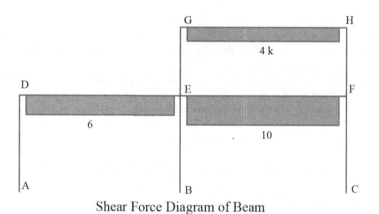

Shear Force Diagram of Beam

Example 15

9.4.6 COLUMN AXIAL FORCES

Determine the axial forces in columns of the building frame of example 11 and draw the corresponding diagrams.

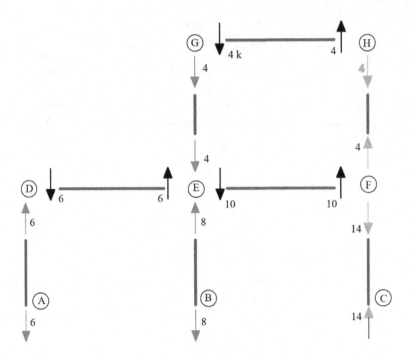

Shear forces of beams must be solved beforehand.

Solution

Shear forces in beams are transferred to the column as axial forces. Here, $\Sigma F_y = 0$ at all joints.

Column EG G_4 kip downward force from beam; therefore, a 4 kip upward force must act at the column. This upward force acts as a tensile for column EG. Same thing applies for joint D.

Column FH: At joint H, there is a 4 kip upward force from the beam; therefore, a 4 kip downward force must act at the column. This downward force acts as a compressive for column FG.

Column BE: At joint E,6 kip upward force from left, 10 kip downward from right, and 4 kip downward from above are present there. To satisfy the equilibrium condition, another 8kip upward force must exists there from below. Alternatively:

$$6 - 10 - 4 + N_{EB} = 0; \ N_{EB} = 8\text{kip}$$

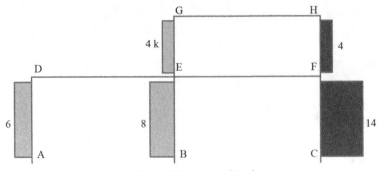

Axial Force Diagram of Columns

9.5 APPROXIMATE ANALYSIS FOR GRAVITY LOAD

9.5.1 ASSUMPTIONS FOR GRAVITY LOAD ANALYSIS

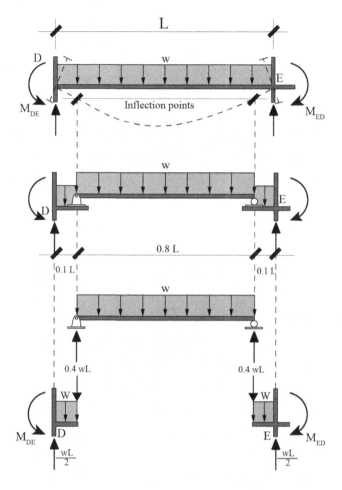

By exact analysis, the inflection points of a fixed beam under uniformly distributed loads are found in the vicinity of 0.01 L distance from each end. At these inflection points, bending moment is zero.

9.5.2 ASSUMPTION OF GRAVITY LOAD ANALYSIS

- The inflection points are located at one-tenth of the span from each end of the girder.
- The girder axial force is zero.

9.5.3 MIDSPAN MOMENT

Based on these assumptions, the length of the simply supported suspended portion of the beam is 0.8 L. Therefore, the maximum bending moment at midspan is:

$$\text{Midspan moment} = \frac{\text{Load} \times \text{Span}^2}{8} = \frac{w(0.8L)^2}{8} = 0.08wL^2$$

9.5.4   END MOMENT

The *end moments* are calculated by summing the moments at support due to the distributed load on the $0.1L$ cantilever segment on the beam plus the moment due to the reaction from the simply supported suspended span.

$$\text{End moment} = (0.1wL)\left(\frac{0.1L}{2}\right) + (0.4wL)(0.1L) = 0.045wL^2$$

End shear:

These are simply given by:

$$\text{End shear} = \frac{wL}{2}$$

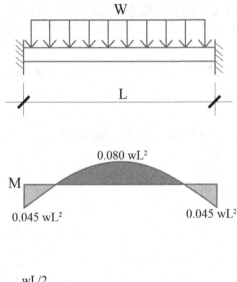

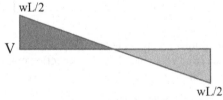

Example 16

Determine the moments in beams of the following building frame and draw the corresponding diagrams.

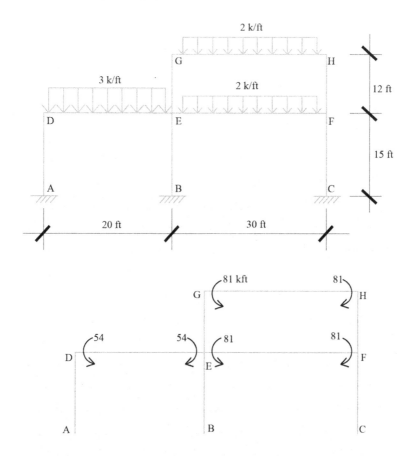

Solution

For each beam, end moments are $0.045\,wL^2$, and midspan moments are $0.080\,wL^2$.

Beam _GH_ and beam _EF_:

$$\text{End moments} = 0.045 \times 2 \times 30^2 = 81\text{k} - \text{ft}$$
$$\text{Mid span moment} = 0.080 \times 2 \times 30^2 = 144\text{k} - \text{ft}$$

Beam _DE_:

$$\text{End moments} = 0.045 \times 3 \times 20^2 = 54\text{k} - \text{ft}$$
$$\text{Mid span moment} = 0.080 \times 3 \times 20^2 = 96\text{k} - \text{ft}$$

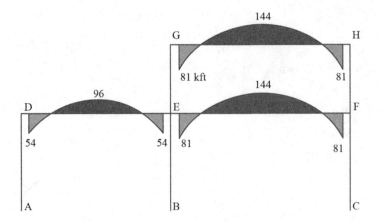

Bending moment diagram of beams.

Example 17

Determine the bending moments in columns of the building frame of example 16, and draw the corresponding diagrams.

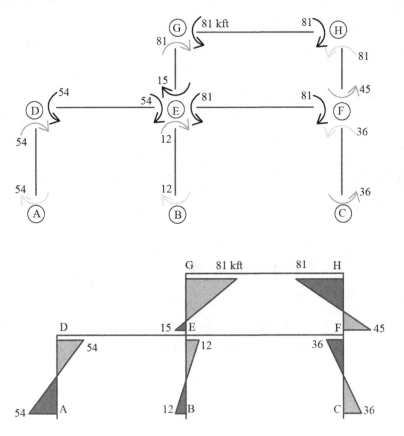

Bending moment of beam must be solved beforehand.

Solution

At each joint, the sum of the moments is zero, that is, $\sum M = 0$.

Joint G: There is a counterclockwise moment of $81\,kft$ acting on the beam; therefore, in the column, there must act a clockwise $81\,ktt$ moment. Alternatively:

$$M_{GE} + M_{GH} = 0;\ M_{GE} - 81 = 0;\ M_{GE} = 81\text{k} - \text{ft}$$

Joint D:

$$M_{DE} + M_{DA} = 0; M_{DA} - 54 = 0; M_{GE} = 54\text{k} - \text{ft}$$

Joint E: Beam moments we solved earlier, $M_{ED} = 54\text{k} - \text{ft}$ and $M_{EF} = 81\text{k} - \text{ft}$. Now we need to determine the column moments M_{EG} and M_{EB}.

$81 - 54 = 27k - ft$ at joint E will be distributed in columns in proportion to their stiffness EI/L. Assuming EI is the same for both columns, using relative stiffness $1/L$, we write:

$$M_{EG} = (81 - 54)\left(\frac{1/12}{1/12 + 1/15}\right) = 15\text{k} - \text{ft}$$

$$M_{EB} = (81 - 54)\left(\frac{1/15}{1/12 + 1/15}\right) = 12\text{k} - \text{ft}$$

Joint F: Similarly as joint E, we write:

$$M_{FH} = 81\left(\frac{1/12}{1/12 + 1/15}\right) = 45\text{k} - \text{ft}$$

$$M_{FC} = 81\left(\frac{1/15}{1/12 + 1/15}\right) = 36\text{k} - \text{ft}$$

Example 18

Determine the shear forces in the columns of the following building frame, and draw the corresponding diagrams.

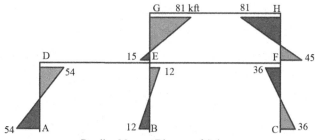

Bending Moment Diagram of Columns

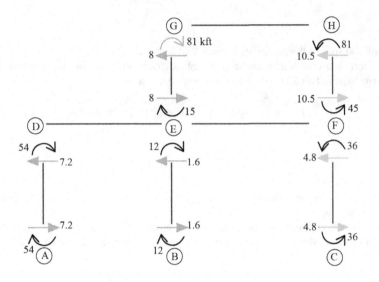

Solution

Column shear forces are found by adding the two end moments and dividing it by the length of the column.

$$V = \frac{M_1 + M_2}{L}$$

Column *EG*:

$$V_{EG} = \frac{81+15}{12} = 8\,\text{kip}$$

Column FH:

$$V_{FH} = \frac{81+45}{12} = 10.5\,\text{kip}$$

Column *AD*:

$$V_{AD} = \frac{54+54}{15} = 7.2\,\text{kip}$$

Remaining shears are found similarly. Note that clockwise shear pair is positive; counterclockwise is negative.

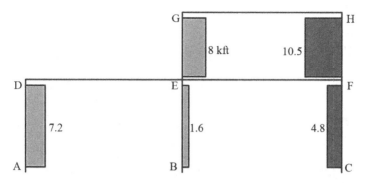

Shear Force Diagram of Columns

Example 19

Determine the shears in beams of the building frame, and draw the corresponding diagrams.

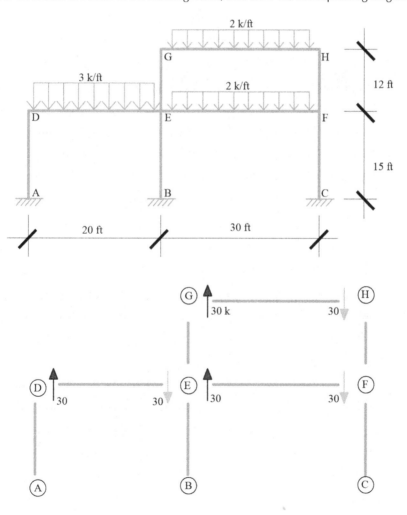

Solution

The shear force in each beam is given by:

$$V = \frac{wL}{2}$$

Beam *GH* and beam *EF*:

$$V_{GH} = V_{EF} = \frac{2 \times 30}{2} = 30 \text{ kip}$$

Beam *DE*:

$$V_{DE} = \frac{3 \times 20}{2} = 30 \text{ kip}$$

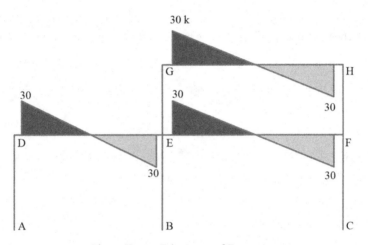

Shear Force Diagram of Beam

Example 20

Determine the axial forces in columns of the building frame of example 16, and draw the corresponding diagrams.

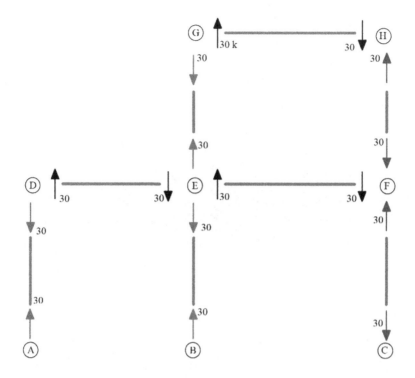

Shear forces of beams must be solved beforehand.

Solution

Shear forces in beams are transferred to the column as axial forces.
Here, $\Sigma F_y = 0$ at all joints.

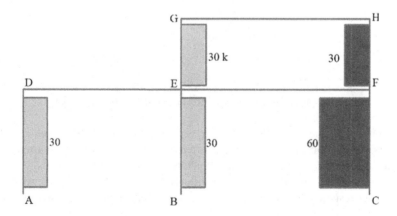

Axial Force Diagram of Columns

9.6   SHEAR WALLS

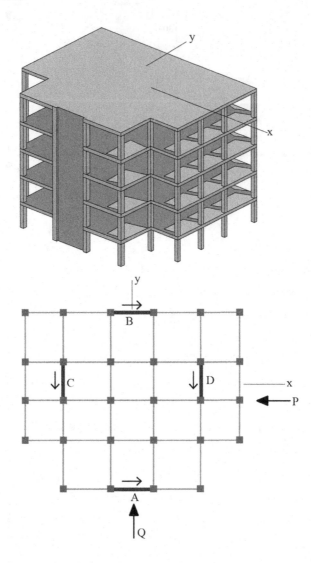

- For tall buildings, it is necessary to provide the adequate stiffness to resist the lateral forces caused by wind and earthquake.
- When reinforced concrete walls with their very large in-plane stiffnesses are placed at certain convenient and strategic locations, often they can be economically used to provide the needed resistance to horizontal loads. Such walls are called shear walls.
- Shear walls are actually deep vertical cantilever beams that provide lateral stability to structures by resisting the in-plane shears and bending moments caused by the lateral forces.
- As the strength of shear walls is almost always controlled by their flexural resistance, their name is something of a misnomer.

- A building that is subjected to horizontal lateral forces, usually from wind or earthquake loads, is shown in the figure.
- If the lateral force comes from the *x* direction, such as *P*, that would be resisted primarily by the shear walls *A* and *D*.
- Lateral load coming from the *y* direction, such as *Q*, will be primarily resisted by shear walls *C* and *D*.

9.6.1 SHEAR WALLS: ARRANGEMENTS

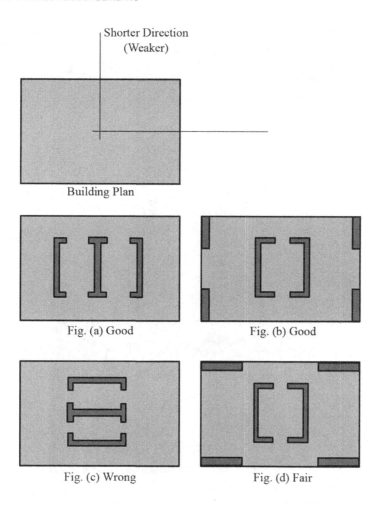

Shorter Direction
(Weaker)

Building Plan

Fig. (a) Good

Fig. (b) Good

Fig. (c) Wrong

Fig. (d) Fair

- A rectangular building plan, as shown in the figure, is inherently weak in the shorter direction and stronger in the longer direction.
- Therefore, lateral stiffness is necessary in shorter direction to avoid failure. Shear walls should be placed primarily in this direction, but they are also necessary in longer direction.
- Figure 9.59(a) and Figure (b): Two possible layout of shear wall. Notice that in both of these figure, the long side of the walls is placed in the shorter direction.

- Figure 9.59(c): An example of wrong placement of the wall, since the long side of the walls is placed in the longer direction of the building.
- Figure 9.59(d): Also acceptable since both sides are strengthened by the walls.
- If the walls are carefully and symmetrically placed in the plan, they will efficiently resist both vertical and lateral loads and do so without interfering substantially with the architectural requirements.
- Reinforced concrete buildings of up to 70 stories have been constructed with shear walls as their primary source of lateral stiffness.

9.6.2   ONE SHELL PLAZA

One Shell Plaza, a 50-story, 715 ft skyscraper located at Houston in Texas, USA.

- Being an example of modern architecture, it was the tallest building in Houston at its completion in 1971.
- The building utilizes the concept of a tube-in-tube structure, which was first proposed by its structural engineer, Fazlur R. Khan.
- The primary lateral resistance of the building is provided by the shear walls that are placed as cores of the building.

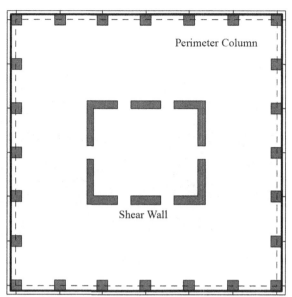

Tube in Tube Structural System

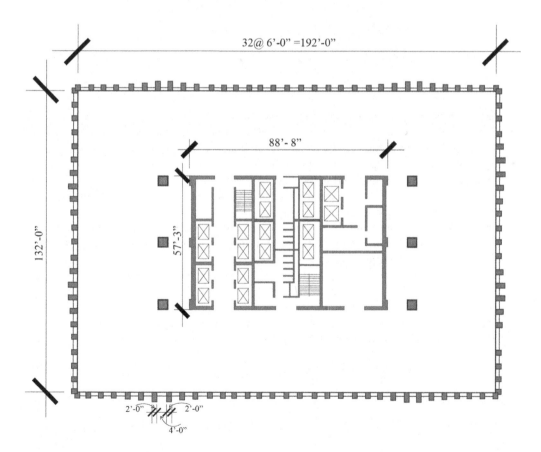

Floor plan of the One Shell Plaza. Notice the arrangement of the shear walls in the shorter direction.

9.6.3 CANTILEVER BEAM ACTION

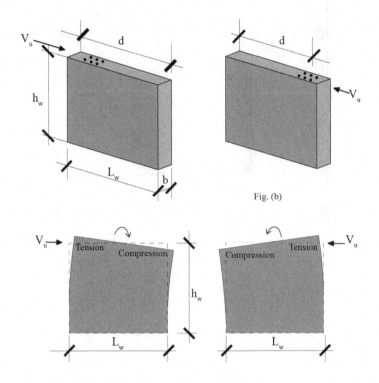

Fig. (b)

- The shear wall is actually a cantilever beam of width $b\ \ell_w$.
- Figure 9.63(a): The wall is being bent from left to right by lateral load V_u, with the result that tensile bars are needed on the left or tensile side.
- Figure 9.63(b): If V_u is applied from the right side, as shown in the figure, tensile bars will be needed on the right-hand end of the wall.
- Thus, it can be seen that a shear wall needs tensile reinforcing on both sides because V_u can come from either direction.
- The depth of the beam from the compression end of the wall to the center of gravity of the tensile bars is estimated to be about 0.8 times the wall length, ℓ_w, as per ACI-318 Section 11.10.4. Therefore:

$$d = 0.8\ell_w$$

- The design shear strength is given by the following equation, which can be used to check the minimum thickness of the wall.

$$V_u = \phi 10\sqrt{f_c'}\,bd$$

9.6.4 STRUCTURAL DESIGN

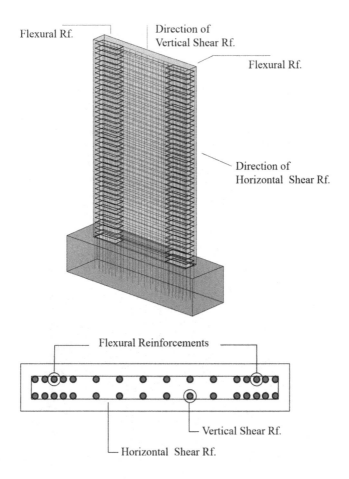

9.6.5 FLEXURAL DESIGN

Reinforcement is provided to resist the moment caused by the lateral force at the base.

- Reinforcements are placed at both sides, as lateral load could come from either direction.
- Same as the rectangular beam design for flexure.

9.6.6 HORIZONTAL SHEAR DESIGN

- Design is the same as the rectangular beam design for shear.
- Maximum spacing shall not be greater than $\ell_w / 5, 3h$ or 18 in.
- Reinforcement ratio shall not be less than 0.0025.

9.6.7 VERTICAL SHEAR DESIGN

- Reinforcement ratio for vertical reinforcing is given by:

$$\rho_v = 0.0025 + 0.5\left(2.5 - \frac{h_w}{\ell_w}\right)$$

- Maximum spacing shall not be greater than $\ell_w/3, 3h$ or 18 in.
- Reinforcement ratio shall not be less than 0.0025.

9.7 SHEAR WALLS DESIGN

Example 21

A shear wall shown in Figure 9.65 is intended to carry a shear force of $V_u = 240$ kip. Find whether its thickness is adequate to carry that force, given that $f_c' = 3000$ psi and $fy = 60,000$ psi.

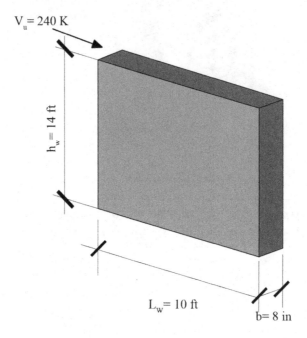

Solution

1. Find the effective depth for the cantilever beam action of the wall.

$$d = 0.8\,\ell_w$$
$$= 0.8 \times (10 \times 12) = 96\,\text{in}$$

2. Maximum shear force is allowed to be carried by the wall according to ACI code:

$$\phi V_n = \phi 10\sqrt{f_c'}\,bd$$
$$= 0.75 \times 10 \times \sqrt{3000} \times 8 \times 96$$
$$= 315,488\,\text{lb} = 315.5\,\text{kip}$$

Since $V_u\,(240\,\text{kip}) < \phi V_n\,(315.5)$, the thickness is adequate.

Answer: Thickness of 8 in is adequate.

Example 22

Design the vertical flexural reinforcing of the shear wall shown in Figure 9.66 if $f'_c = 3000$ psi and $fy = 60,000$ psi.

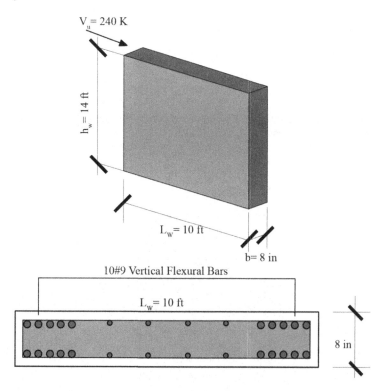

V$_u$= 240 K

h$_w$ = 14 ft

L$_w$= 10 ft

b= 8 in

10#9 Vertical Flexural Bars

L$_w$= 10 ft

8 in

Solution

1. Determine the moment due to the cantilever action at the base of the wall.

$$M_u = V_u h_w = 240 \times 14 = 3360 \text{ k} - \text{ft}$$

2. Nominal flexural resistance factor:

$$R_n = \frac{M_u}{\phi b d^2} = \frac{3360 \times 12}{0.9 \times 8 \times 96^2} = 0.608 \text{ ksi}$$

3. Reinforcement ratio:

$$\rho = \frac{0.85 f'_c}{f_y} \left[1 - \sqrt{1 - \frac{2R_n}{0.85 f'_c}} \right]$$

$$= \frac{0.85 \times 3}{60} \left[1 - \sqrt{1 - \frac{2 \times 0.608}{0.85 \times 3}} \right] = 0.1176$$

4. Amount of steel:

$$A_s = \rho bd = 0.1176 \times 8 \times 96 = 9.03 \text{in}^2$$

5. Use $10\#9\,(10.0\text{in}^2)$ on both sides of the wall, assuming that V_u could come from either direction.

Answer: Use $10\#9\,(10.0\text{in}^2)$ on both sides of the wall.

Example 23

Design horizontal shear reinforcing for the reinforced concrete shear wall shown in Figure 9.67 if $f_c' = 3000\,\text{psi}$ and $fy = 60,000\,\text{psi}$.

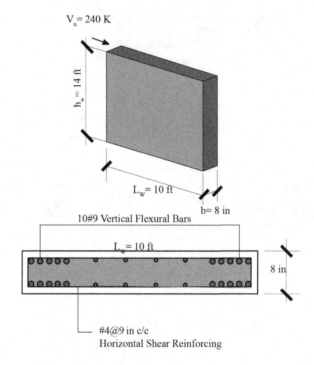

Solution

1. Find the shear capacity of the wall.

$$V_c = 2\sqrt{f_c'}\,bd = 2 \times \sqrt{3000} \times 8 \times 96 = 84,130\,\text{lb} = 84.13\,\text{kip}$$

2. Determine the amount of shear that must be carried by steel.

$$V_s = \frac{V_u - \phi V_c}{\phi} = \frac{240 - 0.75 \times 84.13}{0.75} = 235.87\,\text{kip}$$

3. Determine the spacing of shear reinforcement if #4 bar is used.

$$s = \frac{A_v f_y d}{V_s} = \frac{2 \times 0.20 \times 60 \times 96}{235.87} = 9.77 \text{ in}$$

4. Check maximum spacing.

$$s_{max} = \text{Minimum of} \begin{cases} \ell_w / 5 & = 96/5 = 19.2 \text{in} \\ 3h & = 3 \times 8 = 24 \text{in} \\ 18 \text{ in} & \text{(governs)} \end{cases}$$

5. Provide #4 @ 9 in c/c.
6. Check minimum reinforcing.

$$\rho_h = \frac{A_v}{bs} = \frac{2 \times 0.20}{8 \times 9} = 0.00417$$

Since $\rho_h (0.00417) > \rho_{min} (0.0025)$, selected spacing is adequate.

Answer: Use #4 @ 9 in c/c.

Example 24

Design a vertical shear reinforcing for the reinforced concrete shear wall shown in Figure 9.68 if $f'_c = 3000 \text{psi}$ and $fy = 60,000 \text{psi}$.

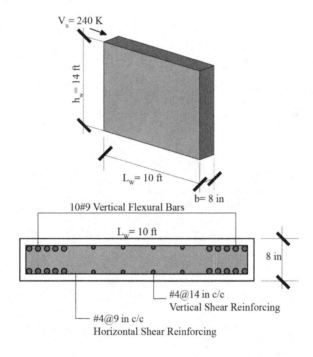

Solution

Horizontal shear reinforcing must be designed earlier, and it was found that $\rho_h = 0.00417$ in example 23.

1. Find the reinforcement ratio for the vertical shear reinforcing.

$$\rho_v = 0.0025 + 0.5\left(2.5 - \frac{h_w}{\ell_w}\right)(\rho_h - 0.0025)$$

$$= 0.0025 + 0.5\left(2.5 - \frac{14}{10}\right)(0.00417 - 0.0025)$$

$$= 0.00342$$

2. Determine the spacing if #4 bar is used.

$$s = \frac{A_v}{\rho b} = \frac{2 \times 0.20}{0.00342 * 8} = 14.6\,\text{in}$$

3. Check maximum spacing.

$$s_{max} = \text{Minimum of} \begin{cases} \ell_w / 3 & = 96/3 = 32\ \text{in} \\ 3h & = 3 \times 8 = 24\ \text{in} \\ 18\ \text{in} & (\text{governs}) \end{cases}$$

5. Provide #4 @ 14 in c/c.

Answer: Use #4 @ 14 in c/c.

9.8 PRESTRESSED CONCRETE

Prestressed members are useful when span-to-depth ratio is very large.

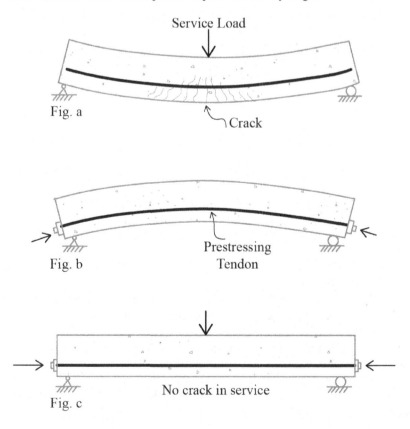

Service Load

Fig. a

Crack

Prestressing
Fig. b   Tendon

No crack in service
Fig. c

- Figure 9.69(a): Regular reinforced concrete members are weak in resisting tension but strong in resisting compression. Crack develops on the tension side due to service loads even though reinforcements are present there.
- Figure 9.70(b): To prevent excessive tension at the service period, concrete is compressed at the construction stage by steel wires known as tendon. The term *pre* refers to this stage, which comes earlier of service period of the beam, and the term stress is referred to the applied compression by the tendon.
- Figure 9.71(c): When this prestressed beam enters into service period, in order to develop cracks, the preapplied compression must be overcome by the tension due to service load.
- As a result, prestressed members deflect less, carry a greater load, or span a greater distance than a conventionally reinforced member of the same size, proportion, and weight.
- There are two types of prestressing: pretensioning and posttensioning.

9.8.1 Pretensioning

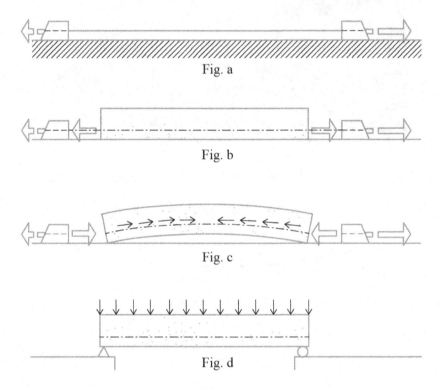

Fig. a

Fig. b

Fig. c

Fig. d

- *Pretensioning* prestresses a concrete member by stretching the reinforcing tendons before the concrete is cast.
- Figure 9.72(a): Steel tendons are first stretched across the casting bed between two abutments until a predetermined tensile force is developed.
- Figure 9.72(b): Concrete is then cast in formwork around the stretched tendons and fully cured. The tendons are placed eccentrically in order to reduce the maximum compressive stress to that produced by bending alone.
- Figure 9.72(c): When the tendons are cut or released, the tensile stresses in the tendons are transferred to the concrete through bond stresses. The eccentric action of the prestressing produces a slight upward curvature or camber in the member.
- Figure 9.72(d): The deflection of the member under loading tends to equalize its upward curvature.

9.8.2 Posttensioning

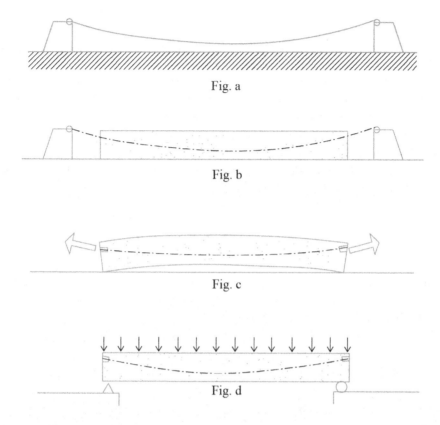

Fig. a

Fig. b

Fig. c

Fig. d

- *Posttensioning* is the prestressing of a concrete member by tensioning the reinforcing tendons after the concrete has set.
- Figure 9.73(a): Unstressed steel tendons, draped inside the beam or slab form, are coated or sheathed to prevent bonding while the concrete is cast.
- Figure 9.73(b): After the concrete has cured, the tendons are clamped on one end and jacked against the concrete on the other end until the required force is developed.
- Figure 9.73(c): The tendons are then securely anchored on the jacking end and the jack removed. After the posttensioning process, the steel tendons may be left unbonded, or they may be bonded to the surrounding concrete by injecting grout into the annular spaces around the sheathed strands.
- Figure 9.73(d): The deflection of the member under loading tends to equalize its upward curvature.

9.8.3 Advantages and Disadvantages of Prestressed Concrete

Advantages

- It is possible with prestressing to utilize the entire cross sections of members to resist loads. Thus, smaller members can be used to support the same loads, or the same size members can be used for longer spans.

- Prestressed members are crack-free under working loads and, as a result, look better and are more watertight, providing better corrosion protection for the steel. Furthermore, crack-free prestressed members require less maintenance and last longer than cracked reinforced concrete members.
- The negative moments caused by prestressing produce camber in the members, with the result that total deflections are reduced.
- Other advantages of prestressed concrete include the following: reduction in diagonal tension stresses, sections with greater stiffnesses under working loads, and increased fatigue and impact resistance compared to ordinary reinforced concrete.

Disadvantages

- Prestressed concrete requires the use of high-strength concretes and steels and the use of more complicated formwork, with resulting higher labor costs.
- Closer quality control is required in manufacture.
- Losses in the initial prestressing forces. When the compressive forces from prestressing are applied to the concrete, it will shorten somewhat, partially relaxing the cables. The result is some reduction in cable tension, with a resulting loss in prestressing forces. Shrinkage and creep of the concrete add to this effect.
- Additional stress conditions must be checked in design, such as the stresses occurring when prestress forces are first applied and then after prestress losses have taken place, as well as the stresses occurring for different loading conditions.
- Cost of end anchorage devices and end beam plates that may be required.

9.8.4 PRESTRESSED BRIDGE GIRDER

9.8.5 POSTTENSIONING ANCHORAGE

9.9   PRESTRESSED BEAM

9.9.1   ELASTIC ANALYSIS DUE TO FLEXURE

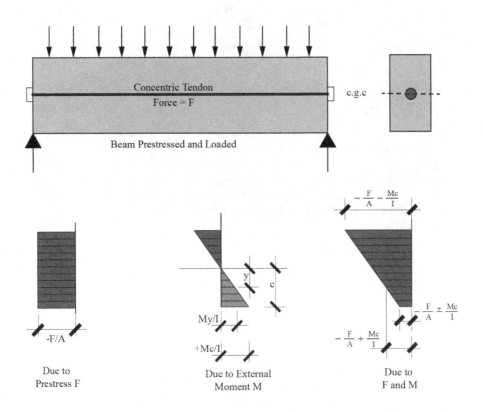

- In its simplest form, let us consider a simple rectangular beam prestressed by a tendon through its centroidal axis (i.e., centroid of concrete, or c.g.c.) and loaded by external loads.
- Due to tensile prestress force F in the tendon, an equal compressive force F is produced in the concrete, which produces a uniform compressive stress of $f = -\dfrac{F}{A}$.
- If M is the external moment at a section due to the load on the beam and the weight of the beam, then the stress at any point across that section due to M is:

$$f = \pm \frac{My}{I}$$

- Thus, the resulting stress distribution is given by:

$$f = -\frac{F}{A} \pm \frac{My}{I}$$

Example 25

Determine the top and bottom fiber stresses at the midsection of the prestressed beam.

Solution

Here, prestressing force $F = 360\,\text{kip} = 360{,}000\,\text{lb}$, eccentricity $e = 6$ in, and $y = 15$ in. Now:

$$A = bh = 20 \times 30 = 600\,\text{in}^2$$
$$I = \frac{bh^3}{12} = \frac{20 \times 30^3}{12} = 45{,}000\,\text{in}^4$$
$$M = \frac{wl^2}{8} = \frac{3 \times 24^2}{8} = 216\,\text{k-ft}$$
$$= 216 \times 12{,}000\,\text{lb-in} = 2592{,}000\,\text{lb-in}$$

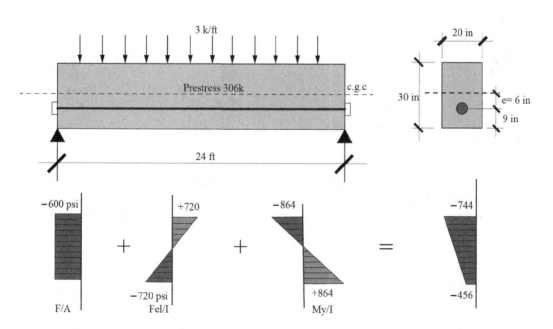

Fiber stresses at midspan:

$$f = \frac{F}{A} \pm \frac{Fey}{I} \pm \frac{My}{I}$$

$$= \frac{-360,000}{600} \pm \frac{-360,000 \times 6 \times 15}{45,000} \pm \frac{2592,000 \times 15}{45,000}$$

$$= -600 \pm 720 \pm 864$$

$$f_{top} = -600 + 720 - 864 = -744 \, psi$$

$$f_{bottom} = -600 - 720 + 864 = -456 \, psi$$

Answer:

$$f_{top} = -744 \, psi, f_{bottom} = -456 \, psi$$

9.9.2 ECCENTRIC TENDON

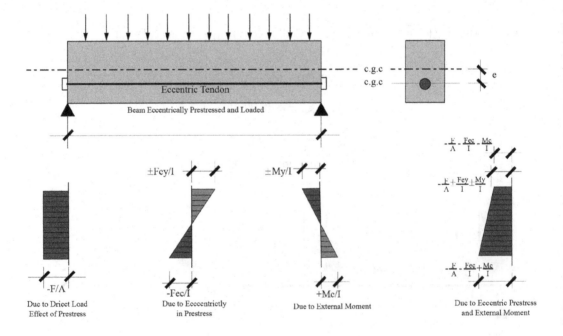

Beam Eccentrically Prestressed and Loaded

Due to Driect Load   Due to Eccentrictly   Due to External Moment   Due to Eccentric Prestress
Effect of Prestress   in Prestress   and External Moment

- Due to eccentrically placed tendon, compressive force F in the concrete now acts at a distance e from the c.g.c., as shown.
- Therefore, a moment of Fe is produced due to eccentric prestress, which produces additional stresses in the section, which is given by:

$$f = \pm \frac{Fe.y}{I}$$

- Thus, the resulting stress distribution is produced due to direct prestress load, eccentricity in prestress, and external moment, which is given by:

$$f = -\frac{F}{A} \pm \frac{Fey}{I} \pm \frac{My}{I}$$

Example 26

Determine the top and bottom fiber stresses at support of the prestressed beam.

Solution

Here, prestressing force $F = 360\,\text{kip} = 360,000\,\text{lb}$, eccentricity $e = 6$ in, and $y = 15$ in. Now:

$$A = bh = 20 \times 30 = 600\,\text{in}^2$$

$$I = \frac{bh^3}{12} = \frac{20 \times 30^3}{12} = 45,000\,\text{in}^4$$

$$M = 0;\ (\textit{Bending moment is Zero at support})$$

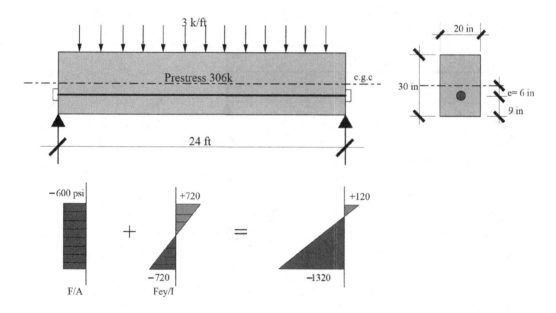

Fiber stresses at support:

$$f = \frac{F}{A} \pm \frac{Fey}{I} \pm \frac{My}{I}$$

$$= \frac{-360,000}{600} \pm \frac{-360,000 \times 6 \times 15}{45,000} \pm 0$$

$$= -600 \pm 720$$

$$f_{top} = -600 + 720 = +120\,\text{psi}$$

$$f_{bottom} = -600 - 720 = -1320\,\text{psi}$$

Answer: $f_{top} = +120$ psi, $f_{bottom} = -1320$ psi

Example 27

Determine the top and bottom fiber stresses at the midsection of the prestressed beam. The prestressing wires are concentric with the concrete section.

Solution

Here, prestressing force $F = 360\,\text{kip} = 360,000\,\text{lb}$, eccentricity $e = 0, y = 15$ in. Now:

$$A = bh = 20 \times 30 = 600\,\text{in}^2$$

$$I = \frac{bh^3}{12} = \frac{20 \times 30^3}{12} = 45,000\,\text{in}^4$$

$$M = \frac{wL^2}{8} = \frac{3 \times 24^2}{8} = 216\,\text{k} - \text{ft}$$

$$= 216 \times 12,000\,\text{lb-in} = 2592,000\,\text{lb-in}$$

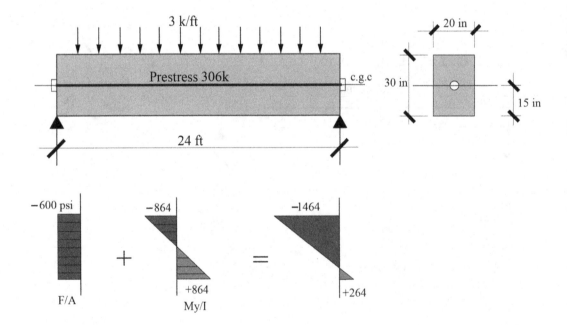

Fiber stresses at midspan:

$$f = \frac{F}{A} \pm \frac{Fey}{I} \pm \frac{My}{I}$$

$$= \frac{-360,000}{600} \pm 0 \pm \frac{2592,000 \times 15}{45,000}$$

$$= -600 \pm 864$$

$$f_{top} = -600 - 864 = -1464\,\text{psi}$$

$$f_{bottom} = -600 + 864 = +264\,\text{psi}$$

Answer: $f_{top} = -1464$ psi, $f_{bottom} = +264$ psi

9.10 PRESTRESSED BEAM

9.10.1 ULTIMATE MOMENT CAPACITY

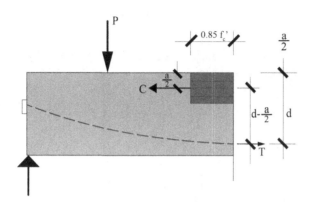

TABLE 9.3

Description of the Symbols

Symbol	Description
f_s	Stress in prestressing steel
f_U	Ultimate stress in prestressing steel
f_c'	Compressive strength of concrete
C	Compressive force in concrete
T	Tensile force in steel
A_S	Area of prestressing steel
ρ	$\rho = A_S / bd$
b	Width of beam
d	Effective depth of concrete
a	Depth of Whitney's stress block

- The analysis of the prestressed beam for ultimate moment capacity is fairly similar to that of reinforced concrete beam. (See "ultimate strength of rectangular beam" of CEE 411 in here.)
- But unlike RC beam, the stresses in steel (f_s) in the prestressed beam must be calculated before an analysis. In RC beam, f_s simply equals to f_y, but in prestress beam, it is given by:

$$f_s = f_u \left(1 - 0.5\rho \frac{f_u}{f_c'}\right)$$

- Compressive force (C) and tensile force (T) are given by:

$$C = 0.85 f_C' ab; \ T = A_s f_s$$

- For equilibrium, C must equal to T. Thus:

$$C = T; \ 0.85 f_C' ab = A_s f_s, \ a = \frac{A_s f_s}{0.85 f_C' b}$$

- The lever arm of C and T force couple is determined by $(d - a/2)$.
- Therefore, the nominal moment capacity is:

$$M_n = C\left(d - \frac{a}{2}\right) = T\left(d - \frac{a}{2}\right) = A_s f_s \left(d - \frac{a}{2}\right)$$

- The ultimate design capacity is:

$$M_u = \phi A_s f_s \left(d - \frac{a}{2}\right)$$

Where ϕ is the strength reduction factor and equals to 0.9.

Example 28

An I-shaped beam is prestressed with effective prestress f_{se} of 160 ksi. Material properties are $f_{pu} = 270$ ksi, $f_C' = 7$ ksi. Determine the ultimate moment capacity of the beam.

9.10.2 REINFORCEMENT RATIO

$$\rho = \frac{A_s}{bd} = \frac{2.75}{18 \times 31.5} = 0.00485$$

Steel stress at ultimate moment capacity:

$$f_s = f_u \left(1 - 0.5\rho \frac{f_u}{f_c'}\right)$$

$$= 270 \left(1 - 0.5 \times 0.00485 \times \frac{270}{7}\right) = 244.75 \, \text{ksi}$$

Check location of neutral axis.
  Compressive area:

$$C = T;$$
$$0.85 f_C' A_C = A_s f_s$$
$$A_c = \frac{A_s f_s}{0.85 f_C'} = \frac{2.75 \times 244.75}{0.85 \times 7} = 113.12 \, \text{in}^2$$

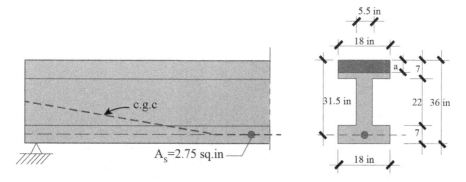

Flange area:

$$A_f = b_f h_f = 18 \times 7 = 126\,\text{in}^2$$

Since $A_c\,(113.12\,\text{in}^2) < A_f\,(126\,\text{in}^2)$, neutral axis is in flange. Therefore, the beam is not a true T-beam.

9.10.3 FIND DEPTH OF NEUTRAL AXIS

$$a = \frac{A_s f_s}{0.85 f_c' b} = \frac{2.75 \times 244.75}{0.85 \times 7 \times 18} = 6.28\ \text{in}$$

9.10.4 DETERMINE CAPACITY

$$M_u = \phi A_s f_s \left(d - \frac{a}{2} \right)$$

$$= 0.9 \times 2.75 \times 244.75 \left(31.5 - \frac{6.28}{2} \right)$$

$$= 17{,}179.2\,\text{k-in} = 1431.6\,\text{k-ft}$$

Answer: 1431.6 kft

9.10.5 T-BEAM

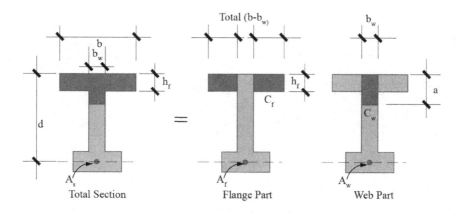

Total Section   Flange Part   Web Part

This part is analogous to the reinforced concrete T-beam analysis. See "T-Beam Analysis" for more details.

$$C_f = 0.85 f_c' \left(b_f - b_w\right) h_f$$
$$C_w = 0.85 f_c' a b_w$$
$$\phi M_n = \phi \left[C_w \left(d - \frac{a}{2}\right) + C_f \left(d - \frac{h_f}{2}\right) \right]$$

Example 29

The same I-shaped beam as example 1, but the steel area is increased to $A_s = 3.67\,\text{in}^2$. Now, determine ϕM_n.

9.10.6   REINFORCEMENT RATIO

$$\rho = \frac{A_s}{bd} = \frac{3.67}{18 \times 31.5} = 0.00647$$

Steel stress at ultimate moment capacity:

$$f_s = f_u \left(1 - 0.5 \rho \frac{f_u}{f_c'}\right)$$
$$= 270 \left(1 - 0.5 \times 0.00647 \times \frac{270}{7}\right) = 236.31\,\text{ksi}$$

9.10.7   CHECK LOCATION OF NEUTRAL AXIS

Compressive area:

$$C = T;\; 0.85 f_c' A_C = A_s f_s$$

$$A_C = \frac{A_s f_s}{0.85 f_c'} = \frac{3.67 \times 236.31}{0.85 \times 7} = 145.76\,\text{in}^2$$

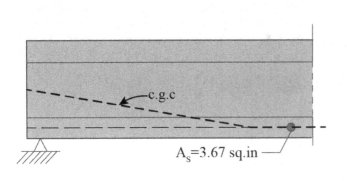

A_s=3.67 sq.in

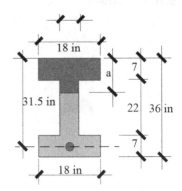

Flange area:

$$A_f = b_f h_f = 18 \times 7 = 126\,\text{in}^2$$

Since $A_C\left(145.76\text{in}^2\right) > A_f\left(126\text{in}^2\right)$, neutral axis is in web. Therefore, the beam is indeed a true T-beam.

9.10.8   FIND DEPTH OF NEUTRAL AXIS

$$a = h_f + \left(A_C - A_f\right)/b_w$$
$$= 7 + \left(145.76 - 126\right)/5.5 = 10.6\,\text{in}$$

Find compressive forces:

$$C_f = 0.85f_C'\left(b_f - b_w\right)h_f$$
$$= 0.85 \times 7 \times \left(18 - 5.5\right) \times 7 = 520.6\,\text{kip}$$
$$C_w = 0.85f_C'ab_w$$
$$= 0.85 \times 7 \times 10.6 \times 5.5 = 346.9\,\text{kip}$$

Determine capacity:

$$\phi M_n = \phi\left[C_w\left(d - \frac{a}{2}\right) + C_f\left(d - \frac{h_f}{2}\right)\right]$$
$$= 0.90\left[346.9\left(31.5 - \frac{10.6}{2}\right) + 520.6\left(31.5 - \frac{7}{2}\right)\right]$$
$$= 21,300k - in = 1775k - ft(Ans.)$$

9.11   VIERENDEEL TRUSS

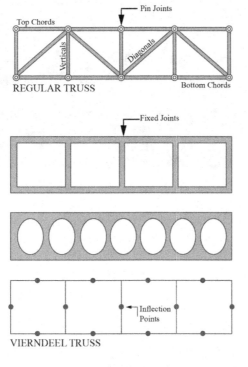

Pin Joints

Top Chords

Verticals

Diagonals

REGULAR TRUSS

Bottom Chords

Fixed Joints

Inflection Points

VIERNDEEL TRUSS

- The members of regular truss are connected by pin joints and have triangular subdivision in their geometry.
- Due to pin connections, the members carry only axial forces (either tension or compression), but no shear or bending moments.
- Unlike regular trusses, Vierendeel trusses do not have any diagonal members, and the vertical web members are rigidly connected to parallel top and bottom chords.
- Therefore, a significant portion of truss action is lost (i.e., forces are no longer transmitted exclusively by axial forces in members).
- Vierendeel trusses are not true trusses because the shear force, which must be transmitted through the top and bottom chords, creates bending moments in these members.
- Vierendeel truss was developed by an engineer, Arthur Vierendeel, in 1896 to serve as the frame of railroad bridges in Belgium.
- Due to the deflected shape of the Vierendeel truss under gravity load, it is seen that double curvature appears on each chord, both horizontals and verticals.
- Therefore, it can be assumed that hinges form at middle points (inflection points) of each member, and bending moments at these points are zero.

Vierendeel bridge at Grammene, Belgium.

Gustav Heinemann Bridge at Berlin, Germany.

Osera de Ebro Rail Bridge, Spain.

Sky bridge, Petronas Twin Towers, Malaysia.

Sky bridge, Petronas Twin Towers, Malaysia.

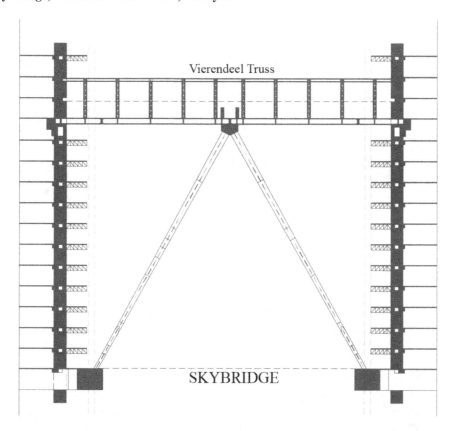

Beinecke Rare Book and Manuscript Library, Connecticut, USA.

Example 30

Determine the shear forces in the horizontal chords of the Vierendeel truss.

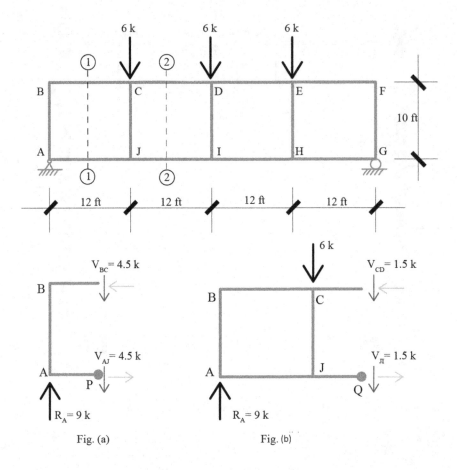

Fig. (a)   Fig. (b)

Solution

Since the loads are symmetric and so is the structure, R_A is equal to half of the total applied load.

$$R_A = \frac{1}{2}(6+6+6) = 9\,\text{kip}$$

9.11.1 SHEAR FORCES IN *BC* AND *AJ*

Figure 9.95(a): Section 1–1 is passed through the center of the first panel. Because the section passes through the points of inflection in the chords, no moments act on the ends of the members at the cut. Assuming the shear is equal in each chord, that is, $V_{BC} = V_A$, we write:

$$\sum F_y = 0 \uparrow^+ \quad R_A - V_{BC} - V_{AJ} = 0$$
$$9 - V_{BC} - V_{BC} = 0; \quad V_{BC} = 4.5\,\text{kip}(\downarrow)$$

Therefore, according to the assumption, $V_{AJ} = 4.5\text{kip}(\downarrow)$.

9.11.2   Shear Forces in CD and μ

Figure 9.95(b): Section 2–2 is passed through the center of the second panel. Assuming the shear is equal in each chord, that is, $V_{CD} = V_{\mu}$, we write:

$$\sum F_y = 0 \uparrow^+ \ R_A - 6 - V_{CD} - V_{1I} = 0$$
$$9 - 6 - V_{CD} - V_{CD} = 0; \ V_{CD} = 1.5\text{kip}(\downarrow)$$

Therefore, according to the assumption, $V_{\mu} = 1.5\,\text{kip}(\downarrow)$.

Note that the shear forces in the top and bottom chords of a panel have equal magnitude and sign.

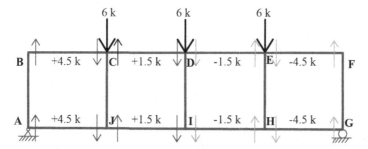

Shear Forces in Top and Bottom Chords

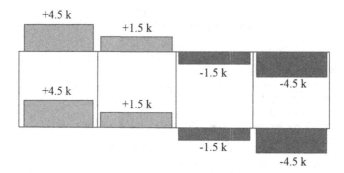

Shear Forces Diagram of Top and Bottom Chords

The shear forces in BC, AJ, CD, and JI were determined earlier. The shear forces in DE, IH, EF, and HG are determined by symmetry, but the sign of these forces will be opposite.

Example 31

Determine the axial forces in the horizontal chords of the Vierendeel truss.

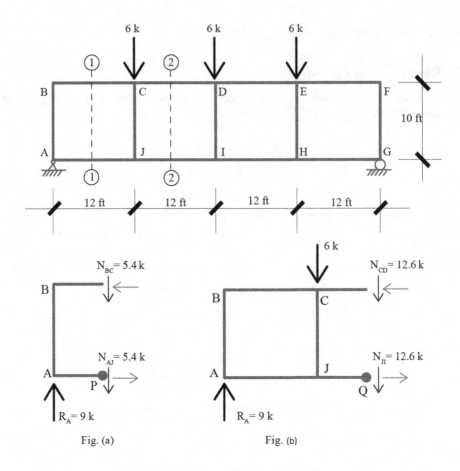

9.11.3 AXIAL FORCES IN *BC* AND *AJ*

Due to gravity loads, the top chords of the Vierendeel truss will always be in compression, and the bottom chords will be in tension.

Figure 9.97(a): Summing moments about the midpoint p (inflection point) of AJ to determine axial force of BC, we write:

$$\sum M_p = 0(\circlearrowleft^+)$$
$$R_A \times 6 - N_{BC} \times 10 = 0$$
$$9 \times 6 - N_{BC} \times 10 = 0$$
$$N_{BC} = 5.4 \text{kip}(\text{C})$$

The equilibrium in the x direction establishes that a tension force of 5.4 kip acts in the bottom chord; therefore, $N_{AJ} = 5.4 \text{kip}(T)$.

9.11.4 AXIAL FORCES IN *CD* AND μ

Figure 9.97(b): Summing moments about the midpoint q (inflection point) of JI to determine axial force of CD, we write:

$$\sum M_q = 0(\circlearrowleft^+)$$

$$R_A \times (12+6) - 6 \times 6 - N_{CD} \times 10 = 0$$

$$9 \times 18 - 6 \times 6 - N_{CD} \times 10 = 0$$

$$N_{CD} = 12.6 \, \text{kip} \, (\text{C})$$

The equilibrium in the x direction establishes that a tension force of 12.6 kip acts in the bottom chord; therefore, $N_{JI} = 12.6 \, \text{kip} \, (\text{T})$.

Note that the axial forces in the top and bottom chords of a panel have equal magnitude but opposite sign.

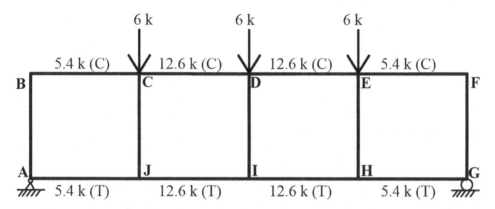

Axial Forces in Top and Bottom Chords

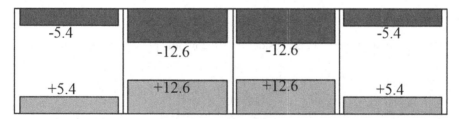

Axial Forces Diagram of Top and Bottom Chords

The axial forces in *BC*, *AJ*, *CD*, and *JI* were determined earlier. The axial forces in *DE*, *IH*, *EF*, and *HG* are determined by symmetry, and they have the same sign of their counterparts.

Example 32

Determine the bending moments in the horizontal chords of the Vierendeel truss.

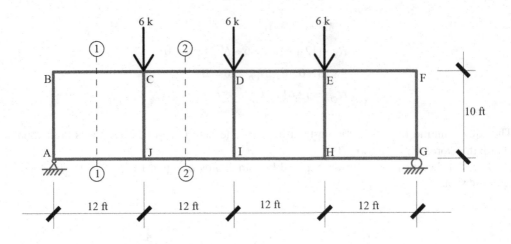

9.11.5   BENDING MOMENT IN *BC*

By taking the *B*1 segment (here "1" is considered as the middle point of *BC*), we write:

$$\sum M_B = 0(\circlearrowleft^+)$$
$$-M_{BC} + V_{BC} \times 6 = 0$$
$$-M_{BC} + 4.5 \times 6 = 0; \; M_{BC} = 27k - ft$$

And now, taking the 1*C* segment, we write:

$$\sum M_C = 0(\circlearrowleft^+)$$
$$V_{BC} \times 6 - M_{CB} = 0$$
$$4.5 \times 6 - M_{CB} = 0; \; M_{CB} = 27k - ft$$

Note that the bending moments on both ends of the same panel (M_{BC} and M_{CD}) are equal in magnitude and have the same direction.

Moreover, they are found simply by multiplying the shear force of that panel (V_{CD}) by half of the panel length, that is, $4.5\text{kip}\times6\text{ft} = 27\text{k}-\text{ft}$ in this case.

9.11.6 BENDING MOMENT IN *CD*

$$M_{CD} = M_{DC} = V_{CD}\times6 = 1.5\times6 = 9\,\text{k-ft}$$

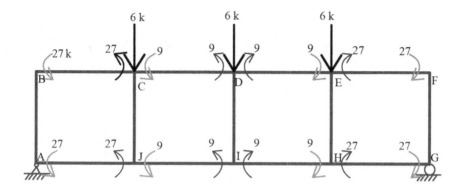

Bending Moment In Top and Bottom Chords

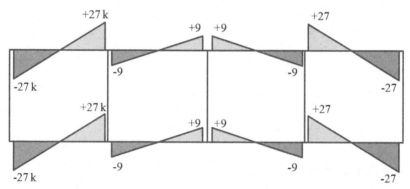

Bending Moment Diagram of Top and Bottom Chords

Example 33

Determine the shear forces in the vertical chords of the Vierendeel truss.

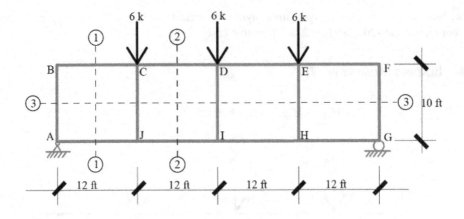

9.11.7 Shear Forces in *AB*

Pass a section 1–1 through midpoint of *BC* and another section 3–3 through midpoint of *AB*. Take the resulting *L*-shaped segment as shown in the following figure.

Now, we write:

$$\sum F_X = 0 \to^+ V_{AB} - N_{BC} = 0$$
$$V_{AB} - 5.4 = 0; \ V_{AB} = 5.4 \, \text{kip}$$

9.11.8 Shear Forces in *JC*

Pass a section 2–2 through the midpoint of *CD*, and another section 3–3 through the midpoint of *AB*. Take the resulting *F*-shaped segment as shown in the following figure.

$$\sum F_X = 0 \to^+ V_{AB} + V_{JC} - N_{CD} = 0$$
$$5.4 + V_{JC} - 12.6 = 0; \ V_{JC} = 7.2 \, \text{kip}$$

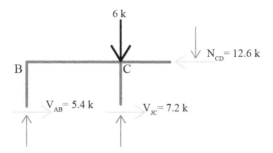

9.11.9 Shear Forces in ID

Pass a section 2–2 through the midpoint of CD, and another section 3–3 through the midpoint of AB. Take the resulting segment as shown in the following figure.

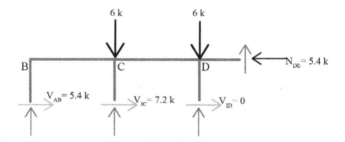

$$\sum F_X = 0 \rightarrow^+ V_{AB} + V_{JC} + V_{ID} - N_{DE} = 0$$
$$5.4 + 7.2 + V_{ID} - 12.6 = 0; \ V_{ID} = 0$$

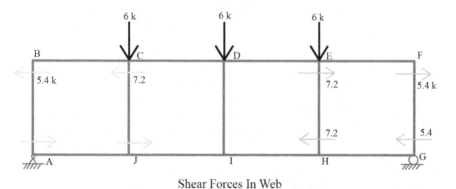

Shear Forces In Web

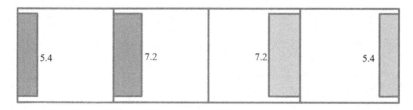

Shear Forces Diagram of Web

Example 34

Determine the axial forces in the vertical chords of the Vierendeel truss.

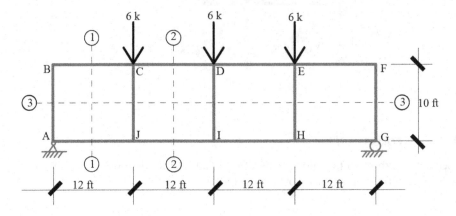

9.11.10   AXIAL FORCES IN *AB*

Pass a section 1–1 through the midpoint of BC, and another section 3–3 through the midpoint of AB. Take the resulting L-shaped segment, as shown in the following figure.

Now, we write:

$$\sum F_y = 0 \uparrow + N_{AB} - V_{BC} = 0$$
$$N_{AB} - 4.5 = 0; \; N_{AB} = 4.5 \, \text{kip}$$

9.11.11   AXIAL FORCES IN *JC*

$CD\,AB$. Take the resulting F-shaped segment as shown in the following figure.

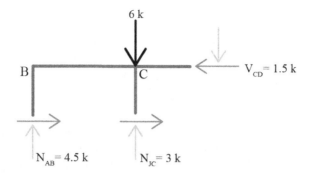

$$\sum F_y = 0 \rightarrow^+ \; N_{AB} + N_{JC} - 6 - V_{CD} = 0$$
$$4.5 + N_{JC} - 6 - 1.5 = 0; \; N_{JC} = 3 \, \text{kip}$$

9.11.12 Axial Forces in *ID*

Pass the section 2–2 and section 3–3, and take the resulting segment, as shown in the following figure.

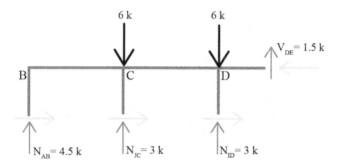

$$\sum F_y = 0 \rightarrow^+ \; N_{AB} + N_{JC} + N_{ID} - 6 - 6 - V_{DE} = 0$$
$$4.5 + 3 + N_{ID} - 6 - 6 + 1.5 = 0; N_{ID} = 3$$

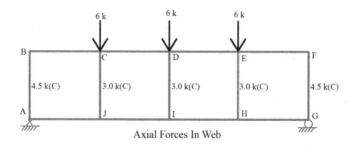

Axial Forces In Web

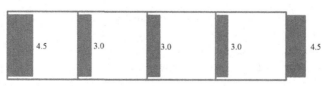

Axial Forces Diagram of Web

Example 35

Determine the bending moments in the horizontal chords of the Vierendeel truss.

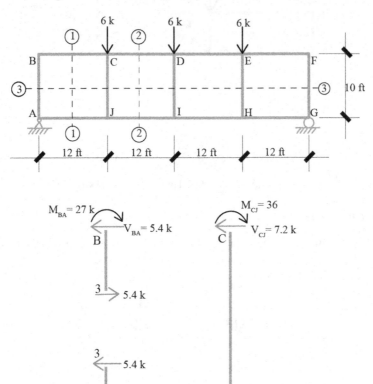

9.11.13 BENDING MOMENT IN *BC*

By taking the *B*3 segment (here, "3" is considered as the middle point of *AB*), we write:

$$\sum M_3 = 0 \circlearrowleft^+ \ M_{BA} - V_{BA} \times 5 = 0$$
$$M_{BA} - 5.4 \times 5 = 0; \ M_{BA} = 27\,\text{k} - \text{ft}$$

And now, taking the *A*1 segment, we write:

$$\sum M_1 = 0(\circlearrowleft^+) \ M_{AB} - V_{AB} \times 5 = 0$$
$$M_{AB} - 5.4 \times 5 = 0; \ M_{AB} = 27\,\text{k} - \text{ft}$$

Note that the bending moments on both ends of the same vertical chord are equal in magnitude and have the same direction.

Moreover, they are found simply by multiplying the shear force of that chord (V_{AB}) by half of its length, that is, $5.4\,\text{kip} \times 5\,\text{ft} = 27\,\text{k} - \text{ft}$ in this case.

9.11.14 BENDING MOMENT IN *JC*

$$M_{JC} = M_{CJ} = V_{CJ} \times 5 = 7.2 \times 5 = 36\, \text{k} - \text{ft}$$

9.11.15 BENDING MOMENT IN *ID*

$$M_{ID} = M_{DI} = V_{ID} \times 5 = 0 \times 5 = 0$$

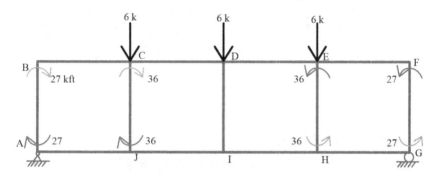

Bending Moments in Vertical Chords

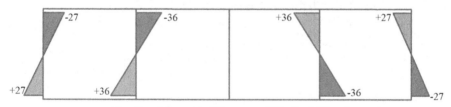

Bending Moments Diagram of Vertical Chords

Example 36

Analyze the horizontal chords of the Vierendeel truss using beam analogy.

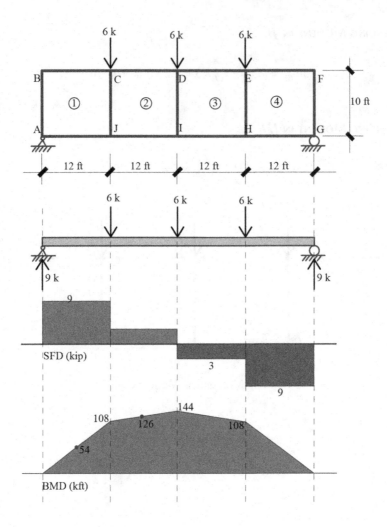

Solution

Imagine a simply supported beam of the same span of the Vierendeel truss, and translate the applied forces on that beam. By analyzing the shear force and bending moment diagrams of this beam, we can determine the internal forces of the Vierendeel truss.

Shear Forces: Numerically equal to half of shear forces of the beam of the same panel.

$$V_{BC} = V_{AJ} = 9/2 = 4.5 \text{ kip}$$
$$V_{CD} = V_{JI} = 3/2 = 1.5 \text{ kip}$$

Bending Moments: These are simply found by multiplying the shear forces of the horizontal chord by half the panel length.

$$M_{BC} = M_{AJ} = 4.5 \times 6 = 27 \text{ k} - \text{ft}$$
$$M_{CD} = M_x = 1.5 \times 6 = 9 \text{ k} - \text{ft}$$

Axial Forces: Determined by dividing the bending moments of the beam at the middle point of each panel by the height of the truss.

$$N_{BC} = N_{AJ} = 54/10 = 5.4 \text{ kip}(C)$$
$$N_{CD} = N_{JI} = 126/10 = 12.6 \text{ kip}(T)$$

Example 37

Analyze the vertical chords of the Vierendeel truss using beam analogy.

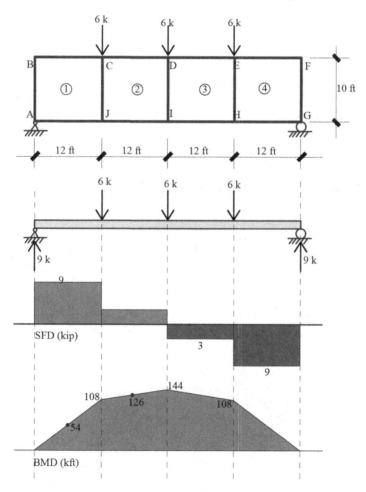

Shear Forces: We need to take the difference in bending moments at the middle point of the adjacent panels of the beam and divide it by the height of the truss.

$$V_{AB} = (54-0)/10 = 5.4 \text{ kip}$$
$$V_{JC} = (126-54)/10 = 7.2 \text{ kip}$$
$$V_{ID} = (126-126)/10 = 0$$

For V_{AB}, we need to take the difference in moments at the middle point of panel 1 ($M = 54\text{k}-\text{ft}$) and its previous panel (which does not exist; therefore, $M = 0$). Now, for V_{JC}, it is panel 2 ($M = 126\text{k}-\text{ft}$) and panel 1 ($M = 54\text{k}-\text{ft}$). Finally, for V_{ID}, it is panel 3 ($M = 126\text{k}-\text{ft}$) and panel 2 ($M = 126\text{k}-\text{ft}$).

Bending Moments: These are simply found by multiplying the shear forces of the chords by half the height of the truss.

$$M_{AB} = 5.4 \times 5 = 27\text{k}-\text{ft}$$
$$M_{JC} = 7.2 \times 5 = 36\text{k}-\text{ft}$$
$$M_{ID} = 0 \times 5 = 0$$

Axial Forces: Determined by taking half of the differences of shear forces in the beam for two adjacent panels.

$$N_{AB} = (0-9)/2 = -4.5\text{kip}(C)$$
$$N_{CJ} = (3-9)/2 = -3\text{kip}(C)$$
$$N_{DI} = (-3-3)/2 = -3\text{kip}(C)$$

9.12   ARCH ANALYSIS

Example 38

Determine the support reactions of the three-hinged arch.

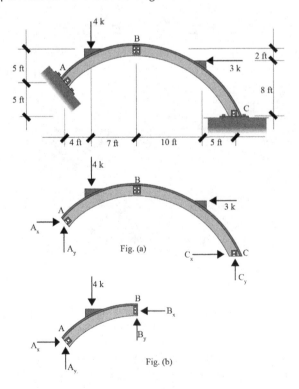

Fig. (a)

Fig. (b)

Solution

Determination of A_x A_y.
   From Figure 9.115(a), by summing the moment at C, we write:

$$\Sigma M_C = 0 \circlearrowleft^+ \quad A_x \times 5 + A_y \times 26 - 4 \times 22 - 3 \times 8 = 0$$

$$5A_x + 26A_y = 112$$

Now, passing a vertical section at hinge B (Figure 9.115b):

$$\Sigma M_B = 0(\circlearrowleft^+) - A_x \times 5 + A_y \times 11 - 4 \times 7 = 0$$
$$-5A_x + 11A_y = 28$$

Solving these two equations simultaneously, we find:

$$A_x = 2.72 \text{kip} (\rightarrow) A_y = 3.78 \text{kip} (\uparrow)$$

Determination of C_X and C_y.
   Now, from Figure 9.115(a), sum the horizontal and vertical forces:

$$\Sigma F_X = 0 \rightarrow^+ A_X - 3 + C_X = 0$$

$$2.72 - 3 + C_X = 0; \ C_X = 0.28 \text{kip} (\rightarrow)$$

$$\Sigma F_y = 0 \uparrow^+ A_y - 4 + C_y = 0$$

$$3.78 - 4 + C_y = 0; \ C_y = 0.22 \text{kip}$$

Example 39

Determine the axial force (thrust), shear force, and bending moment at point P, given that point P is at 4ft right and 2.5ft above from A and slope at P is $\theta = 30°$.

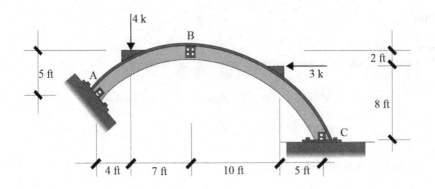

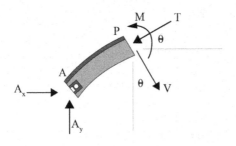

Solution

Passing a section at point P (Figure 9.116b), and summing moments at P, we write:

$$\sum M_P = 0(\circlearrowleft^+) A_y \times 4 - A_x \times 2.5 - M = 0$$
$$3.78 \times 4 - 2.72 \times 2.5 - M = 0;$$
$$M = 8.32 \text{k-ft}$$

Summing horizontal and vertical forces:

$$\sum F_x = 0 \rightarrow^+ \ A_x - T\cos\theta + V\sin\theta = 0$$
$$- T\cos 30 + V\sin 30 \ = -2.72$$
$$\sum F_y = 0\uparrow^+ A_y - T\sin\theta - V\cos\theta = 0$$
$$- T\sin 30 - V\cos 30 = -3.78$$

Solving these two equations simultaneously, we find:

$$T = 4.25\text{kip}(\nearrow)\ V = 1.91\text{kip}(\searrow)$$

9.12.1 THREE-HINGED ARCH, UDL

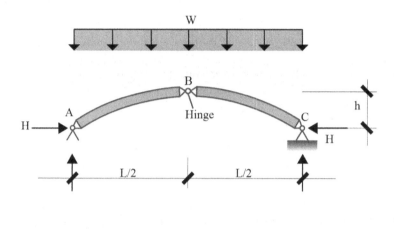

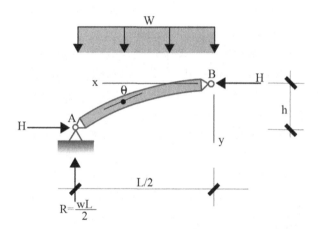

- Many arches carry dead loads that have a uniform or nearly uniform distribution over the span of the structure. For example, the weight per unit length of the floor system of a bridge will typically be constant.
- Consider the symmetric three-hinged arch as shown in Figure 9.117. Because of symmetry, the vertical reactions at supports A and C are $R = wL/2$.
- The horizontal thrust H at the base of the arch can be determine by summing moments about the center hinge at B if the arch is split at the hinge point.

$$\sum M_B = 0U^+ R\left(\frac{L}{2}\right) - \frac{wL}{2}\left(\frac{L}{4}\right) - Hy = 0;\ H = \frac{wL^2}{8y}$$

- An arch may develop bending moments along its axis. But it can be proved that if an arch fits the following parabolic profile, there will be no bending moment or shear force at any section of the arch. Thus, the external loads will be entirely carried by thrusts.

$$y = \left(\frac{4h}{L^2}\right)x^2$$

- The compressive force acting on the arch is known as thrust and is determined by:

$$T = \frac{H}{\cos \theta}$$

- The slope on the arch is found by differentiating the parabolic equation.

$$\tan \theta = \frac{dy}{dx} = \frac{d}{dx}\left[\left(\frac{4h}{L^2}\right)x^2\right] = \frac{8hx}{L^2}; \theta = \tan^{-1}\left(\frac{8hx}{L^2}\right)$$

Example 40

Determine the profile of the parabolic arch, support reactions, horizontal thrust at A, and axial thrust at B and A.

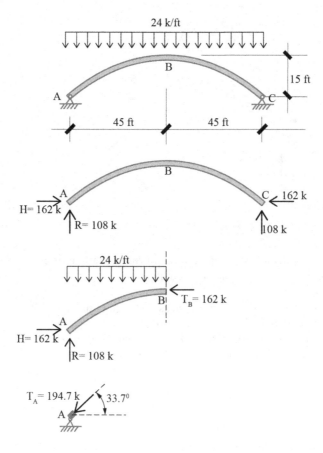

Solution

Though it is a two-hinged arch, the formulas for two-hinged arch are still applicable. Here, arch span $L = 90$ ft and arch height $h = 15$ ft.

9.12.2 ARC PROFILE

$$y = \left(\frac{4h}{L^2}\right)x^2 = \left(\frac{4 \times 15}{90^2}\right)x^2 = \frac{x^2}{135}; \; y = \frac{x^2}{135}$$

9.12.3 SUPPORT REACTIONS

$$R = wL/2 = (2.4 \times 90)/2 = 108 \, \text{kip}$$

9.12.4 HORIZONTAL THRUST AT BASE, A

$$H = \frac{wL^2}{8y} = \frac{2.4 \times 90^2}{8 \times 15} = 162 \, \text{kip}$$

9.12.5 AXIAL THRUST AT B

Passing a vertical section through B and summing horizontal forces:

$$\sum F_X = 0 \rightarrow^+ H - T_B = 0; 162 - T_B = 0; \; T_B = 162 \, \text{kip}$$

9.12.6 AXIAL THRUST AT BASE A

First, we need to determine the slope of arch at base.

$$\theta_A = \tan^{-1}\left(\frac{8hx}{L^2}\right) = \tan^{-1}\left(\frac{8 \times 15 \times 45}{90^2}\right) = 33.7°$$

Now, the axial thrust at base A is:

$$T_A = \frac{H}{\cos\theta} = \frac{162}{\cos 33.7°} = 194.7 \, \text{kip}$$

9.13 SHELL STRUCTURES

9.13.1 SHELLS

- *Shells* are thin curved plate structures typically constructed of reinforced concrete and used for the roofs of buildings.
- They are shaped to transmit applied forces by membrane stresses – the compressive, tensile, and shear stresses acting on the plane of their surfaces.
- A shell can sustain relatively large forces if loads are uniformly applied.

Opera House, Sydney. Architect: Jorn Utzon.

- Because of its thinness, however, a shell has little bending resistance and is unsuitable for concentrated loads.

9.13.2 SINGLE CURVED FORMS: BARREL SHELLS

- Single curved forms, or translational surfaces, are generated by sliding a plane curve along a straight line or over another plane curve.
- Barrel shells are cylindrical shell structures that are example of single curved forms.
- If the length of a barrel shell is three or more times its transverse span, it behaves as a deep beam with a curved section spanning in the longitudinal direction.
- If it is relatively short, it exhibits an archlike action.

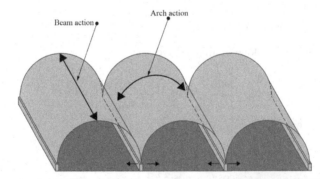

Kimbell Art Museum, USA. Architect: Louis Kahn.

9.13.3 SINGLE CURVED FORMS: BARREL SHELLS

Kimbell Art Museum, USA. Architect: Louis Kahn. Photo Credit: Raihatul Zannah.

9.13.4 SINGLE CURVED FORMS: GROIN VAULTS

Groin or cross vaults are compound vaults formed by the perpendicular intersection of two vaults, forming arched diagonal arrises called groins.

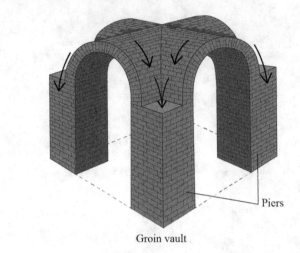

Piers

Groin vault

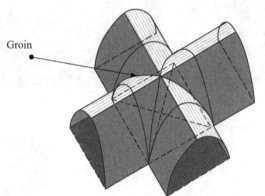

Groin

Interior of Santa Maria Maggiore, Italy.

9.13.5 DOUBLE CURVED FORMS: SPHERICAL DOMES

Rotational surfaces are generated by rotating a plane curve about an axis. Spherical, elliptical, and parabolic dome surfaces are examples of rotational surfaces.

Shell of the Biosphere.

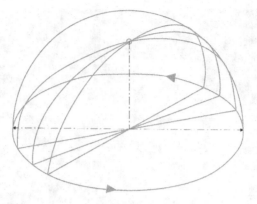

Rotational surfaces are generated by rotating a plane curve about an axis. Spherical, elliptical, and parabolic dome surfaces are examples of rotational surfaces.

Montreal Biosphere, Canada. Architect: Buckminster Fuller.

9.13.6   Double Curved Forms: Hyperbolic Paraboloid

A *hyperbolic paraboloid* is a surface generated by sliding a parabola with downward curvature along a parabola with upward curvature.

The Oceanographic, Spain. Architect: Felix Candela.

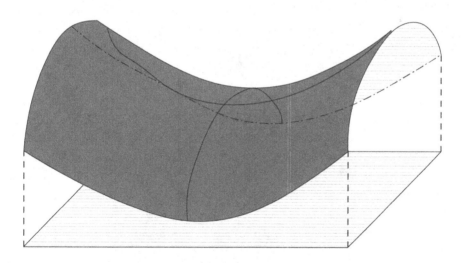

9.13.7   Ruled Surfaces: Hyperboloid

Ruled surfaces are generated by the motion of a straight line. Because of its straight-line geometry, a ruled surface is generally easier to form and construct than a rotational or translational surface.

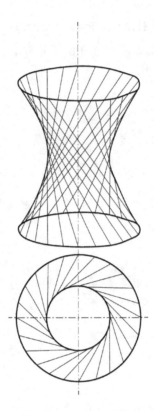

A *one-sheet hyperboloid* is a ruled surface generated by sliding an inclined line segment on two horizontal circles. Its vertical sections are hyperbolas.

Tornado Tower, Qatar.

9.13.8   RULED SURFACES: SADDLE

Saddle surfaces have an upward curvature in one direction and a downward curvature in the perpendicular direction.

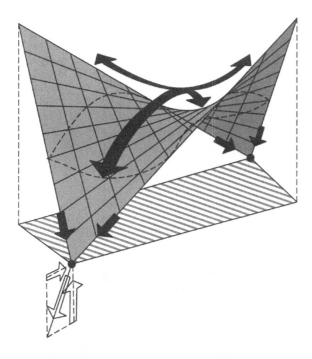

In a saddle-surfaced shell structure, regions of downward curvature exhibit archlike action, while regions of upward curvature behave as a cable structure. If the edges of the surface are not supported, beam behavior may also be present.

Church of St. Aloysius, New Jersey, USA.

9.14 SPHERICAL SHELLS

9.14.1 FORCES IN SPHERICAL SHELLS

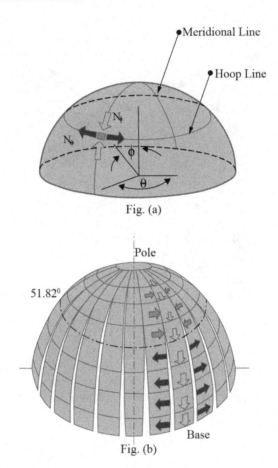

Fig. (a)

Fig. (b)

Two types of forces act on the surface of a spherical shell: meridional forces and hoop forces.

9.14.2 MERIDIONAL FORCES $\left(N_\phi\right)$

These forces act in the direction of the meridian, or latitude, or vertical. Meridional forces are always compressive from pole to base.

$$N_\phi = \frac{Rw}{1+\cos\phi}$$

Here, R is the radius of the dome, w is the distributed load on the dome, and ϕ is the angle measured from the vertical line passing through the pole. For example, for a point on pole, $\phi = 0°$, and for a point on base, $\phi = 90°$.

9.14.3 Hoop Forces (N_θ)

They act in the direction of the longitude, or horizontal. *Hoop forces* are compressive near the pole and tensile near the base.

$$N_\theta = Rw\left(\cos\phi - \frac{1}{1+\cos\phi}\right)$$

It can be proved that N_θ is compressive for all points above $\phi = 51.82°$, and tensile at below of it.

9.14.4 Meridional Stresses (f_ϕ) and Hoop Stresses (f_θ)

Determined by dividing the corresponding forces by the shell thickness (t).

$$f_\phi = \frac{N_\phi}{t}\ f_\theta = \frac{N_\theta}{t}$$

Distribution of meridional and hoop forces:

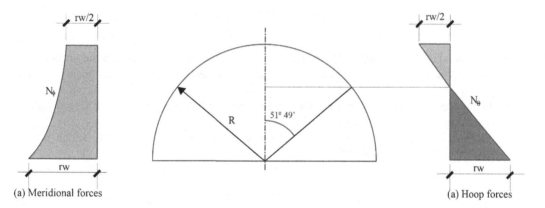

(a) Meridional forces (a) Hoop forces

9.14.5 Tension Ring

A *tension ring* is a planar ring against which the outward thrusts push, causing tension to develop in the ring. Tension force developed in the tension ring is given by:

$$T = RN_\phi\cos\phi\,\sin\phi$$

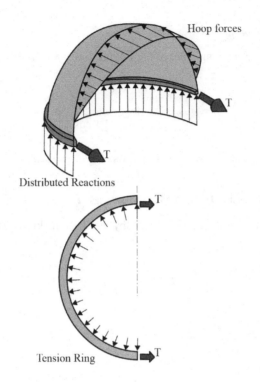

- A tension ring absorbs all the horizontal thrusts involved. When resting on the ground, it also provides a continuous footing for transferring the vertical reaction components to the ground.
- The tension ring must be continuous around the shell base; otherwise, they are no good and severe stress would develop in the shell.
- Alternatively, the ring can be supported on other elements (e.g., columns), which then receive vertical loads only.

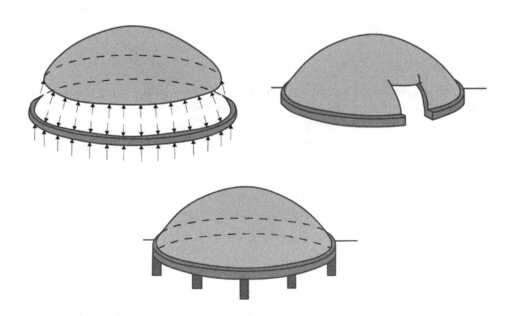

Example 41

Consider the following dome having a 150 ft diameter in plan, 30 ft high from ground, and a thickness of 4 in.

Determine the meridional stress, hoop stress, and ring tension at the base of the shell if there is 25 psf live load.

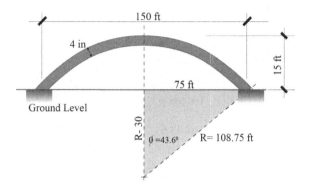

9.14.6 DETERMINE R AND ϕ

From the bottom light gray right triangle as shown in the figure, we write:

$$R^2 = (R-30)^2 + 75^2$$
$$R^2 = R^2 - 2\times30\times R + 30^2 + 75^2$$
$$60R = 6525 \; R = 108.75\,\text{ft}$$

Now, the angle ϕ subtended by the dome to its center:

$$\phi = \sin^{-1}\left(\frac{75}{108.75}\right) = 43.6°$$

9.14.7 CALCULATE LOADS

Dead load of 4 in thick concrete:

$$w_D = t\gamma_{conc.} = \frac{4}{12}\times150\,\text{lb}/\text{ft}^3 = 50\,\text{psf}$$

Therefore, the design load is:

$$w_u = 1.2w_D + 1.6w_L = 50\times1.2 + 1.6\times25 = 100\,\text{psf}$$

9.14.8 MERIDIONAL FORCES AND STRESSES

$$N_\phi = \frac{Rw}{1+\cos\phi} = \frac{180.75\times100}{1+\cos 43.6} = 6307\,\text{lb/ft}$$

$$f_\phi = \frac{N_\phi}{t} = \frac{6307/12}{4} = 131.4\,\text{psi (C)}$$

9.14.9 HOOP FORCES AND STRESSES

$$N_\theta = Rw\left(\cos\phi - \frac{1}{1+\cos\phi}\right)$$

$$= 108.75\times100\left(\cos 43.6 - \frac{1}{1+\cos 43.6}\right) = 1568\,\text{lb/ft}$$

$$f_\theta = \frac{N_\theta}{t} = \frac{1568/12}{4} = 32.7\,\text{psi (C)}$$

9.14.10 RING TENSION

$$T = RN_\phi\cos\phi\sin\phi$$

$$= 108.75\times6307\times\cos 43.6\times\sin 43.6$$

$$= 342,534\,\text{lb} = 342\,\text{kip (T)}$$

Question: Find the amount of reinforcement required to carry the meridional stress, hoop stress, and ring tension, given that $f_C' = 4000$ psi and $f_y = 60\,\text{ksi}$.

9.14.11 REINFORCEMENT TO CARRY MERIDIONAL STRESS

Since meridional stresses are always compressive in dome, we do not need any reinforcement for that, but according to ACI code, a minimum amount of 0.0018 bh reinforcement must be provided.

$$A_{s,\text{min}} = 0.0018\,bh = 0.0018\times12\times4 = 0.0864\,\text{in}^2/\text{ft}$$

If #3 bar is selected for reinforcing, spacing would be:

$$S = \frac{A_{s,\text{provided}}}{A_{s,\text{provided}}} = \frac{0.11}{0.0864}\times12 = 15.27\,\text{in}$$

The maximum spacing allowed is:

$$s_{\text{max}} = \begin{cases} 3h = 3\times4 = 12 \text{ in (governs)} \\ 18 \text{ in} \end{cases}$$

Thus, we cannot provide 15.27 in spacing as calculated, because it exceeds the maximum spacing of 12 inch. Therefore, provide #3 @ 12 in c/c.

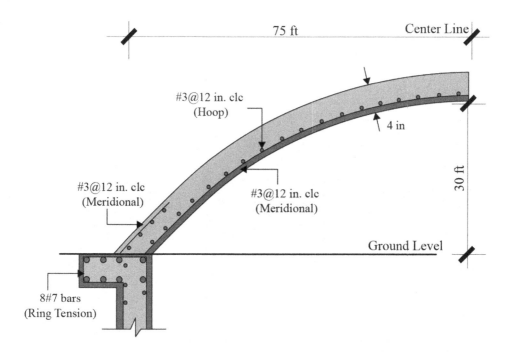

9.14.12 REINFORCEMENT TO CARRY HOOP STRESS

The hoop stress ($f_\theta = 32.7\,\text{psi}$) was found compressive too. Thus, no reinforcement is required here also. Based on the same reasoning as previous, provide the minimum reinforcement amount, that is, #3 @ 12 in c/c.

9.14.13 REINFORCEMENT TO CARRY RING TENSION

The ring tension was found, $T = 342\,\text{kip}$. Now, the amount of steel required:

$$A_s = \frac{T}{\phi f_y} = \frac{342\,\text{kip}}{0.9 \times 60\,\text{ksi}} = 6.33\,\text{in}^2$$

Provide 8 # 7 bars $\left(A_s = 6.32\,\text{in}^2 \right)$.

Note: The provided $A_s = 6.32\,\text{in}^2$ is only $0.01\,\text{in}^2$ less than the required amount of $6.33\,\text{in}^2$, which is acceptable.

Index

Note: Page numbers in *italics* indicate a figure and page numbers in **bold** indicate a table on the corresponding page.

A

admixtures, 300–301
aggregates, 296–297, 299–300, *300*
    coarse, 296–297, 299
    fine, 296, 299–300
    unit weight and, 297
air-entraining, 296, 300
algebraic approach, 105–106
arch analysis, 181–183
    axial force, 523–525
    axial thrust, 527
    horizontal thrust, 527
    left segment, 184–185
    profile, 527
    support reactions, 527
    three-hinged arch, 522–523, 525–527
    whole free body, 183–184
Aspdin's cement, 298
axial force, 41
    column, 467–468, *477*
    internal force diagrams, 72
    Vierendeel trusses, 504

B

β_1 factor, 323–328
beam (cracked), elastic analysis, 313
    bending stresses in RC beam, 313–314
    stress in concrete and steel, 315–317, *316*
    T-beam, 317–320, **318**, *319*
    transformed section, 314, *315*
beam (uncracked), elastic analysis, 308–309
    load, 312
    tensile cracks in RC beam, 309–310
beam analysis, 22–24
    moment distribution method, 224–234
    T-beam, 328, 502
beam deflection
    direct integration method, 110–115
    imaginary beam, 216
    moment area method, 116–123
    real beam, 216
    unit load formula, 216–218
    virtual work method, 215
beam design
    check shear, 422–423
    dead load and load deflection, 422
    load deflection, 422
    moment, 421–422
beam moments, 464–466
beam reactions, 37–40
    fixed supports, 37
    pinned supports, 37
beam shears, 466–467

beam slope deflection, 236
    angular displacement, 237
    equilibrium of joint, 239
    fixed-end moments, 238, 241–242
    load reactions, 239, 242
    moments reactions, 240–241, 243
    reactions, 239, 242
    relative linear displacement, 237–238
beam stiffness method, *see* stiffness method, beam
bearing failure
    plate, 85
    rivet, 86–88
Beinecke Rare Book and Manuscript Library, *507*
bending moment, 43–51, 72, 140, 472, 520–522
    arch analysis, 181–183
    beam moments, 464–466
    beam structure, 78–82
    for a cantilever beam, 77
    column moments, 463–464
    flexural stress, 89–94
    gravity load analysis, 469
    internal force diagrams, 72
    internal force diagrams, commonly used beans, 74
    midspan moment, 469
    two-point loaded beam structure, 75
    uniformly loaded beam structure, 76
    Vierendeel truss, 511, 512, 518
bending stresses, RC beam, 313–314
biaxial bending, 392–393
    capacity, 394–395
    slenderness ratio, 393
    slender/stocky column, 393
biaxial buckling, 150–152
bolt connection, 424
    bearing strength criteria, 425
    bolts required, 428
    demand, 427–428, **428**
    design strength, 427
    edge bolt, tension member, bearing of, 426
    edge bolt on gusset plate, bearing of, 429
    excessive elongation of hole, 425–426
    gusset plate, 427, 429
    hole diameter, 426, 429
    inner bolt on gusset plate, bearing of, 429
    inner bolt on tension member, 427
    shearing on bolt, 427
    shearing on each bolt, 428
    spacing, 429
    tension member, 427
bolt joint, 83
braced core system, 142
brittleness, 57–58
buckling
    column analysis, 128
    compression member, 388

Printed in the United States
by Baker & Taylor Publisher Services